Electromagnetic Instabilities in an Inhomogeneous Plasma

Plasma Physics Series

Series Editor: **Professor E W Laing**, University of Glasgow

Advisory Panel

> **Dr J Lacina**, Czechoslovak Academy of Sciences
> **Professor M A Hellberg**, University of Natal
> **Professor A B Mikhailovskii**, I V Kurchatov Institute of Atomic Energy
> **Professor K Miyamoto**, University of Tokyo
> **Professor H Wilhelmsson**, Chalmers University of Technology

Other books in the series

An Introduction to Alfven Waves
R Cross

MHD and Microinstabilities in Confined Plasma
W M Manheimer and C N Lashmore-Davies

Transition Radiation and Transition Scattering
V L Ginzburg and V N Tsytovich

Radiofrequency Heating of Plasmas
R A Cairns

Plasma Physics via Computer Simulation
C K Birdsall and A B Langdon

Plasma Physics Series

Electromagnetic Instabilities in an Inhomogeneous Plasma

A B Mikhailovskii

I V Kurchatov Institute of Atomic Energy
Moscow

Translation Editor: **Professor E W Laing**

Institute of Physics Publishing
Bristol, Philadelphia and New York

05011401

PHYSICS

©IOP Publishing Ltd 1992

British Library Cataloguing in Publication Data

Mikhailovskii, A. B. (Anatolii Borisovich)
    Electromagnetic instabilities in an inhomogeneous
    plasma. - (Plasma physics series)
    I. Title   II. Series
    530.401

    ISBN 0-7503-0182-1

Library of Congress Cataloging-in-Publication Data are available

Published by IOP Publishing Ltd, a company wholly owned by the Institute of Physics, London, by IOP Publishing Ltd
Techno House, Redcliffe Way, Bristol BS1 6NX, England
335 East 45th Street, New York, NY 10017-3483, USA

US Editorial Office: IOP Publishing Inc., The Public Ledger Buildings, Suite 1035, Independence Square, Philadelphia, PA 19106, USA

Typeset in TEX at IOP Publishing Ltd
Printed in Great Britain by Galliard (Printers) Ltd, Great Yarmouth, Norfolk

Preface

The aim of this book is a comprehensive presentation of the theory of electromagnetic instabilities in a magnetized inhomogeneous plasma, mainly in the classical approximation of straight and parallel magnetic field lines. In addition, the effects of magnetic field curvature are taken into account.

The electromagnetic instabilities are more significant than the well-known electrostatic ones when plasma pressure is finite or large compared with magnetic field pressure. Such plasmas are called finite- or high-β plasmas.

The presentation is mainly given from the general physics viewpoint. Therefore, the book is useful for everyone interested in the current development of plasma theory.

Confinement of finite- or high-β plasmas is one of the main objectives in laboratory controlled fusion investigations. At the same time, such plasmas are widely representative of the state of matter in space. Therefore, the book will facilitate the labour of experts on controlled fusion and space physics who are engaged in electromagnetic instabilities. Keeping in mind this category of readers, in the course of presentation the author mentions the specific problems of laboratory and space physics which led to the notions of instabilities considered.

The study of the instabilities is based on analytical approaches. These approaches are presented in the book in detail. This allows one to show the main physical regularities in the types of instabilities discussed with minimal efforts. Then the reader can reproduce all the results given in the book, and, if necessary, make his/her own contribution in this new branch of plasma theory.

A considerable part of the results given in the book has been obtained by the author and his colleagues. The same concerns the methods and the problem statements. At the same time, the main results in the fields which this book covers which have been obtained

v

by other authors, both Soviet and Western, are also presented.

The book contains reviews, including critical remarks, of publications on every type of instabilities, that should help the reader in orientation on the literature.

I am indebted to Drs V A Pilipenko and O G Onishchenko and my daughter Lyudmila A Mikhailovskaya for their assistance during the preparation of the manuscript.

A B Mikhailovskii
February 1992

Contents

Introduction

The state of the theory of plasma instabilities up to the year 1969 was summarized in [1, 2]. Further development in plasma physics added a new part of plasma theory, the theory of electromagnetic instabilities in a magnetized inhomogeneous plasma. Presenting this here, we assume the reader to be acquainted with the general concepts of electrostatic instabilities in such a plasma as given in [1, 2].

We devote main attention to the approximation of a magnetic field with straight and parallel lines. In addition, we study instabilities for the simplest case of a curvilinear magnetic field: the longitudinally homogeneous shearless field. The study of instabilities specific to the case of a complex geometry of the magnetic field and/or depending on the magnetic shear is beyond the scope of the book. It will be presented in [3].

Using the term *electromagnetic instabilities*, we emphasize their distinction from the electrostatic ones. In studying electrostatic instabilities one neglects magnetic field perturbations and represents the perturbed electric field in the form $\boldsymbol{E} = -\nabla\phi$, where ϕ designates the *scalar potential* which is often also called the *electrostatic potential*. In contrast to this, for electromagnetic instabilities the perturbed magnetic fields are also significant. Thus, these instabilities represent a more general and more complex class of perturbations. As a rule, the *electrostatic approximation* can be justified for a sufficiently small ratio of plasma pressure to magnetic pressure characterized by a dimensionless parameter β. Generally speaking, in the opposite case, i.e., for finite or high β, instabilities have to be studied with allowance for their electromagnetic nature. The *electromagnetic effects* can be important also for vanishing β-values if the plasma contains an additional *high-energy component* or if the perturbations are *very stretched out* along the equilibrium magnetic field lines.

The material of the book is divided into Chapters 1–16.

In Chapter 1 an approach to the theory of collisionless plasma instabilities is presented. This approach is based on the *local dispersion relation* written in terms of the *permittivity tensor of an inhomogeneous plasma*. The latter is calculated from the *Vlasov kinetic equation*. Such a dispersion relation takes into account the *electromagnetic nature of perturbations* and the equilibrium magnetic field inhomogeneity. We then neglect the transverse current due to the *non-stationary transverse electric field* or *to the centrifugal force*. The simplification of the dispersion relation, derived in Chapter 1, for various specific cases of perturbations is the subject of Chapter 2. One of the important results of Chapter 2 is the establishment of the fact that the *low-frequency long-wavelength oscillation branches split* into two types: the *inertialess branches* also called *magnetoacoustic branches*, and the *inertial branches* also known as *Alfven branches* (the term 'oscillation branch' is equivalent to the often used term 'mode'). Such a *splitting of oscillation branches* allows one to analyse the respective types of instabilities separately.

The *inertialess* and *inertial (Alfven) instabilities in collisionless finite-β plasma* with Maxwellian particle distributions are studied in Chapters 3 and 4, respectively. These instabilities are dependent on the magnetic drift of the particles. Therefore they can be called the *magneto-drift instabilities*.

The material concerning the low-frequency instabilities excited by *high-energy particles* is presented in Chapter 5.

The *cyclotron and high-frequency instabilities* specific to *strongly inhomogeneous plasma* are studied in Chapter 6.

Chapter 7 is devoted to the description of electromagnetic perturbations in a *plasma with transverse current* due to the non-stationary transverse electric field or to the centrifugal force. *Instabilities in such a plasma* are discussed in Chapter 8.

From Chapter 9 onwards, *collisions between particles* are taken into account. In Chapter 9 the problem of the *hydrodynamic description of collisional finite-β plasma perturbations* is discussed. In Chapter 10 the *starting equations* for basic types of such perturbations are derived using the *two-fluid approach*. The analysis of *specific types of instabilities in a collisional finite-β plasma* is given in Chapter 11.

Chapters 12–14 both discuss collisionless and collisional instabilities. In Chapter 12 material concerning the *high-β force-free plasma* is summarized. In Chapter 13 we present the theory of the *Kelvin–Helmholtz instabilities*, both the ordinary ones (so-called hydromagnetic or classical) and those associated with the transverse drift-convective motion of particles (*electromagnetic drift Kelvin–Helmholtz instabilities*). In Chapter 14 the *low-β plasma instabilities* are considered.

Throughout Chapters 1–14 we assume the magnetic field lines to be straight and parallel. The role of *magnetic-field line curvature* in finite-β plasma instabilities is discussed in Chapters 15 and 16. We then restrict ourselves to the simplest case of a curvilinear magnetic field—the shearless longitudinally inhomogeneous magnetic field. This field is a particular case of the cylindrical magnetic field. Chapter 15 is devoted to the *general problems of describing the instabilities in the case of such a field. Specific types of these instabilities* are analysed in Chapter 16.

The present state of the theory of electromagnetic instabilities in an inhomogeneous plasma is due to the contributions of a large number of authors. As a rule, different types of instabilities have their own history. For this reason the history of our topic is very extensive and detailed. Therefore, we prefer to present our view of the history within the respective chapters.

We mention applications of the instabilities in the respective chapters also. In addition, the reader can find the names of applied problems, dealing with the instabilities considered, by means of the Index.

1 General Dispersion Relation for a Collisionless Inhomogeneous Plasma

The theory of *electromagnetic instabilities* is more general than that of electrostatic ones, since the total set of the Maxwell equations is taken into account. This extension to *electromagnetic effects* turns out to be important mainly when the plasma pressure is not too small compared with the equilibrium magnetic field pressure. In such finite-β plasmas the equilibrium magnetic field gradient is also comparable to the plasma pressure one. Therefore, the generalization of the theory of the electrostatic instabilities has to take into account *two factors*: the electromagnetic nature of perturbations, on the one hand, and the gradient of the equilibrium magnetic field on the other hand. Naturally, in some particular cases, one of these factors may prove to be insignificant and, moreover, cases are possible where the plasma pressure is rather small, but the electromagnetic part of the perturbations is non-negligible. These situations allow one to have simplified theories.

In this chapter the basic equations of the theory of electromagnetic instabilities in a *collisionless inhomogeneous plasma* are given. The main approximation is that the magnetic field lines are straight and parallel. In addition, we assume that the characteristic length of the perturbations is small in comparison to the characteristic length of the plasma and magnetic field inhomogeneities.

The basic equations for the case of a collisional plasma will be derived in Chapters 9 and 10. The role of the curvature of the magnetic field lines will be studied in Chapters 15 and 16, and some examples of large-scale perturbations will be discussed in Chapters 7, 8 and 13.

We also assume that, besides the *diamagnetic current*, there is no additional transverse current in the equilibrium considered. The basic equations for the case of a plasma with an *additional transverse current* will be given in Chapter 7.

We simplify the problem by using the *WKB approximation* in the form

$$\exp[i \int k_x(x')\mathrm{d}x']$$

where x is the direction of the inhomogeneity. Within our local approximation the *wave number* k_x can be considered to be slowly varying and its derivatives may be neglected. The problem of describing perturbations in the inhomogeneous plasma therefore reduces to the *local dispersion relation*, i.e., to an equation relating the *oscillation frequency* ω with the *wave vector* $\boldsymbol{k}$. The structure of this dispersion relation is discussed in section 1.1.

To find this dispersion relation we use the *permittivity tensor*. To calculate this tensor we must find the *perturbed particle distribution functions*. To do this, knowledge of the *constants of unperturbed particle motion*, the *equilibrium distribution functions* and the *unperturbed particle trajectories* is required. Section 1.2 provides the information about the equilibrium state of an *inhomogeneous plasma* with an *inhomogeneous magnetic field*. In section 1.3 the perturbed distribution functions are derived. In section 1.4 the permittivity tensor is calculated.

Note that the permittivity tensor which we use is somewhat different from that used in the theory of homogeneous plasma instabilities [1.1]. This aspect is discussed in section 1.1.4. The point is that with the usual permittivity tensor the dispersion relation proves to be dependent not only on the tensor itself, but also on its spatial derivatives (see also [1.2, 1.3]). Therefore we use a *modified permittivity tensor*. To derive such a tensor, the relationship between the *perturbed charge density* and two components of the *perturbed current density*, on the one hand, and the *electric field*, on the other hand, is used.

The first study of the instabilities in an inhomogeneous plasma with the help of a permittivity tensor was given in [1.2]. There, the structural specificity associated with the *gradient effects* was explained and the correspondence between the description of the *electrostatic perturbations* and that of *electromagnetic perturbations* was discussed. We should mention, however, that the paper [1.2] was preceded by [1.4, 1.5] where the dispersion relations were already obtained for some particular kinds of perturbations. The study in [1.4] deserves specific attention, since it was the earliest approach to *non-hydromagnetic instabilities* in an inhomogeneous plasma, including electromagnetic plasma instabilities in a straight magnetic field. The approach which we apply in this chapter goes back to [1.2] with further developments given in [1.3, 1.6–1.11].

Note that the above mentioned procedure of deriving the basic equations for inhomogeneous plasma oscillations, based on calculations of the *charge density* and two components of the *perturbed current density*, was also used in [1.12] for perturbations propagating strictly across the equilibrium magnetic field.

We restrict ourselves to the non-relativistic approximation. *Relativistic effects* were taken into account in [1.13] (see also [1.14]).

1.1   General form of the dispersion relation

As mentioned above, we describe the dependence of perturbations on x by means of the wave number $k_x(x)$, neglecting the derivatives of $k_x(x)$ with respect to x. Therefore in our subsequent calculations, k_x will be treated equivalently to the wave vector components k_y and k_z perpendicular to the direction of inhomogeneity. As a result, the phase of a perturbation can be written in the form

$$\exp[i(k_x x + k_y y + k_z z - \omega t)]$$

where ω designates the *oscillation frequency*.

In section 1.1.1 we recall the form of the dispersion relation for a homogeneous plasma. In section 1.1.2 we introduce the potentials of the perturbed electromagnetic field and explain their meaning. In section 1.1.3 we represent the basic electrodynamic equations in terms of these potentials. Finally, in section 1.1.4 we give the general form of the *dispersion relation for an inhomogeneous plasma* and explain the quantities, which characterize the permittivity of such a plasma.

1.1.1   The dispersion relation for a homogeneous plasma

The dispersion relation in a homogeneous plasma is usually represented in the form (see [1.1], equation (1.32))

$$\begin{vmatrix} \epsilon_{11} - N^2\cos^2\theta & \epsilon_{12} & \epsilon_{13} + N^2\cos\theta\sin\theta \\ \epsilon_{21} & \epsilon_{22} - N^2 & \epsilon_{23} \\ \epsilon_{31} + N^2\cos\theta\sin\theta & \epsilon_{32} & \epsilon_{33} - N^2\sin^2\theta \end{vmatrix} = 0. \quad (1.1)$$

Here $N^2 = c^2 k^2/\omega^2$ is the square of the *refractive index*, and $\epsilon_{\alpha\beta}$ (with $\alpha, \beta = 1, 2, 3$) are the components of the *permittivity tensor* determined by the relation

$$\epsilon_{\alpha\beta} = \delta_{\alpha\beta} + 4\pi i\sigma_{\alpha\beta}/\omega \quad (1.2)$$

where $\sigma_{\alpha\beta}$ is the *conductivity tensor* which can be found from the relationship between the perturbed current density j and the perturbed electric field E in the form:

$$j_\alpha = \sigma_{\alpha\beta} E_\beta \qquad (\alpha, \beta = 1, 2, 3). \qquad (1.3)$$

It is assumed that the wave vector k has components in directions 1 and 3 only, so that $k = (k_1, 0, k_3)$. The angle θ is defined by the relation $\theta = \tan^{-1}(k_1/k_3)$. When we choose axis 3 (z-axis) to be in the direction of the equilibrium magnetic field B_0, θ represents the angle between k and B_0.

Considering the perturbations in an inhomogeneous plasma, we shall assume that the wave vector k is approximately perpendicular to B_0, i.e., $\cos\theta \ll 1$. In all other cases, the gradient effects should be unimportant, since we assume the inhomogeneity to be perpendicular to B_0. Under these assumptions equation (1.1) can be transformed as follows. We multiply the third row of matrix (1.1) by $\cos\theta$ and add it to the first row, after that we multiply the third column by $\cos\theta$ and add it to the first one. Then we have

$$\begin{vmatrix} \epsilon_{00} & \epsilon_{02} & \epsilon_{03} \\ \epsilon_{20} & \epsilon_{22} - N^2 & \epsilon_{23} \\ \epsilon_{30} & \epsilon_{32} & \epsilon_{33} - N^2 \end{vmatrix} = 0 \qquad (1.4)$$

with

$$\begin{aligned} \epsilon_{00} &= \epsilon_{11} + 2\epsilon_{13}\cos\theta + \epsilon_{33}\cos^2\theta \equiv k_\alpha \epsilon_{\alpha\beta} k_\beta / k^2 \\ \epsilon_{02} &= -\epsilon_{20} = \epsilon_{12} + \epsilon_{32}\cos\theta \equiv k_\alpha \epsilon_{\alpha 2}/k \\ \epsilon_{03} &= \epsilon_{30} = \epsilon_{13} + \epsilon_{33}\cos\theta \equiv k_\alpha \epsilon_{\alpha 3}/k. \end{aligned} \qquad (1.5)$$

Here, we use the symmetry properties

$$\begin{aligned} \epsilon_{13} &= \epsilon_{31} \\ \epsilon_{12} &= -\epsilon_{21} \\ \epsilon_{23} &= -\epsilon_{32} \end{aligned}$$

(see [1.1], (A.24)). Physically, ϵ_{00} is the plasma *scalar permittivity* (see [1] (A.1.12)). In the transition from (1.1) to (1.4), one neglects terms of order $\cos^2\theta$ compared with unity.

1.1.2   Potentials of the perturbed field
Turning to the perturbations in an inhomogeneous plasma, we present their electric field in the form

$$E = -\nabla\phi - (1/c)\partial A/\partial t \qquad (1.6)$$

where ϕ and $\boldsymbol{A}$ are the *scalar (electrostatic)* and *vector potentials*. Since we use the Coulomb gauge, we have

$$\nabla \cdot \boldsymbol{A} = 0. \tag{1.7}$$

When $k_z \ll k$ holds, (1.7) means approximately that $\nabla \cdot \boldsymbol{A}_\perp = 0$, or, in terms of the wave numbers, $\boldsymbol{k} \cdot \boldsymbol{A} = 0$. Here, the subscript $\perp$ denotes components perpendicular to the magnetic field $\boldsymbol{B}_0$. Under these conditions one quantity A_2 only, given by

$$A_2 = \boldsymbol{k}_\perp \times \boldsymbol{A} \cdot \boldsymbol{e}_z / k \tag{1.8}$$

determines $\boldsymbol{A}_\perp$:

$$A_x = -k_y A_2 / k \qquad A_y = k_x A_2 / k. \tag{1.9}$$

Thus, we describe the electromagnetic field of the perturbations by three quantities: $\phi, A_2, A_3 \equiv A_z$. The quantity A_2 characterizes the components of the perturbed magnetic field $\boldsymbol{B}$ along the equilibrium magnetic field $\boldsymbol{B}_0$,

$$B_z = \mathrm{i} k A_2. \tag{1.10}$$

On the other hand, A_3 describes the perturbed magnetic field $\boldsymbol{B}_\perp$ perpendicular to the direction $\boldsymbol{B}_0$,

$$\boldsymbol{B}_\perp = \mathrm{i} \boldsymbol{k} \times \boldsymbol{e}_3 A_3 \tag{1.11}$$

where $\boldsymbol{e}_3 = \boldsymbol{e}_z$ is the unit vector along $\boldsymbol{B}_0$. According to equation (1.6), in terms of (ϕ, A_2, A_3) the perturbed electric field is

$$\boldsymbol{E} = -\mathrm{i}\boldsymbol{k}\phi + \mathrm{i}(\omega/c)(-\boldsymbol{k} \times \boldsymbol{e}_3 A_2 / k + \boldsymbol{e}_3 A_3). \tag{1.12}$$

Note that the difference between $\boldsymbol{k}_\perp$ and $\boldsymbol{k}$ is important in the first term of the right-hand side of (1.12), although we could neglect this difference in the preceding equations (1.8)–(1.11).

1.1.3 *Electrodynamic equations in terms of* ϕ, A_2, A_3

Describing the perturbations, we start with *Poisson's equation*,

$$k^2 \phi = 4\pi\rho \tag{1.13}$$

and the components of the *Maxwell equations* along the directions $\boldsymbol{A}_\perp$ and $\boldsymbol{B}_0$

$$k^2 A_i = 4\pi j_i / c + \omega^2 A_i / c^2 \qquad (i = 2, 3). \tag{1.14}$$

Here ρ is the *charge density*; $j_3 = j_z$, and j_2 is the *current density* component perpendicular to the wave vector and to the equilibrium magnetic field

$$j_2 = \mathbf{k} \times \mathbf{j} \cdot \mathbf{e}_z / k. \qquad (1.15)$$

Equations (1.13) and (1.15) are our *starting electrodynamic equations*.

1.1.4 *General form of the dispersion relation for an inhomogeneous plasma*

We write the charge density in the form

$$\rho = -\frac{k^2}{4\pi}(\epsilon_{00} - 1)\phi + \frac{k\omega}{4\pi c}(\epsilon_{02}A_2 + \epsilon_{03}A_3). \qquad (1.16)$$

The quantities $\epsilon_{00}, \epsilon_{02}, \epsilon_{03}$ introduced in such a manner are generalizations, for the case of an inhomogeneous plasma, of the quantities denoted by the same symbols which occur in the dispersion relation for a homogeneous plasma (1.4). In the case of a homogeneous plasma these quantities are related to the permittivity tensor $\epsilon_{\alpha\beta}(\alpha, \beta = 1, 2, 3)$ by (1.5). In order to obtain similar relations for the case of an inhomogeneous plasma, we take into account (1.3) and the charge continuity equation and represent the charge density in the form:

$$\rho = \frac{1}{\omega}\left(\mathbf{k} \cdot \mathbf{j} - \mathrm{i}\frac{\partial j_x}{\partial x}\right) = \frac{1}{\omega}\left(k_\alpha \sigma_{\alpha\beta} - \mathrm{i}\frac{\partial \sigma_{x\beta}}{\partial x}\right)E_\beta. \qquad (1.17)$$

Here substituting E_β from (1.12) and allowing for (1.2), we express ρ in terms of $\epsilon_{\alpha\beta}$. Comparing the result with (1.16), we find that

$$\epsilon_{00} = \frac{1}{k^2}\left(k_\alpha \epsilon_{\alpha\beta} - \mathrm{i}\frac{\partial \epsilon_{x\beta}}{\partial x}\right)k_\beta$$

$$\epsilon_{02} = \frac{1}{k^2}[k_\alpha(\epsilon_{\alpha y}k_x - \epsilon_{\alpha x}k_y) - \mathrm{i}\frac{\partial}{\partial x}(\epsilon_{xy}k_x - \epsilon_{xx}k_y)] \qquad (1.18)$$

$$\epsilon_{03} = \frac{1}{k}\left(k_\alpha \epsilon_{\alpha z} - \mathrm{i}\frac{\partial \epsilon_{xz}}{\partial x}\right).$$

Obviously, in contrast to the case of a homogeneous plasma, the quantities $\epsilon_{00}, \epsilon_{02}, \epsilon_{03}$ for the inhomogeneous plasma do include the terms with the derivatives of $\epsilon_{\alpha\beta}$ ($\alpha, \beta = 1, 2, 3$), with respect to x.

Similarly to (1.16), we present the current density components j_2, j_3 in the form

$$j_2 = -\frac{k\omega}{4\pi}\epsilon_{20}\phi + \frac{\omega^2}{4\pi c}[(\epsilon_{22} - 1)A_2 + \epsilon_{23}A_3]$$

$$j_3 = -\frac{k\omega}{4\pi}\epsilon_{30}\phi + \frac{\omega^2}{4\pi c}[\epsilon_{32}A_2 + (\epsilon_{33} - 1)A_3]. \qquad (1.19)$$

Comparing (1.19) with (1.3), we find that, in contrast to (1.18), the quantities ϵ_{ik} with subscripts $i = 2, 3$ and $k = 0, 2, 3$ do not contain the derivatives of $\epsilon_{\alpha\beta}$, so that we have

$$
\begin{aligned}
\epsilon_{20} &= (k_x \epsilon_{y\beta} - k_y \epsilon_{x\beta}) k_\beta / k^2 \\
\epsilon_{30} &= \epsilon_{z\beta} k_\beta / k \\
\epsilon_{22} &= [k_x^2 \epsilon_{yy} + k_y^2 \epsilon_{xx} + k_x k_y (\epsilon_{yx} + \epsilon_{xy})] / k^2 \\
\epsilon_{23} &= (k_x \epsilon_{yz} - k_y \epsilon_{xz}) / k \\
\epsilon_{32} &= (k_x \epsilon_{zy} - k_y \epsilon_{zx}) / k.
\end{aligned} \tag{1.20}
$$

In spite of this asymmetry between the relations for ϵ_{ik} $(i, k = 0, 2, 3)$, and $\epsilon_{\alpha\beta}$ $(\alpha, \beta = 1, 2, 3)$, we shall see that the tensor ϵ_{ik} introduced for an inhomogeneous plasma has the same symmetry properties as the analogous tensor for a homogeneous plasma, i.e.,

$$
\begin{aligned}
\epsilon_{20} &= -\epsilon_{02} \\
\epsilon_{03} &= \epsilon_{30} \\
\epsilon_{32} &= -\epsilon_{23}.
\end{aligned} \tag{1.21}
$$

Substituting (1.16) and (1.19) into (1.13) and (1.14), we arrive at a homogeneous set of equations. Setting its determinant to zero yields the dispersion relation in the form of (1.4).

The dispersion relation for an inhomogeneous plasma can also be presented in the form of (1.1). However, as the following analysis will show, it is more convenient to use a set of somewhat more complicated coefficients $\epsilon'_{\alpha\beta}$, which present themselves as combinations of ϵ_{ik}, instead of $\epsilon_{11}, \epsilon_{12}, \epsilon_{13}$. For our case of a straight magnetic field, these combinations can be found as follows. We add the second row of the matrix (1.4) after its multiplication by $ik_y \kappa_B / k^2$ and the third row of the same matrix after its multiplication by $- \cos \theta$ to the first one. In doing so, we use hereinafter the abbreviation

$$
\kappa_B = \partial \ln B_0 / \partial x. \tag{1.22}
$$

Then, we add the second column multiplied by $-ik_y \kappa_B / k^2$ and the third column multiplied by $- \cos \theta$ to the first one. By this procedure the form of (1.1) for the dispersion relation is obtained, where the unprimed expressions are replaced by the primed ones defined by

$$
\begin{aligned}
\epsilon'_{11} &= \epsilon_{00} - 2i(k_y \kappa_B / k^2) \epsilon_{20} + (k_y \kappa_B / k^2)^2 (\epsilon_{22} - N^2) \\
&\quad - 2i(k_y \kappa_B / k^2) \epsilon_{23} \cos \theta - 2\epsilon_{03} \cos \theta + \epsilon_{33} \cos^2 \theta \\
\epsilon'_{12} &= -\epsilon'_{21} = \epsilon_{02} - i(k_y \kappa_B / k^2)(\epsilon_{22} - N^2) - \epsilon_{32} \cos \theta \\
\epsilon'_{13} &= \epsilon'_{31} = \epsilon_{03} - \epsilon_{33} \cos \theta + i(k_y \kappa_B / k^2) \epsilon_{23}.
\end{aligned} \tag{1.23}
$$

It is clear that in the case of $\kappa_B = 0$, relations (1.23) are the same as (1.5). However, even in this case the forms of ϵ_{ik} and $\epsilon_{\alpha\beta}$ for an inhomogeneous plasma differ, generally speaking, from those for a homogeneous plasma.

Then for simplicity, we will omit the primes in (1.23), keeping in mind that the meaning of these quantities is somewhat different from that for a homogeneous plasma.

1.2   Equilibrium

1.2.1   *Constants of motion*

We assume the equilibrium magnetic field to be *weakly inhomogeneous* so that the characteristic length of the field inhomogeneity is large compared with the *Larmor radius* of the particle. In this case we have the *constant of motion*

$$X = x + v_y/\Omega. \tag{1.24}$$

Here v_y is the component of the particle velocity $\boldsymbol{v}$ along the y-axis, $\Omega = eB_0/Mc$ is the *cyclotron frequency*, where e and M are the electric charge and the mass of the particle, respectively. The quantity X represents the x-coordinate of the particle *guiding centre*. The magnetic field $\boldsymbol{B}_0$ contained in (1.24) is taken at the point of the guiding centre so that Ω is also the constant of motion. The variable quantities in (1.24) are the coordinate x and the velocity v_y.

Equation (1.24) can be obtained, for example, from the following considerations. Let us take into account the fact that the component of the generalized particle momentum, transverse to the magnetic field and the direction of the field inhomogeneity, is a constant of motion in the presence of the slab symmetry considered. Let us denote such a constant of motion divided by the particle mass as

$$V_y \equiv v_y + \int_X^x \Omega(x')\mathrm{d}x' \tag{1.25}$$

where X is a certain, for the time being arbitrary constant. The quantity V_y can be called the *inhomogeneous constant of motion*. Assuming a weak magnetic field inhomogeneity, we replace $\Omega(x')$ by $\Omega(X)$. Choosing $V_y \equiv 0$, we arrive at (1.24).

In the case of a weakly inhomogeneous magnetic field, the *magnetic moment* μ is also a constant of motion, where the quantity μ is determined by

$$\mu = \tilde{v}_\perp^2/2B_0. \tag{1.26}$$

Here $\tilde{\boldsymbol{v}}_\perp = \boldsymbol{v}_\perp - \boldsymbol{V}_d$ is the part of the transverse particle velocity $\boldsymbol{v}_\perp$ associated with its cyclotron rotation, where $\boldsymbol{V}_d = V_d\boldsymbol{e}_y$, $\boldsymbol{e}_y$ is the unit vector along the y-axis, and V_d is the *magnetic-drift velocity* of a particle defined by

$$V_d = \kappa_B\tilde{v}_\perp^2/2\Omega. \tag{1.27}$$

The *particle energy*, i.e., $\mathcal{E} = v^2/2$ (energy per unit mass) is also a constant of motion. In the case of a straight magnetic field, the *transverse particle energy* $\mathcal{E}_\perp = v_\perp^2/2$ and the *longitudinal velocity* v_z can also be regarded as constants of motion.

1.2.2 Equilibrium distribution function

We characterize the equilibrium of each plasma component by the *distribution function* $f_0(\boldsymbol{v},x)$. It is assumed that f_0 is normalized by the condition

$$\int f_0 d\boldsymbol{v} = n_0 \tag{1.28}$$

where n_0 is the equilibrium *number density* of particles of the respective species.

In the case of a *collisionless plasma* the function f_0 can be constructed with the use of the constants of the equilibrium particle motion. We can choose the quantities $\mathcal{E}, v_z, X$ as such constants of motion, so that

$$f_0(\boldsymbol{v},x) = F(\mathcal{E}, v_z, X) \tag{1.29}$$

where F is, generally speaking, an arbitrary function. One of the important examples of the function F is the *isotropic Maxwellian distribution function*

$$F = n_0\left(\frac{M}{2\pi T}\right)^{3/2}\exp\left(-\frac{M\mathcal{E}}{T} + \frac{Mv_zU}{T} - \frac{MU^2}{2T}\right) \tag{1.30}$$

where T and U are the *temperature* and *longitudinal velocity* of the respective plasma component. For the plasma at rest ($U = 0$), we have

$$F = n_0\left(\frac{M}{2\pi T}\right)^{3/2}\exp\left(-\frac{M\mathcal{E}}{T}\right) \tag{1.31}$$

instead of equation (1.30). It is assumed that n_0, T and U in (1.30) and (1.31) are functions of the constant of motion X.

The form of (1.29) is convenient in the case of essentially non-one-dimensional particle velocity distributions, including isotropic ones. If the velocity distribution is anisotropic, it will be more convenient for us to use the *transverse particle energy* $\mathcal{E}_\perp$ instead of the total particle energy $\mathcal{E}$, as the constant of motion. Then, we have the relation

$$f_0(\boldsymbol{v}, x) = F_1(\mathcal{E}_\perp, v_z, X) \qquad (1.32)$$

instead of (1.29), where F_1 is also, generally speaking, an arbitrary function. In particular, for the case of a one-dimensional velocity distribution, the function F_1 has the form

$$F_1(\mathcal{E}_\perp, v_z, X) = (2\pi)^{-1}\delta(\mathcal{E}_\perp)f_\parallel(v_z, X) \qquad (1.33)$$

where $f_\parallel$ is the so-called *one-dimensional distribution function*. Allowing for (1.28), we have the normalization condition for $f_\parallel$:

$$\int f_\parallel \, \mathrm{d}v_z = n_0. \qquad (1.34)$$

One of the examples of the function $f_\parallel$ is the *one-dimensional Maxwellian distribution function*

$$f_\parallel = n_0 \left(\frac{M}{2\pi T_\parallel} \right)^{1/2} \exp\left(-\frac{M(v_z - U)^2}{2T_\parallel} \right). \qquad (1.35)$$

Here, in contrast to (1.30) and (1.31), $T_\parallel$ is the *longitudinal temperature*.

The distribution function f_0 can also be presented as some function F_2 of the constants of motion $\mathcal{E}$, μ, and X

$$f_0(\boldsymbol{v}, x) = F_2(\mathcal{E}, \mu, X). \qquad (1.36)$$

Such a presentation is convenient when the *approximation to a straight magnetic field* is considered as a limiting case of *longitudinally inhomogeneous magnetic fields*. In the latter case the longitudinal velocity and the transverse energy are not constants of motion so that relations (1.29) and (1.32) are invalid.

1.2.3   *Equation of transverse equilibrium*

We assume the inhomogeneity of the plasma, as that of the magnetic field, to be sufficiently weak. Then, from (1.29) we have approximately

$$f_0(x) = F(x) + \frac{v_y}{\Omega} \frac{\partial F(x)}{\partial x}. \qquad (1.37)$$

Similar approximate relations for f_0 can be found from (1.32) and (1.36).

The plasma inhomogeneity results, generally speaking, in an *equilibrium electric current* (*diamagnetic current*) j_{0y} determined by the relation

$$j_{0y} = \sum e \int v_y f_0 \mathrm{d}\boldsymbol{v} \qquad (1.38)$$

where the summation is made over the charge species. Allowing for equations (1.37) and (1.38) we find that

$$j_{0y} = \frac{c}{B_0} \frac{\partial p_\perp}{\partial x} \qquad (1.39)$$

where $p_\perp$ is the *transverse plasma pressure* given by

$$p_\perp = \sum (M/2) \int v_\perp^2 F \mathrm{d}\boldsymbol{v} \qquad (1.40)$$

for f_0 of the form of (1.29). When f_0 has the form of equations (1.32) and (1.35), the quantity $p_\perp$ is given by similar relations with substitutions $F \to F_1$, $F \to F_2$.

The equilibrium current j_{0y} is the reason for the *magnetic-field inhomogeneity*. The value of a *magnetic-field gradient* is determined by the Maxwell equation

$$\partial B_0/\partial x = -4\pi j_{0y}/c. \qquad (1.41)$$

From (1.39) and (1.41) we find the *equation of transverse equilibrium* between the plasma and the magnetic field

$$\frac{\partial}{\partial x}\left(p_\perp + \frac{B_0^2}{4\pi}\right) = 0. \qquad (1.42)$$

Note that, for F_1 of the form of (1.33), we have $p_\perp = 0$ so that $\nabla B_0 = 0$. The magnetic field gradient also vanishes in the case when $p_\perp \neq 0$, $\nabla p_\perp = 0$. This case corresponds to the approximation of a *force-free plasma*.

1.2.4 *Equilibrium trajectories*

The particle motion in an inhomogeneous magnetic field is determined by equations

$$\begin{aligned}
\mathrm{d}\boldsymbol{v}/\mathrm{d}t &= \boldsymbol{v} \times \boldsymbol{\Omega}(\boldsymbol{r}) \\
\mathrm{d}\boldsymbol{r}/\mathrm{d}t &= \boldsymbol{v}
\end{aligned} \qquad (1.43)$$

where $\boldsymbol{\Omega} = \Omega \boldsymbol{e}_0$, $\boldsymbol{e}_0 = \boldsymbol{B}_0/B_0$ and $\boldsymbol{r}$ represents the particle coordinates. We present the *transverse velocity* in the form

$$\begin{aligned}
v_x &= v_\perp \cos\alpha \\
v_y &= v_\perp \sin\alpha.
\end{aligned} \qquad (1.44)$$

Then we have $v_\perp = \text{const}$ from (1.43), while, allowing for (1.24), the function α satisfies the equation

$$\mathrm{d}\alpha/\mathrm{d}t = -\Omega(x) + \kappa_B v_\perp \sin \alpha. \tag{1.45}$$

Here κ_B is taken for $x = X$.

It follows from (1.45) that

$$\alpha = \alpha^{(0)}(t) + \delta\alpha(t) \tag{1.46}$$

where

$$\begin{aligned}
\alpha^{(0)}(t) &= \alpha_0 - \Omega(t - t_0) \\
\delta\alpha(t) &= [\cos \alpha^{(0)}(t) - \cos \alpha_0]\kappa_B v_\perp/\Omega.
\end{aligned} \tag{1.47}$$

Here $\alpha_0 = \tan^{-1}[v_y(t_0)/v_x(t_0)]$, and t_0 is an arbitrary instant of time. Substituting (1.46) into (1.44) and performing the expansion in a series in $\delta\alpha(t)$, from (1.43) we find that

$$\begin{aligned}
x &= x_0 - (v_\perp/\Omega)[\sin \alpha^{(0)}(t) - \sin \alpha_0] \\
y &= y_0 + (v_\perp/\Omega)[\cos \alpha^{(0)}(t) - \cos \alpha_0] + V_\mathrm{d}(t - t_0) \\
z &= z_0 + v_z(t - t_0) \\
v_x &= v_\perp \cos \alpha^{(0)}(t) \\
v_y &= v_\perp \sin \alpha^{(0)}(t) + V_\mathrm{d} \\
v_z &= \text{const.}
\end{aligned} \tag{1.48}$$

Here $x_0 = x(t_0)$, $y_0 = y(t_0)$, $z_0 = z(t_0)$.

The set (1.48) describes the *cyclotron rotation* of the particle across the field lines with frequency Ω and with *Larmor radius* $v_\perp/\Omega$, the *uniform motion along the field lines* with velocity v_z and the *drift across the field* with velocity V_d (*magnetic-drift velocity*).

1.3   Perturbed distribution function

In order to find the *perturbed distribution function* of some charge species designated by f we start with the perturbed *Vlasov kinetic equation* for this species

$$\frac{\partial f}{\partial t} + \boldsymbol{v} \cdot \nabla f + \boldsymbol{v} \times \boldsymbol{\Omega} \cdot \frac{\partial f}{\partial \boldsymbol{v}} = -\frac{e}{M}\left(\boldsymbol{E} + \frac{1}{c}\boldsymbol{v} \times \boldsymbol{B}\right) \cdot \frac{\partial f_0}{\partial \boldsymbol{v}}. \tag{1.49}$$

We solve this equation by the standard *method of integration along trajectories* described, in particular, in: [1], section 7.2; [2], section 1.6; [1.1], Appendix. The solution can be presented in the form

$$f = -\frac{e}{M} \int_{-\infty}^{t} \left(E + \frac{1}{c} v \times B \right) \cdot \frac{\partial f_0}{\partial v} dt'. \qquad (1.50)$$

The functions of coordinates and velocities in the integrand are taken for $r = r(t', r_0, v_0)$, $v = v(t', r_0, v_0)$ determined by the relations in (1.48).

Recall that, in our assumptions, the perturbed fields E and B depend on r and t as $\exp(ikr - i\omega t)$ and are expressed in terms of ϕ, A_2, A_3 by relations (1.10)–(1.12).

In the case where f_0 is of the form of (1.29) its derivative can be written as

$$\frac{\partial f_0}{\partial v} = v \frac{\partial F}{\partial \mathcal{E}} + e_z \frac{\partial F}{\partial v_z} + \frac{e_y}{\Omega} \frac{\partial F}{\partial X}. \qquad (1.51)$$

Here we have taken relation (1.24) into account.

Allowing for the discussion above, and following [1, 2], we find that

$$f = \frac{e}{M} \left[\phi \Lambda + \frac{1}{c} \left(\frac{A_2 \cos \chi}{\Omega} \frac{\partial F}{\partial X} + A_3 \Phi \right) - \omega G I \right]. \qquad (1.52)$$

Here

$$\Lambda = \partial F / \partial \mathcal{E} \qquad \Phi = \partial F / \partial v_z \qquad (1.53)$$

$$G = \frac{\partial F}{\partial \mathcal{E}} + \frac{k_z}{\omega} \frac{\partial F}{\partial v_z} + \frac{k_y}{\omega \Omega} \frac{\partial F}{\partial X} \qquad (1.54)$$

$$I = \sum \zeta_n \exp[i\xi \sin(\alpha - \chi) - in(\alpha - \chi)] \qquad (1.55)$$

$$\times [\phi J_n + \frac{1}{c} (iv_\perp J'_n + V_d J_n \cos \chi) A_2 - \frac{v_z}{c} J_n A_3] \equiv \sum I_n$$

$$\zeta_n = (\omega - k_z v_z - \omega_d - n\Omega)^{-1} \qquad (1.56)$$

where $\xi = k_\perp v_\perp / \Omega$, $J_n = J_n(\xi)$ is Bessel's function, J'_n is the derivative of J_n with respect to the argument, $\omega_d = k_y V_d$ is the so-called *magnetic-drift frequency* of the particle, and $\chi = \tan^{-1}(k_y/k_x)$. The summation in (1.55) is performed over all the integer values of n.

When f_0 is of the form of (1.32), we obtain instead of (1.52)

$$f = \frac{e}{M} \left[\phi \Lambda_1 + \frac{1}{c} \left(\frac{A_2 \cos \chi}{\Omega} \frac{\partial F_1}{\partial X} + A_3 \Phi_1 \right) - \omega G_1 I \right]. \qquad (1.57)$$

Here

$$\Lambda_1 = \frac{\partial F_1}{\partial \mathcal{E}_\perp} \qquad \Phi_1 = \frac{\partial F_1}{\partial v_z} - v_z \frac{\partial F_1}{\partial \mathcal{E}_\perp} \qquad (1.58)$$

$$G_1 = \frac{\partial F_1}{\partial \mathcal{E}_\perp} + \frac{k_z}{\omega}\left(\frac{\partial F_1}{\partial v_z} - v_z \frac{\partial F_1}{\partial \mathcal{E}_\perp}\right) + \frac{k_y}{\omega\Omega}\frac{\partial F_1}{\partial X}. \qquad (1.59)$$

Similarly, when f_0 is of the form of (1.36), we have

$$f = \frac{e}{M}\left(\phi\Lambda_2 + \frac{v_\perp \sin(\alpha - \chi)}{cB_0}A_2\frac{\partial F_2}{\partial \mu} - \omega\sum_n G_{2n}I_n\right) \qquad (1.60)$$

where

$$\Lambda_2 = \partial F_2/\partial \mathcal{E} \qquad (1.61)$$

$$G_{2n} = \frac{\partial F_2}{\partial \mathcal{E}} + \frac{n\Omega}{B_0\omega}\frac{\partial F_2}{\partial \mu} + \frac{k_y}{\omega\Omega}\frac{\partial F_2}{\partial X}. \qquad (1.62)$$

Subsequently, the functions of the argument X will be expanded in a series in v_y/Ω (cf. equation (1.37)).

In addition, we note that the terms with $A_2\cos\chi$ in (1.52), (1.55), and (1.57) turn out to be insignificant in problems involving *small-scale perturbations* (these terms can be of importance when studying *large-scale perturbations*, when the WKB approximation is invalid).

1.4   Permittivity of an inhomogeneous plasma

With the help of (1.52) (or (1.57), (1.60)) we calculate the *perturbed charge density* ρ and the components of the *perturbed current density* j_2 and j_3, using the relations

$$\rho = \sum e \int f \mathrm{d}v$$
$$j_i = \sum e \int v_i \mathrm{d}v \qquad (i = 2,3). \qquad (1.63)$$

Here $v_2 \equiv \boldsymbol{e}_z \cdot \boldsymbol{k} \times \boldsymbol{v}/k$, and $v_3 = v_z$. On the other hand, ρ and j_i are related to the *permittivity tensor* components $\epsilon_{ik}(i,k = 0,2,3)$ by means of (1.16) and (1.19). Allowing for the above-said and using (1.52), we find that for f_0 of the form of (1.29), these components are given by

$$\epsilon_{00} = 1 - \sum \frac{4\pi e^2}{Mk^2} \left\langle \Lambda - \omega G \sum_n \zeta_n J_n^2 \right\rangle$$

$$\epsilon_{02} = -\epsilon_{20} = i \sum \frac{4\pi e^2}{Mk\omega} \left\langle v_\perp G \sum_n \zeta_n J_n J_n' \right\rangle$$

$$\epsilon_{03} = \epsilon_{30} = -\sum \frac{4\pi e^2}{Mk\omega} \left\langle v_z \left(\Lambda - \omega G \sum_n \zeta_n J_n^2 \right) \right\rangle$$

$$\epsilon_{22} = 1 + \sum \frac{4\pi e^2}{M\omega} \left\langle v_\perp^2 G \sum_n \zeta_n J_n'^2 \right\rangle \qquad (1.64)$$

$$\epsilon_{32} = -\epsilon_{23} = i \sum \frac{4\pi e^2}{M\omega} \left\langle v_z v_\perp G \sum_n \zeta_n J_n J_n' \right\rangle$$

$$\epsilon_{33} = 1 + \sum \frac{4\pi e^2}{M\omega^2} \left\langle v_z \left(\Phi + v_z \omega G \sum_n \zeta_n J_n^2 \right) \right\rangle.$$

The summation sign without index represents a summation over all the charge species. The symbol $\langle \ldots \rangle$ means the integration over the *transverse velocity modulus* and over the *longitudinal velocities*

$$\langle \ldots \rangle = 2\pi \int (\ldots) v_\perp \, dv_\perp \, dv_z. \qquad (1.65)$$

Using (1.60), one can find that, for f_0 of the form of (1.32), the expressions for ϵ_{ik} are given by (1.64) with the following substitutions

$$G \to G_1 \qquad \Lambda \to \Lambda_1 \qquad \Phi \to \Phi_1. \qquad (1.66)$$

When f_0 is of the form of (1.36), the relations in (1.64) are modified, according to (1.60), in the following manner. First, the substitutions are done in all the components ϵ_{ik}

$$G \sum_n \to \sum_n G_{2n}. \qquad (1.67)$$

Second, Λ is replaced by Λ_2, and the term with Φ in ϵ_{33} is omitted. Third, an additional term appears in the component ϵ_{22} so that

$$\epsilon_{22} = 1 + \sum \frac{4\pi e^2}{M\omega} \left\langle v_\perp^2 \left(\frac{1}{2B_0\omega} \frac{\partial F_2}{\partial \mu} + \sum_n G_{2n} \zeta_n J_n'^2 \right) \right\rangle. \qquad (1.68)$$

Note that the terms of order $(nk_y/k^2)\partial G/\partial x$ in relations (1.64) for ϵ_{00}, ϵ_{03}, and ϵ_{33} are neglected as compared to G. These terms can be significant in the case of *high-frequency long-wavelength perturbations* (see sections 1.7 and 2.3 of [2] and also section 7.5.2 of this book in detail).

2 Simplified Dispersion Relations

The *general dispersion relation* (1.4) with the *permittivity tensor* (1.64) is extremely complicated. Therefore, we are forced to study only various limiting cases of perturbations described by some simplified variations of (1.4). Deriving such simplified dispersion relations is the goal of the present chapter.

In section 2.1 we deal with perturbations whose *frequencies* are small compared with the *ion cyclotron frequency*, called *low-frequency perturbations*, with respect to ions. Then we discuss those perturbations whose *wave length* is small compared with the *ion Larmor radius*, called *short-wavelength perturbations*, with respect to ions. The *dispersion relations* for such perturbations with $k_z = 0$ and $k_z > \omega/v_{Te}$ are derived in sections 2.2 and 2.3. Finally, in section 2.4 the case of a low-β plasma is considered.

Note that one of the important results of this chapter is the *splitting* of the *general dispersion relation* for *low-frequency long-wavelength perturbations* into two mutually independent dispersion relations. One of them corresponds to the *Alfven* or *inertial perturbations*, while the second corresponds to *magnetoacoustic* and some other perturbations called, in general, *inertialess perturbations*. Such a splitting for arbitrary angles between k and B_0 was revealed in [1.8], and that splitting for the case $k \cdot B_0 = 0$ was shown in [1.6]. This considerably facilitates the analysis of the low-frequency long-wavelength perturbations. In addition, as shown in [2.1, 2.2], the splitting is revealed under the *hydrodynamic description* of perturbations (see also Chapters 9 and 10). Moreover, according to [2.3] (see also [2.4]), the same holds in the case of *large-scale perturbations* which cannot be described by the WKB approach, at least at $k \cdot B_0 = 0$. Therefore, one should treat the results of papers which do not show the splitting with care.

In section 2.1 we use the results of [1.6–1.11, 2.5–2.7].

The *short-wavelength perturbations* with $k_z = 0$ discussed in section 2.2 can be of importance in the study of *low-frequency instabilities* where $\omega \ll \Omega_i$, for *ion-cyclotron instabilities* where $\omega \simeq \Omega_i$, and for *high-frequency ones* where $\omega \gg \Omega_i$.

The main results of section 2.2 were obtained in [2.8, 2.9].

The problem of short-wavelength perturbations with $\omega \ll k_z v_{Te}$ discussed in section 2.3 was originally studied in [2.10] for the case of a *low-β plasma*. Such perturbations in the case of a *finite-β plasma* were considered in [2.11]. The dispersion relation for the *electromagnetic perturbations* of a low-β plasma, section 2.4, has been found in [1.2] (see also [1.3, 2.10]). The kinetic theory of low-frequency drift instabilities in a *relativistic finite-β plasma* was presented in [2.12].

2.1 Permittivity and dispersion relations for low-frequency long-wavelength perturbations

The dispersion relation of the form of (1.1) proves to be more convenient than that of (1.4) for the analysis of *low-frequency perturbations*. In this connection, it is necessary to obtain expressions for the *permittivity tensor* components $\epsilon_{\alpha\beta}$ ($\alpha, \beta = 1, 2, 3$), starting with equations (1.23) and (1.64).

We perform this procedure using the *equilibrium equation* (1.42) and the *quasineutrality condition*. In addition, we assume the equilibrium longitudinal electric current to vanish. Restricting ourselves to f_0 of the form of (1.29), we find for the case of *low-frequency* ($\omega \ll \Omega$) *perturbations* that

$$
\begin{aligned}
\epsilon_{11} &= 1 - \sum \frac{4\pi e^2}{Mk^2}\left\langle G\left[\left(1 - J_0^2\right)\left(1 - \frac{k_z v_z}{\omega}\right)\right.\right. \\
&\quad \left.\left. - \frac{\omega_d}{\omega}\left(1 + J_0^2 - \frac{4J_0 J_1}{\xi^2}\right) - \zeta_0 \frac{\omega_d^2}{\omega}\left(\frac{2J_1}{\xi} - J_0\right)^2\right]\right\rangle \\
\epsilon_{12} &= -\epsilon_{21} = -\mathrm{i}\sum \frac{4\pi e^2}{M\omega}\left\langle \frac{v_\perp^2}{2\Omega}G\left[\frac{2J_0 J_1}{\xi} - 1\right.\right. \\
&\quad \left.\left. - \frac{2J_1}{\xi}\left(\frac{2J_1}{\xi} - J_0\right)\zeta_0\omega_d\right]\right\rangle \\
\epsilon_{13} &= \epsilon_{31} = -\sum \frac{4\pi e^2}{Mk\omega}\left\langle v_z G\left[1 - J_0^2 - \zeta_0\omega_d J_0\left(J_0 - \frac{2J_1}{\xi}\right)\right]\right\rangle.
\end{aligned}
$$
(2.1)

Using (1.64), we obtain that in the *low-frequency approximation* the remaining components $\epsilon_{\alpha\beta}$ contained in (1.1) are equal to

$$\epsilon_{22} = 1 + \sum \frac{4\pi e^2}{M\omega} \left\langle v_\perp^2 G \left(\zeta_0 J_1^2 - \frac{1}{\Omega^2 \zeta_0} \sum_{n\neq 0} \frac{J_n'^2}{n^2} \right) \right\rangle$$

$$\epsilon_{23} = -\epsilon_{32} = i \sum \frac{4\pi e^2}{M\omega} \langle v_z v_\perp \zeta_0 J_0 J_1 \rangle \qquad (2.2)$$

$$\epsilon_{33} = 1 + \sum \frac{4\pi e^2}{M\omega^2} \left\langle v_z \left(\Phi + v_z \omega G \zeta_0 J_0^2 \right) \right\rangle.$$

In section 2.1.1 we shall give the expressions for the *permittivity tensor* in the case of *long-wavelength perturbations*, $k_\perp \rho \ll 1$, where $\rho = v_\perp / \Omega$ is the *Larmor radius*. (We denote the Larmor radius by the same letter as the charge density since both notations are generally adopted. This will not lead to misunderstandings since the corresponding quantities are used in different sections.)

In section 2.1.2 we consider the zeroth approximation in $k_\perp \rho$ and show that in this approximation the above-mentioned *splitting* of the general dispersion relation into two dispersion relations holds. In section 2.1.3 we give the dispersion relations and the expressions for the tensor $\epsilon_{\alpha\beta}$ for the case of low-frequency long-wavelength perturbations in the approximation $k_\perp \rho \to 0$. In section 2.1.4, expressions (2.1) and (2.2) are simplified in the approximation of a *force-free plasma*. In section 2.1.5 the expressions for $\epsilon_{\alpha\beta}$ relevant to the problem of low-frequency long-wavelength perturbations of a plasma with an *inhomogeneous velocity profile* are given.

Since the dispersion relation obtained in section 2.1.2 for the Alfven perturbations does not contain the effects of *resonant interaction* between waves and particles, in section 2.1.6 we derive the generalization of this dispersion relation taking into account small terms of order $k_\perp^2 \rho_i^2$ describing such an interaction.

2.1.1   *Permittivity for long-wavelength low-frequency perturbations*

In the case when the transverse wavelength of the perturbations is large compared with the Larmor radius of the particles, $\xi \ll 1$, the permittivity tensor $\epsilon_{\alpha\beta}$ $(\alpha, \beta = 1, 2, 3)$ determined by relations (2.1) and (2.2) takes the form

$$\epsilon_{11} = \epsilon_{11}^{(0)} \equiv 1 - \frac{4\pi e^2}{M\Omega^2} \left\langle \frac{v_\perp^2}{2} G \left(1 - \frac{k_z v_z}{\omega} - \frac{1}{2} \frac{\omega_{\mathrm{d}}}{\omega} \right) \right\rangle$$

$$\epsilon_{12} = i \frac{4\pi e^2 k_\perp^2}{M\omega\Omega^3} \left\langle \frac{v_\perp^4}{16} G \left(3 + \zeta_0 \omega_{\mathrm{d}} \right) \right\rangle$$

$$\epsilon_{13} = -\frac{4\pi e^2 k_\perp}{M\omega\Omega^2} \left\langle \frac{v_z v_\perp^2}{8} G \left(2 + \zeta_0 \omega_{\mathrm{d}} \right) \right\rangle$$

$$\epsilon_{22} = \sum \frac{4\pi e^2 k_\perp^2}{M\omega\Omega^2} \left\langle \frac{v_\perp^4}{4} G\zeta_0 \right\rangle + \epsilon_{11}^{(0)} \tag{2.3}$$

$$\epsilon_{23} = i \sum \frac{4\pi e^2 k_\perp}{M\omega\Omega} \left\langle \frac{v_z v_\perp^2}{2} G\zeta_0 \right\rangle$$

$$\epsilon_{33} = 1 + \sum \frac{4\pi e^2}{M\omega^2} \left\langle v_z \left(\Phi + v_z \omega G\zeta_0 \right) \right\rangle.$$

In this approximation the components ϵ_{11}, ϵ_{12}, and ϵ_{13} are due only to the contribution of the ions (the ion index in the expressions for these components is omitted for simplicity). Components ϵ_{22}, ϵ_{23}, and ϵ_{33} have contributions from both the ions and the electrons as represented by the summation sign. We note also that only the leading terms in the parameter ζ^2 are kept in each component $\epsilon_{\alpha\beta}$.

The term $\epsilon_{11}^{(0)}$ in the expression for ϵ_{22} is important for the problem of excitation of the *fast magnetoacoustic waves* by *high-energy particles*. Such a problem is discussed in Chapter 5. This term will be neglected until Chapter 5. The term 1 in ϵ_{33} is significant only in the case of a *highly rarefied plasma*. As a rule, this term will also be neglected.

2.1.2 Splitting of the general dispersion relation in the approximation $k_\perp \rho \to 0$

Let us present the dispersion relation (1.1) in the form

$$D_1^{(0)} D_2 + \Delta = 0 \tag{2.4}$$

where

$$D_1^{(0)} = \epsilon_{11}^{(0)} - c^2 k_z^2 / \omega^2 \tag{2.5}$$

$$D_2 = \begin{vmatrix} \epsilon_{22} - N^2 & \epsilon_{23} \\ \epsilon_{32} & \epsilon_{33} \end{vmatrix} \tag{2.6}$$

and Δ denotes the remaining terms of the left-hand side of (1.1). Allowing for (2.3), we conclude that $D_1^{(0)}$ is of zeroth order in $k_\perp$, D_2 is of second order in $k_\perp$, while Δ is composed of terms of orders $k_\perp^4$ and $k_\perp^6$. Consequently, neglecting the small corrections of orders $k_\perp^2$ and $k_\perp^4$, the dispersion relation splits into

$$D_1^{(0)} = 0 \tag{2.7}$$

$$D_2 = 0. \tag{2.8}$$

Equation (2.7) corresponds to the *inertial perturbations* (the *Alfven perturbations*) while (2.8) corresponds to the *inertialess ones*.

In the case of an *isotropic plasma at rest* with a *Maxwellian velocity distribution* of the particles, i.e., when F has the form of (1.31), the components $\epsilon_{\alpha\beta}$ in (2.5) and (2.6) are given by

$$\epsilon_{11}^{(0)} = \frac{c^2}{c_{\mathrm{A}}^2}\left(1 - \frac{\omega_{pi}^*}{\omega} - \frac{\omega_{\mathrm{Di}}}{\omega} + \sigma\frac{\omega_{pi}^*\omega_{\mathrm{Di}}}{\omega^2}\right)$$

$$\epsilon_{22} = -\sum \frac{k_\perp^2}{k_y^2 \kappa_B^2 \lambda_{\mathrm{D}}^2 \omega}\hat{l}\langle\omega_{\mathrm{d}}^2\zeta_0\rangle$$

$$\epsilon_{23} = -\epsilon_{32} = -i\sum\frac{k_\perp}{k_y \kappa_B \lambda_{\mathrm{D}}^2 \omega}\hat{l}\langle v_z\omega_{\mathrm{d}}\zeta_0\rangle$$

$$\epsilon_{33} = -\sum\frac{1}{\lambda_{\mathrm{D}}^2 \omega}\hat{l}\langle v_z^2\zeta_0\rangle.$$

(2.9)

Here

$$c_{\mathrm{A}}^2 = \frac{B_0^2}{4\pi M_i n_0} \qquad\qquad \sigma = \frac{1 + 2\eta_i}{1 + \eta_i}$$

$$\omega_{pi}^* = \frac{ck_y}{e_i n_0 B_0}\frac{\partial p_i}{\partial x} \qquad \eta_i = \frac{\partial \ln T_i}{\partial \ln n_0}$$

$$\omega_{\mathrm{Di}} = \frac{cT_i k_y}{e_i B_0}\frac{\partial \ln B_0}{\partial x} \qquad \lambda_{\mathrm{D}}^2 = \frac{T}{4\pi e^2 n_0}.$$

(2.10)

Note that c_{A} is the *Alfven velocity*, ω_{pi}^* is the so-called *ion pressure-gradient drift frequency*, ω_{Di} is the *average ion magnetic-drift frequency*, and λ_{D} is the *Debye length*.

The operator $\hat{l}$ is defined by

$$\hat{l} = 1 - \frac{k_y T}{M\omega\Omega}\left(\frac{\partial \ln n_0}{\partial x} + \frac{\partial T}{\partial x}\frac{\partial}{\partial T}\right).$$

(2.11)

The symbol $\langle\ldots\rangle$ means averaging over the Maxwellian distribution

$$\langle\ldots\rangle = \left(\frac{M}{2\pi T}\right)^{1/2}\frac{M}{T}\int_{-\infty}^{\infty}\exp\left(-\frac{Mv_z^2}{2T}\right)\mathrm{d}v_z$$

$$\times \int_0^\infty\exp\left(-\frac{Mv_\perp^2}{2T}\right)(\ldots)v_\perp\mathrm{d}v_\perp.$$

(2.12)

The subscript i marks quantities corresponding to the ions.

2.1.3   *Dispersion relations for low-frequency long-wavelength perturbations with $k_z = 0$*

One of the simplest types of perturbations is that with $k_z = 0$. Similarly to section 2.1.2, we consider such perturbations to be of *low frequency* and *long wavelength*. Then instead of two dispersion relations (2.7) and (2.8), we have

$$\epsilon_{11}^{(0)} = 0 \tag{2.13}$$

$$\epsilon_{22} - N^2 = 0 \tag{2.14}$$

$$\epsilon_{33} = 0. \tag{2.15}$$

The last two equations can be found from (2.8) and (2.6) using the fact that $\epsilon_{23} = 0$ for $k_z = 0$.

When $k_z = 0$, the components $\epsilon_{11}^{(0)}$, ϵ_{22}, and ϵ_{33} are also simplified. Thus, according to (2.3), for f_0 of the form of (1.29) they are equal to

$$\epsilon_{11}^{(0)} = 1 - \frac{4\pi e^2}{M\Omega^2}\left\langle \frac{v_\perp^2}{2}\left(\frac{\partial F}{\partial \mathcal{E}} + \frac{k_y}{\omega\Omega}\frac{\partial F}{\partial x}\right)\left(1 - \frac{\omega_d}{2\omega}\right)\right\rangle$$

$$\epsilon_{22} = \sum \frac{4\pi e^2 k_\perp^2}{M\omega\Omega^2}\left\langle \frac{v_\perp^4}{4(\omega - \omega_d)}\left(\frac{\partial F}{\partial \mathcal{E}} + \frac{k_y}{\omega\Omega}\frac{\partial F}{\partial x}\right)\right\rangle \tag{2.16}$$

$$\epsilon_{33} = \sum \frac{4\pi e^2}{M\omega^2}\left\langle v_z\left[\frac{\partial F}{\partial v_z} + \frac{\omega v_z}{\omega - \omega_d}\left(\frac{\partial F}{\partial \mathcal{E}} + \frac{k_y}{\omega\Omega}\frac{\partial F}{\partial x}\right)\right]\right\rangle.$$

When F is of the form of (1.31), the first equation of (2.16) yields the former result of equation (2.9) for $\epsilon_{11}^{(0)}$, while (2.9) as well as (2.16) yield the following expressions for ϵ_{22} and ϵ_{33}:

$$\epsilon_{22} = -\sum \frac{k_\perp^2}{k_y^2\kappa_B^2\lambda_D^2\omega}\hat{i}\left\langle \frac{\omega_d^2}{\omega - \omega_d}\right\rangle \tag{2.17}$$

$$\epsilon_{33} = -\sum \frac{1}{\lambda_D^2\omega}\hat{i}\left(\frac{T}{M}\left\langle\frac{1}{\omega - \omega_d}\right\rangle\right). \tag{2.18}$$

Here, in contrast to (2.12), the symbol $\langle\ldots\rangle$ means averaging only over transverse velocities

$$\langle\ldots\rangle = \frac{M}{T}\int_0^\infty \exp\left(-\frac{Mv_\perp^2}{2T}\right)(\ldots)v_\perp dv_\perp. \tag{2.19}$$

Equations (2.13) and (2.14) for the case of Maxwellian velocity distributions with $T_e = T_i$ were first obtained by Tserkovnikov [1.4]. The perturbations of type (2.13) themselves represent a limiting case of the Alfven perturbations. In this connection, these perturbations as well as those of a more general type described by the dispersion relations (2.7) with $D_1^{(0)}$ of the form of (2.5) are the *Alfven perturbations*. The perturbations of type (2.14) are called *Tserkovnikov's perturbations* or *magneto-drift waves* with $k_z = 0$. Equation (2.15) corresponds to the perturbations with the electric field transverse to the wave vector ($\boldsymbol{E} \parallel \boldsymbol{B}_0 \perp \boldsymbol{k}$). These perturbations are called the *transverse drift waves*. For the case of a *Maxwellian low-β plasma* these perturbations were originally studied in [2.5] in the approximation $\omega_\mathrm{d} = 0$ and in [1.12] for $\omega_\mathrm{d} \neq 0$.

2.1.4 *Permittivity of a force-free plasma*

The *force-free high-β plasma* is an important particular kind of magnetized inhomogeneous plasma. We describe plasma as force-free if $\nabla p_\perp = 0$. In this case, according to (1.42), $\nabla B_0 = 0$ and, according to (1.39), $\boldsymbol{j}_0 = 0$. Naturally, these conditions can be satisfied for arbitrary β, but we shall be interested mainly in the case where $\beta \gg 1$. This case can be realized in systems where the plasma is confined with the help of a *dense blanket of neutral gas*, whereas the magnetic field is used only to reduce the *transport coefficients*.

According to (1.27), when $\nabla B_0 = 0$, the particle magnetic drift vanishes ($V_\mathrm{d} = 0$). As a result, the permittivity tensor $\epsilon_{\alpha\beta}$ can be substantially simplified. In addition, we assume the plasma to be Maxwellian and the relative *temperature gradients* of electrons and ions to be the same, $\partial \ln T_\mathrm{e}/\partial \ln T_\mathrm{i} = 1$. Then additional simplifications in the tensor $\epsilon_{\alpha\beta}$ can be performed using the condition $\partial \ln T/\partial \ln n_0 = -1$ arising from the requirement $\nabla p_\perp = 0$.

In particular, for *low-frequency perturbations*, $\omega \ll \Omega$, we find from (2.1) and (2.2), with the above assumptions and allowing for (1.31), that

$$
\epsilon_{11} = \sum \hat{l}^0 \frac{\omega_p^2}{\Omega^2} \frac{1 - I_0 \exp(-z)}{z}
$$

$$
\epsilon_{12} = -\mathrm{i} \sum \hat{l}^0 \frac{\omega_p^2}{\omega\Omega}(I_0 - I_1)\exp(-z)
$$

$$
\epsilon_{13} = 0
$$

$$
\epsilon_{22} = 2\mathrm{i}\pi^{1/2} \sum \hat{l}^0 \frac{\omega_p^2}{\omega^2} xWz(I_0 - I_1)\exp(-z) \qquad (2.20)
$$

$$
\epsilon_{23} = \mathrm{i} \sum \hat{l}^0 \frac{k_\perp}{k_z} \frac{\omega_p^2}{\omega\Omega}(1 + \mathrm{i}\pi^{1/2}xW)(I_0 - I_1)\exp(-z)
$$

$$
\epsilon_{33} = \sum \hat{l}^0 \frac{2\omega_p^2}{\Omega^2} x^2(1 + \mathrm{i}\pi^{1/2}xW)I_0 \exp(-z).
$$

Here

$$
\hat{l}^0 A(n_0, T) \equiv A - \frac{\omega_T}{\omega}\left(\frac{\partial}{\partial T}(TA) - n_0\frac{\partial A}{\partial n_0}\right) \qquad (2.21)
$$

where A is some function of density and temperature, and ω_T is the so-called *temperature-gradient drift frequency* defined by

$$\omega_T = \frac{ck_y}{eB_0}\frac{\partial T}{\partial x}. \tag{2.22}$$

Note that, when $\nabla(n_0 T) = 0$, we have $\omega_T = -\omega_n$ where ω_n is the so-called *density-gradient drift frequency* defined by

$$\omega_n = \frac{ck_y T}{eB_0 n_0}\frac{\partial n_0}{\partial x}. \tag{2.23}$$

The remaining designations in (2.20) are: $\omega_p^2 = 4\pi e^2 n_0/M$ is the square of the *plasma frequency* of corresponding particle species, $z = k_\perp^2 T/M\Omega^2$, $I_0(z)$ and $I_1(z)$ are the Bessel functions of imaginary argument of first kind, $x = \omega/\mid k_z \mid v_T$, $v_T = (2T/M)^{1/2}$ is the *thermal velocity*, and $W(x)$ is the error function for a complex argument (the *Kramp function*) defined by

$$W(x) = \exp(-x^2)\left(1 + 2\mathrm{i}\pi^{-1/2}\int_0^x \exp(t^2)\mathrm{d}t\right). \tag{2.24}$$

Note that $W(x)$ is related to the so-called *plasma dispersion function* $Z(x)$ by the expression (see [1], equation (2.46))

$$Z(x) = \mathrm{i}\pi^{1/2}xW(x). \tag{2.25}$$

Equations (2.20) were obtained in [2.6]. They will be used in Chapter 12.

2.1.5 *Permittivity of a plasma flow with an inhomogeneous velocity profile*

Now we assume that a plasma moves *along* the magnetic field and its *longitudinal velocity* is *inhomogeneous*. In the case of an isotropic Maxwellian velocity distribution, such a plasma is characterized by a distribution function of the form of (1.30) with $U' \equiv dU/dx \neq 0$. We shall find the expressions for the permittivity tensor components $\epsilon_{\alpha\beta}$ assuming for simplicity that the gradients of density, temperature and magnetic field vanish, $\nabla n_0 = \nabla T = \nabla B_0 = 0$.

Let us restrict ourselves to the case of *long-wavelength perturbations*. Then the dispersion relation reduces to equations (2.7) and (2.8) so that only the components $\epsilon_{11}^{(0)}$, ϵ_{22}, ϵ_{23}, and ϵ_{33} are necessary for us. Starting with (2.3) and allowing for (1.30), we find that, under these assumptions

$$\epsilon_{11}^{(0)} = 1 + \frac{c^2}{c_A^2}\left(1 + \frac{k_y k_z T_i U'}{M_i \Omega_i \tilde{\omega}^2}\right)$$

$$\epsilon_{22} = \sum \frac{\omega_p^2}{\omega^2} \frac{k_\perp^2 v_T^2}{\Omega^2}\left[(1-\alpha)Z(\tilde{x}) - \alpha\right]$$

$$\epsilon_{23} = i\sum \frac{\omega_p^2 k_\perp}{\omega\Omega k_z}(1-\alpha)(1 + Z(\tilde{x})) \qquad (2.26)$$

$$\epsilon_{33} = \sum \frac{M\omega_p^2}{Tk_z^2}(1-\alpha)(1 + Z(\tilde{x})).$$

Here $\alpha = k_y U'/k_z \Omega$, $\tilde{x} = (\omega - k_z U)/\mid k_z \mid v_T$, $\tilde{\omega} = \omega - k_z U$.

The relations in (2.26) were obtained in [2.7]. They will be used in Chapter 13.

2.1.6 Allowance for the interaction of Alfven perturbations with resonant particles

In contrast to (2.8), equation (2.7) does not contain the terms corresponding to *resonant interaction* between waves and particles. In order to take into account such an interaction we should, first, augment ϵ_{11} by terms of order ξ^2 and, second, allow for the contribution in (2.4) by Δ. Then (2.7) is replaced by

$$D_1^{(0)} + \epsilon_{11}^{(1)} + \Delta^{(0)}/D_2 = 0 \qquad (2.27)$$

where $\epsilon_{11}^{(1)}$ is a correction to ϵ_{11} of order ξ^2, and $\Delta^{(0)}$ denotes the leading terms of expansion of Δ in powers of $k_\perp^2$.

For subsequent calculations of the growth rate due to resonant interaction, we need to know only the imaginary part of $\epsilon_{11}^{(1)}$. According to (2.1), it is equal to

$$\mathrm{Im}\,\epsilon_{11}^{(1)} = \frac{4\pi e^2}{M\omega\Omega^2}\mathrm{Im}\left\langle \mathcal{E}_\perp G\zeta_0\omega_d^2\frac{\xi^2}{32}\right\rangle. \qquad (2.28)$$

Note: the right-hand side of this equation represents the contribution due to the ions; the ion indices are here omitted for simplicity; as before, $\mathcal{E}_\perp = v_\perp^2/2$.

We find the expression for $\Delta^{(0)}$, taking into account equation (2.7), in the zeroth approximation in ξ^2

$$\Delta^{(0)} = -C_1 D_2 - C_2. \qquad (2.29)$$

Here

$$C_1 = \frac{k_\perp^2}{k_z^2 \epsilon_{33}} \left(\epsilon_{11}^{(0)} + \frac{k_z}{k_\perp} \epsilon_{13} \right)^2 \qquad (2.30)$$

$$C_2 = -\frac{k_\perp^2}{k_z^2 \epsilon_{33}} \left[\left(\epsilon_{11}^{(0)} + \frac{k_z}{k_\perp} \epsilon_{13} \right) \epsilon_{23} + \frac{k_z}{k_\perp} \epsilon_{12} \epsilon_{33} \right]^2. \qquad (2.31)$$

Then equation (2.27) reduces to

$$D_1^{(0)} + \epsilon_{11}^{(1)} - C_1 - C_2/D_2 = 0. \qquad (2.32)$$

Using (2.32), we find the general expression for the *growth rate* $\gamma \equiv$ Im ω of the *resonant interaction* between the *Alfven perturbations* and the particles

$$\gamma = -\frac{1}{\partial D_1^{(0)}/\partial \omega} \mathrm{Im}\, (\epsilon_{11}^{(1)} - C_1 - C_2/D_2). \qquad (2.33)$$

When $k_z = 0$, equation (2.33) is replaced by

$$\gamma = -\frac{1}{\partial D_1^{(0)}/\partial \omega} \mathrm{Im}\, \left(\epsilon_{11}^{(1)} + \frac{\epsilon_{12}^2}{\epsilon_{22} - N^2} \right). \qquad (2.34)$$

Relations (2.33) and (2.34) will be used in Chapters 4 and 5.

2.2 Dispersion relations for short-wavelength perturbations with $k_z = 0$

Let us now consider how the *general dispersion relations* will be simplified if the *transverse wavelength* of perturbations is *small* compared with the ion Larmor radius, $k_\perp \rho_i \gg 1$, but *larger* than or *comparable* with the electron one, $k_\perp \rho_e \lesssim 1$. The perturbation frequencies are assumed to be *small* compared with the electron cyclotron frequency, $\omega \ll \Omega_e$, but *arbitrary* with respect to the ion one. Therefore, the dispersion relations given below can be used in the cases of *low-frequency perturbations* where $\omega \ll \Omega_i$, *ion cyclotron perturbations* where $\omega \simeq \Omega_i$, and *high-frequency* ones where $\omega \gg \Omega_i$.

We restrict ourselves to perturbations with $k_z = 0$. In this case equation (1.1) *splits* into two dispersion relations, one of which corresponds to perturbations with $\boldsymbol{E} \perp \boldsymbol{B}_0$ and the second to those with $\boldsymbol{E} \parallel \boldsymbol{B}_0$. We shall consider their structure separately.

2.2.1 *Perturbations with $\boldsymbol{E} \perp \boldsymbol{B}_0$*

In the case where $k_z = 0$ and $\boldsymbol{E} \perp \boldsymbol{B}_0$ the dispersion relation (1.1) reduces to

$$\begin{vmatrix} \epsilon_{11} & \epsilon_{12} \\ \epsilon_{21} & \epsilon_{22} - N^2 \end{vmatrix} = 0. \tag{2.35}$$

Let us give the expressions for $\epsilon_{\alpha\beta}$ contained in (2.35) assuming the electrons to be *Maxwellian*. Starting with (1.64) and using (1.23), we find (as mentioned previously, the primes at $\epsilon_{\alpha\beta}$ have been omitted for simplicity)

$$\epsilon_{11} = \epsilon_{11}^{(0)} + \epsilon_{11}^{(1)} + \epsilon_{11}^{(2)} + \delta\epsilon_{11}. \tag{2.36}$$

Here

$$\epsilon_{11}^{(0)} = -\frac{4\pi e^2}{M_i k^2} \int \left(\frac{\partial F_{\perp i}}{\partial \mathcal{E}_\perp} + \frac{k_y}{\omega \Omega_i} \frac{\partial F_{\perp i}}{\partial x} \right) \left(1 - \frac{\omega_{di}}{\omega} \right) v_\perp dv_\perp$$

$$\epsilon_{11}^{(1)} = \frac{1}{k^2 \lambda_{De}^2} \hat{l}_e \left[1 - I_0 \exp(-z_e) - \frac{\omega_{De}}{\omega} \left(1 - \frac{2I_1}{z_e} \exp(-z_e) \right) \right]$$

$$\epsilon_{11}^{(2)} = -\frac{1}{k^2 \lambda_{De}^2 \omega} \hat{l}_e \left\langle \frac{\omega_{de}^2}{\omega - \omega_{de}} \left(J_0 - \frac{2J_1}{\xi} \right)^2 \right\rangle$$

$$\delta\epsilon_{11} = \frac{4\pi e^2 \omega}{M_i k^2} \sum_{n=-\infty}^{\infty} \int \left(\frac{\partial F_{\perp i}}{\partial \mathcal{E}_\perp} + \frac{k_y}{\Omega_i \omega} \frac{\partial F_{\perp i}}{\partial x} \right) \frac{J_n^2 d\mathcal{E}_\perp}{\omega - \omega_{di} - n\Omega_i}$$

$$\tag{2.37}$$

where $F_{\perp i} \equiv F_{\perp i}(\mathcal{E}_\perp) = \int F_i dv_z$ is the *ion distribution function* over the *transverse energies* (per unit mass), $\mathcal{E}_\perp = v_\perp^2/2$, $I_0(z_e)$ and $I_1(z_e)$ are the Bessel functions of the imaginary argument of the first kind, $z_e = k_\perp^2 T_e / M_e \Omega_e^2$, J_0 and J_1 are the usual Bessel functions of argument $\xi = k_\perp v_\perp / \Omega_e$ and J_n is the Bessel function of argument $\xi_i = k_\perp v_\perp / \Omega_i$. The remaining notations are clear from the above presentation. In particular, the values ω_{de} and ω_{di} are clear from section 1.3, and ω_{Di} from (2.10).

The component ϵ_{12} has the form

$$\epsilon_{12} = \epsilon_{12}^{(0)} + \epsilon_{12}^{(1)}. \tag{2.38}$$

Here

$$\epsilon_{12}^{(0)} = i \frac{4\pi e_e c n_e}{B_0 \omega} \left(1 - \frac{\omega_{pi}^*}{\omega} \right)$$

$$\epsilon_{12}^{(1)} = \frac{i}{k_y \kappa_B \lambda_{De}^2} \hat{l}_e \left[\frac{\omega_{De}}{\omega} \left(\frac{2}{z_e} I_1 \exp(-z_e) - 1 \right) \right. \tag{2.39}$$

$$\left. + \left\langle \frac{\omega_{de}}{\omega - \omega_{de}} \frac{2J_1}{\xi} \left(\frac{2J_1}{\xi} - J_0 \right) \right\rangle \right]$$

where $\omega_{pi}^* = (ck_y/e_i B_0 n_0)\partial p_{\perp i}/\partial x$ (cf. equation (2.10)), and $p_\perp$ is determined by (1.40).

Finally,

$$\epsilon_{22} = -\frac{1}{\kappa_B^2 \lambda_{De}^2 \omega} \hat{l}_e \left\langle \frac{\omega_{de}^2}{\omega - \omega_{de}} \frac{4J_1^2}{\xi^2} \right\rangle. \qquad (2.40)$$

We take into account the *ion-cyclotron effects* only in ϵ_{11} since $2J_n' J_n/\xi$ and $4J_n'^2/\xi^2$ are small compared with J_n^2 for large arguments of the ion Bessel functions (see equations (1.64)). We also note that, when the *cyclotron effects* described by $\delta\epsilon_{11}$ are ignored, the above expressions for $\epsilon_{\alpha\beta}$ can be found from equations (2.1) and (2.2).

When the transverse wavelength is large compared with the electron Larmor radius, $k\rho_e \ll 1$, the expressions for $\epsilon_{11}^{(1)}, \epsilon_{11}^{(2)}, \epsilon_{12}^{(1)}$, and ϵ_{22} are reduced to

$$
\begin{aligned}
\epsilon_{11}^{(1)} &= \frac{1}{k^2 \lambda_{De}^2} \hat{l}_e \left[z_e \left(1 - \frac{\omega_{De}}{\omega} \right) \right] \\
\epsilon_{11}^{(2)} &= -\frac{1}{k^2 \lambda_{De}^2} \hat{l}_e \left\langle \frac{\xi^4}{64} \frac{\omega_{de}^2}{\omega - \omega_{de}} \right\rangle \\
\epsilon_{12}^{(1)} &= \frac{i}{k_y \kappa_B \lambda_{De}^2} \hat{l}_e \left(-\frac{3}{2} z_e \frac{\omega_{De}}{\omega} + \left\langle \frac{\xi^2}{8} \frac{\omega_{de}}{\omega - \omega_{de}} \right\rangle \right) \\
\epsilon_{22} &= -\frac{1}{\kappa_B^2 \lambda_{De}^2 \omega} \hat{l}_e \left\langle \frac{\omega_{de}^2}{\omega - \omega_{de}} \right\rangle.
\end{aligned}
\qquad (2.41)
$$

Note that in this approximation, $\epsilon_{11}^{(1)}$ coincides with $\epsilon_{11}^{(0)}$ determined by (2.9) to an accuracy of the replacement of electron indices by ion indices (quantitatively, the ratio of these values is of order M_e/M_i). The relationship between $\epsilon_{11}^{(2)}$ and $\epsilon_{12}^{(1)}$ determined by (2.41), on the one hand, and $\epsilon_{11}^{(1)}$ and ϵ_{12} determined by (2.28) and (2.3) on the other hand, is the same. The component ϵ_{22} in (2.41) corresponds to the electron part of ϵ_{22} in the fourth equation of (2.3). Note also that, in the case of *short-wavelength perturbations* discussed, the parts of $\epsilon_{\alpha\beta}$ equivalent to those in the theory of *long-wavelength perturbations* are characterized by the *shifted superscripts*: $(0) \to (1)$, $(1) \to (2)$. This is due to the fact that in the long-wavelength approximation the leading terms in ϵ_{11} and ϵ_{12} prove to be cancelled under summation of the electron and ion contributions while in the short-wavelength approximation there is no such cancellation.

When the *ion velocity distribution* is *Maxwellian*, the expressions for $\epsilon_{11}^{(0)}, \delta\epsilon_{11}, \epsilon_{12}^{(0)}$ have the form

$$\epsilon_{11}^{(0)} = \frac{1}{k^2 \lambda_{Di}^2} \hat{l}_i \left(1 - \frac{\omega_{Di}}{\omega} \right)$$

$$\delta\epsilon_{11} = -\frac{\omega}{k^2 \lambda_{Di}^2} \hat{l}_i \sum_{n=-\infty}^{\infty} \left\langle \frac{J_n^2}{\omega - \omega_{di} - n\Omega_i} \right\rangle \qquad (2.42)$$

$$\epsilon_{12}^{(0)} = -\frac{i}{k_y \kappa_B \lambda_{Di}^2} \hat{l}_i \frac{\omega_{di}}{\omega}.$$

In the case of such a distribution the expression for ϵ_{22} in the *long-wavelength approximation* (with respect to electrons) can be presented in the form

$$\epsilon_{22} = N^2 - (\kappa_B \lambda_{Di})^{-2} \hat{l}_i \omega_{Di}/\omega - (k/\kappa_B) I^{(e)} \qquad (2.43)$$

where

$$I^{(e)} = (k\kappa_B \lambda_{De}^2)^{-1} \hat{l}_e \left\langle \omega_{de}/(\omega - \omega_{de}) \right\rangle. \qquad (2.44)$$

When the ion distribution is *arbitrary*, (2.43) is replaced by

$$\epsilon_{22} = N^2 + \frac{1}{\kappa_B^2 \lambda_{De}^2} \frac{\omega_{De}}{\omega} \left(1 - \frac{\omega_{pi}^*}{\omega} \right) - \frac{k}{\kappa_B} I^{(e)}. \qquad (2.45)$$

In the case of a *loss-cone distribution*, i.e., when $F_{\perp i}(0) = 0$, the expression for $\epsilon_{11}^{(0)}$ is essentially modified. In this case it is convenient to present this quantity in the form

$$\epsilon_{11}^{(0)} = -(k\lambda_{De})^{-2} \omega_{ne}/\omega. \qquad (2.46)$$

The expression for $\delta\epsilon_{11}$ can be yet more simplified if the *transverse wavelength is small compared with the ion Larmor radius*, $\xi_i \gg 1$. In this case we can consider $J_n^2(\xi_i) \approx (\pi |\xi_i|)^{-1}$ so that (2.37) yields

$$\delta\epsilon_{11} = \frac{4e^2 |\Omega_i| \omega}{M_i k^3} \sum_{n=-\infty}^{\infty} \int \left(\frac{\partial F_{\perp i}}{\partial \mathcal{E}_\perp} + \frac{k_y}{\Omega_i \omega} \frac{\partial F_{\perp i}}{\partial x} \right) \frac{dv_\perp}{\omega - \omega_{di} - n\Omega_i}.$$

$$(2.47)$$

In the approximation of a *low-β plasma* ($\beta \to 0, \omega_{di} \to 0$) and under a *Maxwellian ion velocity distribution* with $\nabla T = 0$, equation (2.47) reduces to

$$\delta\epsilon_{11} = -\omega(k\lambda_{Di})^{-2} \hat{l}_i (2\pi z_i)^{-1/2} \sum_n (\omega - n\Omega_i)^{-1} \qquad (2.48)$$

whereas in the case of a *loss-cone distribution*, when $F_{\perp i}(0) = 0$, we have

$$\delta\epsilon_{11} = \frac{\omega_{pi}\omega\,|\,\Omega_i\,|}{\pi k^3}\left(\langle v_\perp^{-3}\rangle + \frac{k_y\kappa_n}{\omega\Omega_i}\langle v_\perp^{-1}\rangle\sum_n(\omega - n\Omega_i)^{-1}\right). \quad (2.49)$$

Here $\omega_{pi}^2 = 4\pi e^2 n_0/M_i$ is the square of the *ion plasma frequency*,

$$\begin{aligned}
\langle v_\perp^{-3}\rangle &= \int (F_{\perp i}/v_\perp^2)\mathrm{d}v_\perp/n_0 \\
\langle v_\perp^{-1}\rangle &= \int (F_{\perp i})\mathrm{d}v_\perp/n_0.
\end{aligned} \qquad (2.50)$$

In obtaining (2.49), we considered that only the density is inhomogeneous so that $\kappa_n = \partial\ln n_0/\partial x$.

If β is not too small, ω_{di} can be sufficiently large, $\omega_{di} \gtrsim \Omega_i$. In this case the ion contribution to (2.47) is due to a *large number of n-terms in the series*. Then we can approximately write

$$(\omega - \omega_{di} - n\Omega_i)^{-1} \approx -\frac{i\pi}{|\,\Omega_i\,|}\delta\left(n - \frac{\omega - \omega_{di}}{\Omega_i}\right) \qquad (2.51)$$

and can change from *summation* over n to *integration* in equation (2.47). As a result, we find from (2.48) and (2.49), respectively, that

$$\delta\epsilon_{11} = \frac{i\pi^{1/2}\omega}{k^3\lambda_{Di}^2 v_{Ti}}\left(1 - \frac{\omega_{ni}}{\omega}\right) \qquad (2.52)$$

$$\delta\epsilon_{11} = -i\frac{\omega_{pi}^2\omega}{k^3}\left(\langle v_\perp^{-3}\rangle + \frac{k_y\kappa_n}{\omega\Omega_i}\langle v_\perp^{-1}\rangle\right). \qquad (2.53)$$

When β is sufficiently low and ω_{di} is unimportant, equations (2.52) and (2.53) can also be obtained from (2.48) and (2.49) if the imaginary part of the frequency is considered to be large enough, $\gamma \gtrsim \Omega_i$.

Finally, relations (2.49) and (2.53) can be simplified if the ion *transverse velocity distribution* is assumed to be a *δ-function*

$$F = n_0 v_0^{-1}\delta(v_\perp - v_0) \qquad (2.54)$$

with the velocity v_0 independent of the coordinate x. In this case the above-mentioned relations mean, respectively

$$\delta\epsilon_{11} = \frac{\omega_{pi}^2\omega\,|\,\Omega_i\,|}{k^3 v_0^3}\left(1 + \frac{\omega_{ni}}{\omega}\right)\sum_n(\omega - n\Omega_i)^{-1} \qquad (2.55)$$

$$\delta\epsilon_{11} = -i(kv_0)^{-3}\omega_{pi}^2\omega(1 + \omega_{ni}/\omega) \qquad (2.56)$$

where $\omega_{ni} = k_y \kappa_n v_0^2/\Omega_i$.

Note that assuming the velocity distribution of the ions to be a δ-function, we reinforce the effects of their *thermodynamical non-equilibrium*. Therefore, the stabilization criteria obtained in this approximation are sufficient for all ion velocity distributions.

The above relations were demonstrated in [2.8, 2.9]. They will be used in Chapter 6.

2.2.2 Perturbations with $E \parallel B_0$

As follows from (1.1), the perturbations with $k_z = 0$ and $E \parallel B_0$ are described by the dispersion relation

$$\epsilon_{33} - N^2 = 0. \tag{2.57}$$

According to (1.64), ϵ_{33} can be presented in the form

$$\epsilon_{33} = \epsilon_{33}^{(0)} + \delta\epsilon_{33}. \tag{2.58}$$

Here $\epsilon_{33}^{(0)}$ corresponds to the *low-frequency* part of ϵ_{33} obtained by ignoring the *ion-cyclotron effects*. This quantity is determined by the right-hand side of the last equation of (2.2) for $k_z = 0$, i.e., by

$$\epsilon_{33}^{(0)} = \sum \frac{4\pi e^2}{M\omega^2}\left\langle v_z\left[\frac{\partial F}{\partial v_z} + \frac{\omega v_z}{\omega - \omega_d}\left(\frac{\partial F}{\partial \mathcal{E}} + \frac{k_y}{\omega\Omega}\frac{\partial F}{\partial x}\right)J_0^2\right]\right\rangle. \tag{2.59}$$

The quantity $\delta\epsilon_{33}$ corresponds to the *cyclotron contribution* of the ions. According to (1.64)

$$\delta\epsilon_{33} = \frac{4\pi e^2}{M_i\omega}\left\langle v_z^2\left(\frac{\partial F_i}{\partial \mathcal{E}} + \frac{k_y}{\omega\Omega_i}\frac{\partial F_i}{\partial x}\right)\sum_{n\neq 0}\frac{J_n^2}{\omega - \omega_{di} - n\Omega_i}\right\rangle. \tag{2.60}$$

When the electron and ion distributions are *Maxwellian*, relations (2.59) and (2.60) mean (cf. equation (2.9))

$$\epsilon_{33}^{(0)} = -\sum \frac{1}{\lambda_D^2\omega}\hat{i}\left\langle\frac{v_z^2 J_0^2}{\omega - \omega_d}\right\rangle$$
$$\delta\epsilon_{33} = -\frac{1}{\lambda_{Di}^2\omega}\hat{i}\left\langle\sum_{n\neq 0}\frac{v_z^2 J_n^2}{\omega - \omega_{di} - n\Omega_i}\right\rangle. \tag{2.61}$$

The results of this section will be used in section 6.6.

2.3 Oblique short-wavelength perturbations

2.3.1 Low-frequency perturbations

Let us discuss how the *general dispersion relation* is simplified in the case where $\omega \ll \Omega_i, 1/\rho_i \ll k_\perp \ll 1/\rho_e$, and $\omega/k_z v_{Te} \ll 1$, assuming the plasma to be *Maxwellian*.

Taking into account the small terms of order $\omega/k_z v_{Te}$, we present the *permittivity tensor* ϵ_{ik} ($i, k = 0, 2, 3$) defined by (1.64) in the form

$$\epsilon_{ik} = \epsilon_{ik}^{(0)} + \mathrm{i}\pi^{1/2}\frac{\omega}{|k_z| v_{Te}}c_{ik} \qquad (2.62)$$

where the terms $\epsilon_{ik}^{(0)}$ are leading while those with c_{ik} are correctional. We find $\epsilon_{ik}^{(0)}$ and c_{ik} using (1.64). As a result, we have the following expressions for $\epsilon_{ik}^{(0)}$:

$$
\begin{aligned}
\epsilon_{00}^{(0)} &= \frac{1}{k^2 \lambda_{Di}^2}\left(1 + \frac{T_i}{T_e}\right) \\
\epsilon_{02}^{(0)} &= 0 \\
\epsilon_{03}^{(0)} &= \frac{1}{k_z k \lambda_{De}^2}\left(1 - \frac{\omega_{ne}}{\omega}\right) \\
\epsilon_{22}^{(0)} &= 0 \\
\epsilon_{23}^{(0)} &= \frac{\mathrm{i}\omega_{pe}^2}{\Omega_e \omega}\frac{k}{k_z}\left(1 - \frac{\omega_{pe}^*}{\omega}\right) \\
\epsilon_{33}^{(0)} &= \frac{1}{k_z^2 \lambda_{De}^2}\left[1 - \frac{\omega_{ne}}{\omega} - \frac{\omega_{De}}{\omega}\left(1 - \frac{\omega_{pe}^*}{\omega}\right)\right].
\end{aligned}
\qquad (2.63)
$$

Here $\omega_{pe}^*, \omega_{De}$, and ω_{ne} are determined by equations (2.10) and (2.23) with the substitutions $p_i \to p_e, (T_i, T_e) \to T_e, (e_i, e) \to e_e$.

Similarly,

$$
\begin{aligned}
c_{00} &= \frac{\chi_1}{k_\perp^2 \lambda_{De}^2} & c_{03} &= \frac{1}{kk_z \lambda_{De}^2}\left(\chi_1 - \frac{\omega_{De}}{\omega}\chi_2\right) \\
c_{02} &= \mathrm{i}\frac{\omega_{pe}^2 \chi_2}{\Omega_e \omega} & c_{23} &= \mathrm{i}\frac{\omega_{pe}^2 k_\perp}{\Omega_e \omega k_z}\left(\chi_2 - 2\frac{\omega_{De}}{\omega}\chi_3\right) \\
c_{22} &= \beta_e N^2 \chi_3 & c_{33} &= \frac{1}{k_z^2 \lambda_{De}^2}\left(\chi_1 - 2\frac{\omega_{De}}{\omega}\chi_2 + 2\frac{\omega_{De}^2}{\omega^2}\chi_3\right)
\end{aligned}
\qquad (2.64)
$$

where

$$\chi_1 = 1 - \frac{\omega_{ne}}{\omega}\left(1 - \frac{\eta_e}{2}\right)$$

$$\chi_2 = 1 - \frac{\omega_{ne}}{\omega}\left(1 + \frac{\eta_e}{2}\right)$$

$$\chi_3 = 1 - \frac{\omega_{ne}}{\omega}\left(1 + \frac{3}{2}\eta_e\right)$$

$$\eta_e \equiv \frac{\partial \ln T_e}{\partial \ln n_0}$$

$$\beta_e = \frac{8\pi n_0 T_e}{B_0^2}. \qquad (2.65)$$

Substituting (2.62)–(2.65) into (1.4) we arrive at the dispersion relation

$$D^{(0)} + D^{(1)} = 0. \qquad (2.66)$$

Here

$$D^{(0)} = \left(1 - \frac{\omega_{ne}}{\omega}\right)\left(1 - \frac{\omega_{ni}}{\omega}\right) + \frac{\beta}{2}\left(1 - \frac{\omega_{pe}^*}{\omega}\right)\left(1 - \frac{\omega_{pi}^*}{\omega}\right)$$
$$- \frac{c^2 k_\perp^2 k_z^2 \lambda_{De}^2}{\omega^2}\left(1 + \frac{T_i}{T_e}\right) \qquad (2.67)$$

$$D^{(1)} = i\pi^{1/2}\frac{\omega}{|k_z|v_{Te}}\frac{T_i}{T_e}\left[\frac{\chi_1}{1 + T_e/T_i}\left(2 + \frac{T_e}{T_i} - \frac{\omega_{ne}}{\omega}\right)^2\right.$$
$$+ \beta_e\left(2 + \frac{T_e}{T_i} - \frac{\omega_{ne}}{\omega}\right)\left(1 - \frac{\omega_{pi}^*}{\omega}\right)\chi_2$$
$$\left.+ \frac{\beta_e^2}{2}\left(1 + \frac{T_e}{T_i}\right)\left(1 - \frac{\omega_{pi}^*}{\omega}\right)^2\chi_3\right] \qquad (2.68)$$

where $\beta = \beta_e + \beta_i$, $\beta_i = 8\pi n_0 T_i/B_0^2$.

The dispersion relation (2.66) will be used in section 6.6.

2.3.2   *Perturbations with $\omega \simeq \Omega_i$*

When $\omega \simeq \Omega_i$, the terms with the *ion-cyclotron resonance* can also be significant in the dispersion relation (2.66). Using (2.47), we find that, allowing for these terms, the dispersion relation is of the form

$$D^{(0)} + D^{(1)} + D^{(2)} = 0 \qquad (2.69)$$

where $D^{(0)}$ and $D^{(1)}$ are given by (2.67) and (2.68), and

$$D^{(2)} = \frac{i}{2^{3/2}}\frac{(1 - \omega_{ne}/\omega)^2}{1 + T_e/T_i}\hat{I}_i\frac{1}{k\rho_i}\frac{\omega}{|\omega_{Di}(\omega - n\Omega_i)|^{1/2}}\exp\left(-\frac{\omega - n\Omega_i}{\omega_{Di}}\right). \qquad (2.70)$$

It is assumed that $(\omega - n\Omega_i)/\omega_{Di} > 0$.

Relations (2.69) and (2.70) will be used in section 6.7.

2.4 Electromagnetic perturbations in a low-β plasma

For the case where $\beta \ll 1$ the perturbations can be studied in the approximation

$$\boldsymbol{E} = -\nabla\phi - (\boldsymbol{e}_z/c)\partial A_z/\partial t \qquad (2.71)$$

i.e., in the approximation $A_2 = 0$ (see section 1.1.2). Then the dispersion relation (1.4) reduces to the form

$$\begin{vmatrix} \epsilon_{00} & \epsilon_{03} \\ \epsilon_{30} & \epsilon_{33} - N^2 \end{vmatrix} = 0. \qquad (2.72)$$

This equation corresponds to the so-called *approximation of low-β plasma*.

Generally speaking, when (2.72) is used, the term with ω_d should also be omitted in the expressions for ζ_n (see equation (1.56)) since at $\omega \simeq \omega^*$ the ratio ω_D/ω is of the order of β. (Hereafter ω^* represents a *characteristic drift (gradient) frequency*.) Then, according to (1.64), the components ϵ_{ik} in (2.72) take the form

$$\epsilon_{00} = 1 - \sum \frac{4\pi e^2}{Mk^2}\left\langle \frac{\partial F}{\partial \mathcal{E}} - \omega G \sum_n \frac{J_n^2}{\omega - n\Omega - k_z v_z}\right\rangle$$

$$\epsilon_{03} = \epsilon_{30} = -\sum \frac{4\pi e^2}{Mk\omega}\left\langle v_z\left(\frac{\partial F}{\partial \mathcal{E}} - \omega G \sum_n \frac{J_n^2}{\omega - n\Omega - k_z v_z}\right)\right\rangle$$

$$\epsilon_{33} = 1 + \sum \frac{4\pi e^2}{M\omega^2}\left\langle v_z\left(\frac{\partial F}{\partial v_z} + v_z\omega G \sum_n \frac{J_n^2}{\omega - n\Omega - k_z v_z}\right)\right\rangle. \qquad (2.73)$$

In the case of *low-frequency perturbations*, $\omega \ll \Omega_i$, it is convenient to present the expressions of (2.73) in the following form (cf. equations (1.5) and (1.23)):

$$\epsilon_{00} = \epsilon_\perp + \epsilon_\parallel \cos^2\theta \qquad \epsilon_{03} = \epsilon_{30} = \epsilon_\parallel \cos\theta \qquad \epsilon_{33} = \epsilon_\parallel \qquad (2.74)$$

where

$$\epsilon_\perp = 1 - \sum \frac{4\pi e^2}{Mk^2}\left\langle G(1 - J_0^2)\left(1 - \frac{k_z v_z}{\omega}\right)\right\rangle$$

$$\epsilon_\parallel = 1 + \sum \frac{4\pi e^2}{M\omega^2}\left\langle v_z\left(\frac{\partial F}{\partial v_z} + v_z\omega G\frac{J_0^2}{\omega - k_z v_z}\right)\right\rangle. \qquad (2.75)$$

The dispersion relation (2.72) is then written in the form

$$\epsilon_\parallel[1 - (\omega/ck_z)^2\epsilon_\perp] + (k_\perp/k_z)^2\epsilon_\perp = 0. \qquad (2.76)$$

When the *electron* and *ion* distributions are *Maxwellian*, equations (2.75) yield

$$\epsilon_\perp = 1 + \sum (k\lambda_D)^{-2}\hat{l}[1 - I_0(z)\exp(-z)]$$
$$\epsilon_\parallel = 1 + \sum (k_z\lambda_D)^{-2}\hat{l}\{[1 + Z(\omega/\mid k_z \mid v_T)]I_0(z)\exp(-z)\}. \tag{2.77}$$

When $z_i \ll 1$, the quantity $(k_\perp/k_z)^2\epsilon_\perp$ is of the order of $(k_\perp\rho_i)^2$ compared with separate terms in $\epsilon_\parallel$. Therefore, when small terms of order $(k_\perp\rho_i)^2$ are ignored, equation (2.76) *splits* into two dispersion relations:

$$\epsilon_\parallel \equiv \epsilon_{33} = 0 \tag{2.78}$$

$$\epsilon_\perp - N^2\cos^2\theta = 0. \tag{2.79}$$

The results of this section will be used in Chapter 14.
Note also that, when $N^2 \to \infty$, equation (2.72) yields

$$\epsilon_{00} = 0. \tag{2.80}$$

This dispersion relation corresponds to the *electrostatic approximation*.
The analysis of (2.80) was performed in [2].

3   Inertialess Magneto-drift Instabilities in a Collisionless Finite-β Plasma

In this chapter the *inertialess perturbations* described by the dispersion relation (2.8) are studied.

We restrict ourselves to the analysis of different limiting cases of this dispersion relation. The simplest case is that with $k_z = 0$ when equation (2.8) splits into two dispersion relations: (2.14) and (2.15). Equation (2.14) describes *Tserkovnikov's perturbations* while (2.15) describes the *transverse drift waves*. In section 3.1 Tserkovnikov's perturbations are studied. In section 3.2 we consider the transverse drift waves allowing for corrections with $k_z \neq 0$. Then we analyse essentially *oblique perturbations*: with $v_{Ti} \ll \omega/k_z \ll v_{Te}$ (section 3.3), and with $k_z \simeq \omega_{Di}/v_{Ti}$ (section 3.4). The results are summarized in section 3.5.

The magnetic drift of particles plays an important role in all the kinds of perturbations discussed in this chapter. In this connection, the relevant instabilities are called the *magneto-drift instabilities*. Thereby, we emphasize their difference from the so-called *drift instabilities in a low-β plasma* when the magnetic drift is insignificant.

Following [1.4], Tserkovnikov's perturbations were discussed in [1.5]. There the results of [1.4] concerning these perturbations have been reproduced by means of the *method of drift kinetic equation*. (In [1.4], as well as in this book, the *Boltzmann–Vlasov kinetic equation* was used.) Moreover, in contrast to [1.4], where the velocity distribution of the particles was considered to be Maxwellian, in [1.5] the dispersion relation was obtained for an arbitrary distribution function. In addition, an instability in a plasma with a *monoenergetic transverse distribution* of particles was discussed in [1.5].

It should be noted that in both [1.4] and [1.5] the geometry of a *cylindrical plasma column* with a *longitudinal current* and *without a longitudinal magnetic field* was considered. In such a case the magnetic field is curvilinear and the field lines have the form of concentric circles. (We shall discuss this geometry in Chapters 15 and 16.) It had been difficult to understand from [1.4, 1.5] whether the results of these papers have any meaning when the curvature of the magnetic field is vanishing. Moreover, following [1.4], the paper [3.1] had appeared, which was based on the *two-fluid magnetohydrodynamic equations* and evidently showed that inertialess perturbations are sensitive to the *magnetic-field curvature*. The important point is that in both [1.5] and [3.1] the coefficient proportional to the curvature occurs in the dispersion relation for inertialess perturbations. This question had been clarified in [1.6] where it had been shown that the dispersion relation by Tserkovnikov [1.4] holds in the approximation of straight magnetic field lines. The authors of [1.12] considered the geometry of straight field lines and reproduced the respective results of [1.4] for the case $\nabla T = 0$. Tserkovnikov [1.4] assumed the electron and ion temperature to be equal, $T_e = T_i$. The generalization of this analysis for the case of a plasma with $T_e \neq T_i$ was made in [3.2].

The point of special consideration was clarification of the *hydrodynamic meaning* of Tserkovnikov's waves. This consideration was started in [3.1] where it had been shown that in a collisional plasma there are so-called *magneto-drift entropy waves*, physically similar to Tserkovnikov's waves. Other papers dealing with the hydrodynamic approach will be mentioned in Chapter 11.

The transverse drift waves were discussed in the above-mentioned papers [1.6, 1.12, 2.5] and also in [3.3–3.10].

In [2.5] (see also [1.3]) the instability arising under the *intersection* of a *transverse drift wave branch* with the *ion-cyclotron harmonic branches*, corresponding to the *transverse drift-cyclotron instability*, was considered. These studies were continued in [3.8] where one further type of transverse drift-cyclotron instability was discovered. However, the above intersection of oscillation branches is possible only in a *plasma with finite electron Larmor radius*, i.e., in one with a sufficiently steep inhomogeneity. Therefore, we delay the discussion of the above instabilities until Chapter 6.

In [1.12] and then in [3.3], the excitation of the transverse drift waves due to the *magneto-drift resonance* with ions in a plasma with a homogeneous temperature was studied. In [3.4] the *quasilinear and three-wave processes* accompanying the instability were considered.

In [1.6] it was noted that the instability of transverse drift waves discussed in [1.12] can be *suppressed by magnetic-field curvature* (see in detail section 16.2).

The magneto-drift resonance between transverse drift waves and electrons is possible in the presence of an electron temperature gradient. Then both a stabilization of these waves and their additional destabilization can take place. The role of the *temperature gradient* in the problem of transverse drift waves was examined in [3.2, 3.7]. In [3.6] the case of an *almost force-free high-β plasma* was considered (see Chapter 12 about instabilities in such a plasma). In [3.9] an instability of the transverse drift waves in inhomogeneous *counterstreaming plasma flows* was studied. In [3.10] the *temperature anisotropy* was taken into account.

The study of *oblique inertialess waves* ($k_z \neq 0$) in a finite-β Maxwellian plasma was begun in [1.8]. (Early on, in [1.3, 1.7, 2.10] the electromagnetic instabilities in a low-β plasma related to the oblique waves were examined. We shall consider them in Chapter 14.) The results of [1.8] are presented in sections 3.3 and 3.4.

The line of [1.8] has been continued in [2.2, 3.2, 3.11, 3.12]. In [3.11] instabilities with $\omega \ll \mid k_z \mid v_{Ti}$ in a plasma with $\nabla T \neq 0$ were discussed. In [3.12] the *force-free high-β plasma* was considered (see Chapter 12). In [2.2], by means of the *two-fluid description*, the *oblique magnetoacoustic perturbations* were analysed. (We shall discuss such an approach in Chapters 9–11.) The author of [3.2] has generalized some of the results of [1.8] for the case of a plasma with $\nabla \ln T_e \neq \nabla \ln T_i$.

The analysis of inertialess perturbations shows that for a Maxwellian velocity distribution (and in the absence of high-energy particle components), they grow in time only if there is a *temperature gradient* in addition to the *density gradient*. Only the transverse drift waves are the exception. They grow also when $\nabla T = 0$. However, their growth rate is small, proportional to the ratio of the electron to ion mass. Therefore, having in mind the *thermonuclear applications* of the theory of inertialess instabilities, we can conclude that such instabilities, roughly speaking, are unimportant for systems in which the plasma pressure is comparable with the magnetic field pressure, $\beta \simeq 1$, and the temperature gradient is negligibly small compared with the density gradient.

The above-mentioned conditions are realized in *linear theta-pinches* where plasma is detached from the walls, see e.g., [3.13, 3.14]. According to [3.13, 3.14], in the *finite-β regime there is no evidence of microinstabilities* (the diffusion coefficient is close to classical). This corresponds quite well with the theoretical predictions of both the inertialess instabilities discussed in this chapter and in Chapter 11 and the Alfven ones discussed in Chapters 4 and 11. However, the analysis of instabilities in a plasma with *negligibly small pressure* [2] predicts *very significant instabilities* for a plasma with an arbitrary temperature gradient, including the so-called *drift* or *universal instability* [3.15, 3.16] for the case when $\nabla T = 0$. Experiments with very low-pressure plasma (see, e.g., [3.17, 3.18]) also testify to a significant role of instabilities. Thus, both theory and experiment indicate that a plasma with a small temperature gradient is unstable for small β but stable for finite β (see in detail in Chapter 14). At the same time, the theory predicts very significant instabilities for a plasma with $\beta \simeq 1$ and large temperature gradient. However, laboratory verification of this prediction, evidently, is still absent.

The theory of instabilities in an inhomogeneous finite-β plasma is widely used for the interpretation of *collective processes* in *space plasma*. The notions of various types of instabilities are then discussed. Here we shall touch upon space applications of instabilities considered in the present chapter only. Really, we should only mention here the transverse drift wave instabilities since the role of magnetoacoustic instabilities under space conditions have been only briefly analysed (see [3.2], however). In this statement, we do not refer to the magnetoacoustic instabilities in a *two-energy component plasma* since those instabilities will be considered elsewhere (see Chapter 5).

The transverse drift waves, discussed in section 3.2 and originally studied in [1.12], were the first of the electromagnetic drift-type instabilities among those considered in connection with space problems. In [3.3] it was suggested that the *acceleration of electrons* along the Earth's magnetic field lines and *auroral bombardment* can be due to this instability and to the electric field $\boldsymbol{E} \parallel \boldsymbol{B}_0$ generated as a result of it. Various particular cases of the transverse drift waves predicted theoretically in [1.12, 2.5, 3.7, 3.8], were considered in [3.19, 3.20] as a possible reason for the generation of *plasma turbulence* with frequencies $\omega \lesssim \Omega_i$ in the *tangential discontinuities of the solar wind*.

3.1   Tserkovnikov's magneto-drift instabilities

We will now study the *low-frequency long-wavelength perturbations* in a collisionless inhomogeneous *finite-β Maxwellian plasma* with $k_z = 0$ and $\boldsymbol{E} \perp \boldsymbol{B}_0$. Such perturbations are described by the dispersion relation (2.14) with ϵ_{22} of the form of (2.17). In analysing equation (2.14), we restrict ourselves to the case $T_e = T_i \equiv T$.

Remembering definition (2.11) for the operator $\hat{l}$, we calculate the action of this operator in (2.17). Introducing the dimensionless frequency $s \equiv \omega / |\omega_{Di}|$, we reduce (2.14) to the form

$$F(s) \equiv -2 + (s + \Gamma)\Phi(s) + (-s + \Gamma)\Phi(-s) = 0. \qquad (3.1)$$

Here

$$\Phi(s) = \int_0^\infty \frac{\exp(-x)\,\mathrm{d}x}{s+x} \qquad \Gamma = \frac{\eta-1}{\eta(1+\beta/2)+\beta/2}. \qquad (3.2)$$

For $\mid s \mid \gg 1$

$$F(s) = \frac{2(2-\Gamma)}{s^2} + \frac{12(4-\Gamma)}{s^4} + \dots. \qquad (3.3)$$

It can be seen from (3.1) and (3.3) that for $\Gamma < 2$, i.e., for

$$\eta < -1 \qquad (3.4)$$

an instability holds with $\mathrm{Re}\,\omega = 0$ and

$$\gamma = \mid \omega_{\mathrm{Di}} \mid [-12/(1+\beta)(1+\eta)]^{1/2}. \qquad (3.5)$$

Evidently, equation (3.5) is valid only when $\mid 1+\eta \mid \ll 1$.

For $\mid s \mid \ll 1$ we have

$$F(s) = -2\Gamma \ln s - 2(1+C\Gamma) + i\pi\Gamma \qquad (3.6)$$

where $C \approx 0.577$ is the Euler constant. In this case and under the condition $\Gamma > 0$, i.e.,

$$\eta > 1 \qquad (3.7)$$

we find that $\mathrm{Re}\,\omega = 0$ and

$$\gamma = \mid \omega_{\mathrm{Di}} \mid \exp[-(1+\beta)/(\eta-1)+C]. \qquad (3.8)$$

This expression is valid if $\eta - 1 \ll 1$.

The above limiting cases allow one to make a general conclusion: an *inhomogeneous finite-β plasma is unstable if*

$$\left| \frac{\partial \ln T}{\partial \ln n_0} \right| > 1. \qquad (3.9)$$

It is clear that, when η moves away from the instability boundary, the *growth rate* becomes of the order of the *magnetic-drift frequency*

$$\gamma \simeq \mid \omega_{\mathrm{Di}} \mid. \qquad (3.10)$$

It is also clear that the instability discussed is possible for both $\beta \gtrsim 1$ and $\beta \ll 1$. However, when β decreases, the growth rate becomes small since $\omega_{\mathrm{D}} \sim \beta$ so that for small β

$$\gamma \simeq \beta\omega^*. \qquad (3.11)$$

Here, as above, ω^* is the characteristic gradient frequency.

Tserkovnikov [1.4] analysed dispersion relation (3.1) by means of *Nyquist's method* (see [1], section 2.7). He has found the instability condition

$$0 < \Gamma < 2. \tag{3.12}$$

It can be seen that this condition is the same as (3.9).

Since the characteristic frequencies of the perturbations discussed are of the order of the *magnetic-drift frequency*, we call the respective oscillation branches *Tserkovnikov's magneto-drift branches*. We have considered these branches for such values of η where their frequencies are purely imaginary. In deviating from the instability boundary into the stability region, their frequencies become real. Since the dispersion relation (3.1) is symmetric with respect to the change of the sign of the frequency, one can talk about *two oscillation branches with different frequency signs*. The notion of two branches of *magneto-drift perturbations* with $k_z = 0$ is also justified by the analysis of limiting cases of large and small frequencies using equations (3.3) and (3.6), respectively.

3.2   Transverse drift waves

In addition to the Tserkovnikov branches, for $k_z = 0$ there is also a branch of *transverse drift waves*. The electric field of these waves is directed along the equilibrium magnetic field, $\boldsymbol{E} \parallel \boldsymbol{B}_0$. Such waves are described by dispersion relation (2.15). For a Maxwellian velocity distribution of particles, the expression for ϵ_{33} is given by (2.18).

Let us consider (2.15) and (2.18) under the assumptions $\beta \ll 1$, $\omega \gg \omega_D$. Then

$$\epsilon_{33} = \epsilon_{33}^{(e)} + i \operatorname{Im} \epsilon_{33}^{(i)} \tag{3.13}$$

where

$$\epsilon_{33}^{(e)} = -\frac{\omega_{pe}^2}{\omega^2}\left(1 - \frac{\omega_{pe}^*}{\omega}\right) \tag{3.14}$$

$$\operatorname{Im} \epsilon_{33}^{(i)} = \pi \frac{M_e}{M_i} \frac{\omega_{pe}^2}{\omega \mid \omega_{Di}\mid}\left(1 + \frac{\partial \ln T_i}{\partial \ln B_0} - \frac{\omega_{ni}}{\omega}\right) \exp\left(-\frac{\omega}{\omega_{Di}}\right). \tag{3.15}$$

It is assumed that $\omega/\omega_{Di} > 0$. In the contrary case $\operatorname{Im} \epsilon_{33}^{(i)} = 0$.

Assuming the right-hand side of (3.14) to vanish, we find the real part of the oscillation frequency

$$\operatorname{Re}\omega = \omega_{pe}^* \tag{3.16}$$

that corresponds to the branch of the transverse drift waves.

We find the growth rate γ assuming $\gamma/\mathrm{Re}\,\omega \ll 1$. In this approximation we have a general expression

$$\gamma = -\frac{\mathrm{Im}\,\epsilon_{33}^{(i)}}{\partial \mathrm{Re}\,\epsilon_{33}^{(e)}/\partial\omega}. \tag{3.17}$$

Then, using (3.14) and (3.15), we obtain

$$\gamma = \pi\frac{M_e}{M_i}\left(1 + \frac{2}{\beta}\frac{\eta}{1+\eta}\right)\frac{\omega_{pe}^{*2}}{|\omega_{Di}|}\exp\left(-\frac{2}{\beta}\frac{T_e}{T_i}\right). \tag{3.18}$$

In order of magnitude

$$\frac{\gamma}{\mathrm{Re}\,\omega} \simeq \frac{M_e}{M_i\beta}\exp\left(-\frac{2}{\beta}\right). \tag{3.19}$$

It can be seen that, in contrast to the Tserkovnikov magneto-drift waves discussed in section 3.1, the *transverse drift waves are unstable even when* $\nabla T = 0(\eta = 0)$. It should be mentioned that the growth rate of the transverse drift waves is small as M_e/M_i. In addition, the growth rate of (3.18) and (3.19) contains also the smallness of order $\exp(-2/\beta)$. But this smallness is absent if the above results are extrapolated into the region $\beta \simeq 1$.

Relation (3.17) has an *energetic sense* (cf. [1], section 2.2). This relation can also be presented in the form of *the oscillation energy balance*:

$$dW/dt = -\mathrm{Re}\,\sigma\,|\,E_z\,|^2\,. \tag{3.20}$$

Here W is the *oscillation energy* defined by

$$W = \omega\frac{\partial\mathrm{Re}\,\epsilon_{33}}{\partial\omega}\frac{|\,E_z\,|^2}{8\pi} \tag{3.21}$$

and σ is the real part of the conductivity

$$\mathrm{Re}\,\sigma = \frac{\omega}{4\pi}\mathrm{Im}\,\epsilon_{33}^{(i)}. \tag{3.22}$$

It can be seen from (3.14) and (3.21) that the energy of transverse drift waves is negative, $W < 0$. Their excitation is due to the fact that the ions take away energy from the waves (cf. [1], section 2.3).

These results concern not only perturbations with $k_z = 0$ but also those with finite but fairly small k_z, so that $k_z v_{Te}/\omega^* \ll (M_e/M_i)^{1/2}$. In the contrary case, when

$$(M_e/M_i)^{1/2} \ll k_z v_{Te}/\omega^* \ll 1 \tag{3.23}$$

expression (3.16) for the oscillation frequency remains in force while the growth rate is now defined by the terms of the dispersion relation (2.8) proportional to k_z^2. When $\beta \ll 1$ and $\omega \gg \omega_D$, one can use, instead of the full dispersion relation (2.8), the following approximate one:

$$\epsilon_{33} + i(i\epsilon_{23})^2 \mathrm{Im}\,\epsilon_{22}/N^4 = 0. \tag{3.24}$$

According to (2.9), one should substitute here the following expressions for $i\epsilon_{32}$ and $\mathrm{Im}\,\epsilon_{22}$:

$$
\begin{aligned}
i\epsilon_{23} &= \frac{\omega_{pe}^2}{\omega\Omega_e}\frac{k_\perp k_z T_e}{M_e\omega^2}\left(1 - \frac{\omega_{ne}}{\omega}(1+2\eta)\right) \\
\mathrm{Im}\,\epsilon_{22} &= \pi\frac{\beta_i}{2}N^2\frac{\omega^3}{|\,\omega_{Di}\,|^3}\left(1 + \frac{2}{\beta}\frac{\eta}{1+\eta}\right)\exp\left(-\frac{\omega}{\omega_{Di}}\right).
\end{aligned}
\tag{3.25}
$$

Then we find that the growth rate is equal to

$$\gamma = 2\pi\frac{\beta_e\beta_i}{\beta^3}\left(\frac{T_e}{T_i}\right)^3\frac{k_z^2 T_e\eta^2}{M_e\,|\,\omega_{pe}^*\,|}\left(1 + \frac{2}{\beta}\frac{\eta}{1+\eta}\right)\exp\left(-\frac{2}{\beta}\frac{T_e}{T_i}\right). \tag{3.26}$$

Instability takes place if

$$\eta > 0 \qquad \text{or} \qquad \eta < -1. \tag{3.27}$$

As in the case where $k_z = 0$, this instability is explained by excitation of *negative energy waves* due to ion absorption.

For other results of transverse drift wave theory see [1.12, 2.1, 2.5, 3.8–3.10, 3.19–3.21].

3.3 Instabilities with finite k_z

In this section we consider the perturbations with $\omega/v_{Te} \ll k_z \ll \omega/v_{Ti}$ (perturbations with finite k_z).

When the longitudinal wave number k_z increases from $k_z \ll \omega/v_{Te}$ up to $k_z \gtrsim \omega/v_{Te}$, the *magneto-drift oscillation branches* considered in section 3.1 *intermix* with the branch discussed in section 3.2. In addition, when $k_z \gtrsim \omega/v_{Te}$, the electron longitudinal thermal motion *broadens* the electron magneto-drift resonance, $\omega = \omega_d$. This is the reason for the *disappearance of one of the two magneto-drift oscillation branches* discussed in section 3.1. Therefore, for $\omega/v_{Te} \ll k_z \ll \omega/v_{Ti}$, we should expect the presence of two oscillation branches: the electron drift and magneto-drift ones.

Subsequent analysis justifies this qualitative consideration. Thereby, one of the main purposes of this analysis is the elucidation of the picture of oscillation branches in the above-mentioned range of the longitudinal wave number k_z. For this purpose we shall consider the dispersion relation following from (2.8) in the approximation $\omega/k_z v_{Te} \to 0$, $k_z v_{Ti}/\omega \to 0$. Such a dispersion relation has the form

$$\hat{l}_e \left(\frac{\omega_{De}}{\omega} \right) \hat{l}_e + \frac{T_e}{T_i} \hat{l}_i \left(\left\langle \frac{\omega_{di}}{\omega - \omega_{di}} \right\rangle \right) \hat{l}_e \left(1 - \frac{\omega_{De}}{\omega} \right) = 0. \qquad (3.28)$$

(If nothing stands to the right of the operator $\hat{l}$, one should consider $\hat{l} \equiv \hat{l}(1)$.)

We shall also study the interaction of the above-mentioned oscillation branches with the resonant electrons and ions, i.e., the possible instabilities associated with these oscillation branches. Equation (3.28) takes into account only the *ion resonance*. Therefore, in the course of our presentation, we shall modify equation (3.28) by adding small imaginary terms with the *electron resonance*.

3.3.1 Electron drift wave branch

Assuming $\beta \ll 1$, we find that (3.28) has the solution

$$\omega = \omega_{ne}. \qquad (3.29)$$

This solution corresponds to the *electron drift waves* which are well-known in the theory of low-β plasma instabilities (see [2], sections 3.2 and 3.3).

Allowing for the interaction of *resonant electrons* with these waves leads to the conclusion that they are damped for $\eta > 0$ and grow for $\eta < 0$. Their growth rate is of the order of (see [2], section 3.2)

$$\gamma \simeq - \frac{\omega_{ne}^2}{|k_z| \, v_{Te}} \eta. \qquad (3.30)$$

As before, considering β to be small, we shall study the role of finite-β effects in perturbations of the type of (3.29). The main effect associated with a finite β is the resonant interaction of ions with the waves. We find from (3.28) that the waves can be excited due to such an interaction with the growth rate

$$\gamma = - \pi \eta^2 \frac{T_e}{T_i} \frac{\omega_{ne}^2}{|\omega_{Di}|} \frac{1 + (1 - \eta)T_i/T_e + 2\eta/\beta(1 + \eta)}{[1 + (1 + \eta)T_i/T_e]^2}$$
$$\times \exp\left(- \frac{2}{\beta} \frac{T_e}{T_i(1 + \eta)} \right). \qquad (3.31)$$

It can be seen that the ions resonantly interact with the oscillations only for $\eta > -1$. The waves are damped for arbitrary positive η and also for $-\beta(1 + T_i/T_e)/2 < \eta < 0$. The excitation is possible only for

$$-1 < \eta < -\beta(1 + T_i/T_e)/2. \qquad (3.32)$$

Let us compare this result with (3.27) which shows that the $\omega = \omega_{pe}^*$ waves (for $\omega \gg k_z v_{Te}$) are excited for $\eta > 0$. As mentioned in section 3.2, the energy of these waves is negative, $W < 0$. In contrast to this, in the case of (3.29) we have

$$\mathrm{Re}\,\epsilon_{33} = (k_z \lambda_{De})^{-2}(1 - \omega_{ne}/\omega) \qquad (3.33)$$

so that $W > 0$. Therefore, in both cases the ions take away energy from the waves, resulting in the excitation of the waves in the first case, and in damping in the second case (this corresponds to the general results of [1], section 2.3).

Let us now consider a plasma with $\beta \gg 1$. Assuming $\omega \ll (\omega_{Di}, \omega_{De})$, we reduce (3.28) to the relation

$$\hat{l}_e \omega_{De} = \frac{i\pi\omega^2}{4} \frac{T_e}{T_i} \hat{l}_i \left(\frac{1}{|\omega_{Di}|} \right) \frac{\hat{l}_e}{(1 + T_e/T_i)^2}. \qquad (3.34)$$

Hence it follows that $\mathrm{Re}\,\omega$ is defined by (3.16) while

$$\gamma = \frac{\pi\eta^2}{(1+\eta)^2} \left(\frac{2}{\beta} \right)^2 \frac{|\omega_{pe}^*|}{(1 + T_i/T_e)^2} \left(1 + \frac{T_i}{T_e} \frac{1-\eta}{1+\eta} \right). \qquad (3.35)$$

If $T_i = T_e$, the waves are unstable only when

$$\eta < -1. \qquad (3.36)$$

When $T_i > T_e$, the growth rate can be positive even for $\eta > 0$, namely, for

$$\eta > (1 + T_e/T_i)/(1 - T_e/T_i). \qquad (3.37)$$

Note also that the root of (3.29) is a precise solution of (3.28) for all β if $\nabla T = 0$. This follows from the fact that for $\nabla T = 0$ the operator $\hat{l}_e$ defined by (2.11) reduces to

$$\hat{l}_e = 1 - \omega_{ne}/\omega. \qquad (3.38)$$

Therefore, it is clear that for $\nabla T = 0$ the solution of (3.16) coincides with (3.29) while the growth rate (3.35) vanishes in this case.

Using this and expanding in a series in η, we find, by means of (3.28), that for $|\eta| \ll 1$ and arbitrary β, the electron drift waves are characterized by the frequency

$$\omega = \omega_{ne}\left(1 + \eta\frac{\beta/2}{1+\beta/2} - \frac{\eta^2}{4}\frac{\beta}{(1+\beta/2)^3\langle\lambda\zeta_{0i}\rangle}\right). \tag{3.39}$$

Here $\lambda = M_i v_\perp^2/2T$, $\zeta_{0i} = (\omega - \omega_{di})^{-1}$. For simplicity we take $T_e = T_i \equiv T$.

Taking into account (3.29) and the condition $\omega_{ne}/\omega_{Di} > 0$, we find

$$\mathrm{Im}\,\langle\lambda\zeta_{0i}\rangle = -\pi\lambda\delta(\omega - \lambda\omega_{Di}) < 0. \tag{3.40}$$

It then follows from (3.39) that $\mathrm{Im}\,\omega < 0$, i.e., the waves are damped for small $|\eta|$. It is clear from the above that this damping is due to the interaction of waves with resonant ions. In the particular cases of small and large β, the expression for the growth rate following from (3.39) is in accordance with (3.31) and (3.35).

3.3.2 Magneto-drift oscillation branch

Let us consider (3.28) for $\omega \ll \omega_{Di}$. In this case we have approximately

$$\left\langle\frac{\omega_{di}}{\omega - \omega_{di}}\right\rangle = -\left(1 + \frac{\omega}{\omega_{di}}\ln\frac{\omega_{di}}{\omega} + i\pi\,\mathrm{sgn}\,\omega_{Di}\right). \tag{3.41}$$

Substituting (3.41) into (3.28), we find

$$\omega = -\omega_{Di}\exp\left\{\frac{T_i}{T_e}\frac{1+\eta}{1-\eta}\left[\frac{\beta}{2}\left(1 + \frac{T_e}{T_i}\right) + \frac{1}{(1+\eta)^{1/2}}\right]\right\}. \tag{3.42}$$

Here we have used

$$\ln(\omega_{Di}/\omega) + i\pi\,\mathrm{sgn}\,\omega_{Di} = \ln(-\omega_{Di}/\omega).$$

Since we assumed $\omega \ll \omega_{Di}$, equation (3.42) is strictly valid only if

$$0 < \eta^2 - 1 \ll 1. \tag{3.43}$$

However, this equation is qualitatively valid also for $\eta^2 - 1 \simeq 1$ so that the frequency of equation (3.42) is of the order of the ion magnetic-drift frequency ω_{Di}. Therefore, the oscillation branch discussed is called the *ion magneto-drift branch*. However, one should bear in mind that the signs of ω and ω_{Di} are opposite, $\omega/\omega_{Di} < 0$, so that our terminology is quite conventional.

Since $\omega/\omega_{Di} < 0$, the resonant interaction of ions with the branch (3.42) is absent, and, to obtain the growth rate, we should take into account the interaction of this branch with electrons. Using equation (2.8), we add the terms of order $\omega/k_z v_{Te}$ into (3.28). Then (3.42) is replaced by

$$\omega = \omega^{(0)} \exp\left(-i\frac{\beta}{2}\pi^{1/2}\frac{\omega_{ne} - \omega_{Te}/2}{|k_z| v_{Te}}\frac{T_i}{T_e}\frac{1+\eta}{1-\eta}\right) \tag{3.44}$$

where $\omega^{(0)}$ is the right-hand side of (3.42). It follows from (3.44) that the waves are damped for $\eta > 1$ and excited for

$$\eta < -1. \tag{3.45}$$

According to the above, at the validity limit of (3.42), i.e., for $\eta^2 - 1 \simeq 1$, the oscillation frequency is of the order of the ion magnetic-drift frequency, $|\omega| \simeq \omega_{Di}$. Therefore, one can expect that, for $|\eta| \gg 1$, the ion magneto-drift branch will be characterized by a frequency larger than ω_{Di}. To make sure of this, let us consider (3.28) for $\omega \gg \omega_{Di}$ and $|\eta| \gg 1$. Neglecting the terms of order ω_{Di}/ω, we have

$$\left\langle \frac{\omega_{di}}{\omega - \omega_{di}} \right\rangle \approx \frac{\omega_{Di}}{\omega}. \tag{3.46}$$

Substituting (3.46) into (3.28), assuming $\beta \ll 1$, $\omega \ll (\omega_{Te}, \omega_{Ti})$ and taking into account $|\eta| \gg 1$, we arrive at the quadratic equation for ω:

$$\omega^2 - \omega\omega_{ne} - \omega_{Te}\omega_{Di}/(1 + T_i/T_e) = 0. \tag{3.47}$$

When $\eta \ll 2/\beta$, the larger root of this equation yields the branch in (3.29) while the smaller one is

$$\omega = -\eta\omega_{Di}/(1 + T_i/T_e). \tag{3.48}$$

This ω is large compared with ω_{Di} just at $|\eta| \gg 1$. On the other hand, at the validity limit, i.e., when $|\eta| \simeq 1$, it follows from (3.47) that $\omega \simeq \omega_{Di}$, i.e., we have the same as follows from (3.42). This circumstance allows one to refer the waves of type (3.42) and (3.47) to just the same branch.

If the resonant electrons are taken into account, using (2.8), we obtain the following equation instead of (3.47):

$$\omega^2\left(1 + i\pi^{1/2}\frac{\omega_{Te}}{2|k_z| v_{Te}}\right) - \omega\omega_{ne} - \frac{\omega_{Di}\omega_{pi}^*}{1 + T_i/T_e} = 0. \tag{3.49}$$

Thus, the growth rate of the oscillations of type (3.48) is

$$\gamma = \pi^{1/2}\eta(\mathrm{Re}\,\omega)^2/\mid k_z \mid v_{Te} \tag{3.50}$$

where $\mathrm{Re}\,\omega$ is given by (3.48). It is evident that this branch is unstable for large positive η

$$\eta \gg 1. \tag{3.51}$$

Note also that, for such η and $\beta \ll 1$, the electron drift branch is damped due to interaction with resonant electrons (cf. (3.50) with (3.30)).

3.4   Instabilities with large k_z

Now we consider perturbations with large k_z implying that, for such k_z, the finiteness of parameter $k_z v_{Ti}/\omega$ is important. This means that the ion longitudinal thermal motion is assumed to be significant.

Generalization of equation (3.28) to the case of arbitrary $k_z v_{Ti}/\omega$ is the following equation obtained by using relations (2.8), (2.6), and (2.9):

$$2 + i\pi^{1/2}\hat{l}_i\left(\frac{\omega}{\mid k_z \mid v_{Ti}}\langle W\rangle\right) - i\pi^{1/2}\frac{\beta}{2\omega_{Di}^2}\hat{l}_i\left(\omega_{Di}^2\frac{\omega}{\mid k_z \mid v_{Ti}}\langle \lambda^2 W\rangle\right)$$
$$+ \frac{\pi\beta}{4}\frac{\omega^2}{\omega_{Di}^2}\left\{\hat{l}_i\left(\frac{\langle W\rangle}{\mid k_z \mid v_{Ti}}\right)\hat{l}_i\left(\frac{\omega_{Di}^2}{\mid k_z \mid v_{Ti}}\langle \lambda^2 W\rangle\right)\right.$$
$$\left. - \left[\hat{l}_i\left(\frac{\omega_{Di}}{\mid k_z \mid v_{Ti}}\langle \lambda W\rangle\right)\right]^2\right\} = 0. $$

$$\tag{3.52}$$

Here $W(x)$ is defined by (2.24) where

$$x = (\omega - \omega_{Di}\lambda)/\mid k_z \mid v_{Ti} \tag{3.53}$$

and $\lambda = M_i v_\perp^2/2T$. The symbol $\langle \ldots \rangle$ means averaging only over the transverse velocities, this symbol is defined in equation (2.19). For simplicity, in obtaining (3.52), we restricted ourselves to the case $T_e = T_i = T$.

In the low-β limit, when the terms with β and ω_{Di} are unimportant, equation (3.52) reduces to

$$2 + i\pi^{1/2}\hat{l}_i\left(\frac{\omega}{\mid k_z \mid v_{Ti}}W\frac{\omega}{\mid k_z \mid v_{Ti}}\right) = 0. \tag{3.54}$$

This equation describes the well-known *ion temperature-gradient drift instability* (see [2], section 3.2) originally studied in [3.22–3.24]. In the discussed *long-wavelength approximation*, such an instability takes place for $\eta < 0$ and $\eta > 2$. (In the case of perturbations with $k_\perp \rho_i \simeq 1$, the boundary of this instability in a plasma with positive η is displaced to $\eta = 0.96$, [3.24], see also [1.3].) The results following from (3.54) for $\omega \gg k_z v_{Ti}$, $|\eta| \to \infty$ and for $\omega \ll k_z v_{Ti}$, $\eta \approx 2$ are also well-known. In the first case, equation (3.54) reduces to the cubic equation:

$$\omega^3 + \omega_{Ti} k_z^2 T/M_i = 0. \tag{3.55}$$

It describes the hydrodynamic variant of the ion temperature-gradient drift instability. In the second case, it follows from (3.54) that

$$\omega = k_z^2 v_{Ti}^2/\omega_{ni} - i\pi^{1/2}|k_z|v_{Ti}(1 - \eta/2). \tag{3.56}$$

This corresponds to the *kinetic instability* for $\eta > 2$.

The present analysis has two goals. First, we shall study how equations (3.55) and (3.56) are modified when accounting for the effects of order β in the approximation $\beta \ll 1$ (section 3.4.1). Second, we shall analyse what equation (3.52) yields for a plasma with $\beta \gg 1$ (section 3.4.2).

3.4.1 *Influence of finite β on the ion temperature-gradient drift instability*

Assuming $\beta \ll 1$, $k_z v_{Ti} \ll \omega$, and $\nabla n_0 = 0$ in (3.52), we arrive at the following generalization of equation (3.55):

$$\omega^3 + \omega\omega_{Ti}\omega_{Di}/2 + \omega_{Ti}k_z^2 T/M_i = 0. \tag{3.57}$$

We calculate the discriminant of this cubic equation and find that all three roots are real (instability is absent) if

$$k_z v_{Ti}/\omega_{Ti} < 2^{1/2}3^{3/4}\beta^{3/4}. \tag{3.58}$$

When (3.58) is a strong inequality, the roots of (3.57) are

$$\omega_{1,2} = \pm\beta^{1/2}\omega_{Ti}/2 \tag{3.59}$$

$$\omega_3 = -k_z^2 T/M_i\omega_{Di}. \tag{3.60}$$

Note that the roots $\omega_{1,2}$ can also be obtained from (3.47).

Equations (3.58)–(3.60) show a stabilization of perturbations with sufficiently small k_z. One can consider such a stabilization as a weakening of the coupling between the temperature-gradient effects and the longitudinal motion of the particles due to the ion magnetic drift.

We now assume in equation (3.52) that $(\omega, \omega_{\mathrm{Di}}) \ll k_z v_{T\mathrm{i}}$, $\omega_{n\mathrm{i}} \gg k_z v_{T\mathrm{i}}$, $\eta \approx 2$. As before, β is assumed to be small but non-vanishing. We then obtain instead of (3.57) (cf. (3.56)):

$$\omega = k_z^2 v_{T\mathrm{i}}^2 / \omega_{n\mathrm{i}} + (\pi - 3)\omega_{\mathrm{Di}} - i\pi^{1/2} \mid k_z \mid v_{T\mathrm{i}}(1 - 4\beta - \eta/2). \quad (3.61)$$

It can be seen that in this approximation the instability boundary is displaced into the smaller η

$$\eta = 2(2 - 4\beta) \qquad \beta \ll 1. \quad (3.62)$$

Thus, for perturbations of the type discussed, the finite-β effect plays a destabilizing role: there is an instability in a plasma with the smaller temperature gradient. This effect is similar to the displacement of the instability boundary into the region of smaller η for finite $k_\perp$ (see [2], section 3.2).

3.4.2 Suppression of the ion temperature-gradient drift instability in a plasma with $\beta \gg 1$

We now consider a plasma with $\beta \gg 1$. In this case $\omega_{\mathrm{Di}} \gg (\omega_{n\mathrm{i}}, \omega_{T\mathrm{i}})$. Therefore only perturbations with $\omega \ll \omega_{\mathrm{Di}}$ are of interest for the theory of instabilities. In the opposite case the gradient terms ($\sim \nabla n_0, \nabla T$) disappear from the dispersion relation while, according to the general theorem of [1], section 2.6, a homogeneous Maxwellian plasma is stable. In saying this, we do not imply the case of a force-free (or almost force-free) plasma, $\nabla p \approx 0$, which will be considered separately in Chapter 12.

Perturbations with $k_z v_{T\mathrm{i}} \ll \omega \ll \omega_{\mathrm{Di}}$ were discussed in section 3.3. Therefore, we should consider only the perturbations with

$$\omega \ll (k_z v_{T\mathrm{i}}, \omega_{\mathrm{Di}}). \quad (3.63)$$

In this case the terms with ω in the function for W in equation (3.52) can be neglected. The terms of this equation which are quadratic with respect to W can be omitted since they remain with the small weighting factor of order $(\omega/k_z v_{T\mathrm{i}})^2$. Omitting also other small terms, we obtain from (3.52) an equation which is linear in ω:

$$\omega = -\frac{4i}{\pi^{1/2}\beta} \frac{\mid k_z \mid v_{T\mathrm{i}}}{\langle \lambda^2 W \rangle} + \frac{k_y T}{M_{\mathrm{i}}\Omega_{\mathrm{i}}} \frac{\partial}{\partial x} \ln \left(\frac{n_0 \omega_{\mathrm{Di}}}{v_{T\mathrm{i}}} \langle \lambda^2 W \rangle \right) \quad (3.64)$$

where $W = W(-\lambda \omega_{\mathrm{Di}} / \mid k_z \mid v_{T\mathrm{i}})$.

When $k_z v_{Ti} \gg \omega_{Di}$, solution (3.64) corresponds to the *aperiodically damped perturbations* of homogeneous plasma with decay rate (cf. [1.1], equation (6.38))

$$\gamma = -2 \mid k_z \mid v_{Ti}/\pi^{1/2}\beta. \tag{3.65}$$

If the temperature gradient vanishes, $\eta = 0$, equation (3.64) reduces to the form

$$\omega = -\frac{4i}{\pi^{1/2}\beta}\frac{\mid k_z \mid v_{Ti}}{\langle \lambda^2 W \rangle} + \omega_{ni}. \tag{3.66}$$

Hence it is clear that a plasma with $\nabla T = 0$ is stable for large β (since $\operatorname{Re} W > 0$ for all values of the argument). The role of an inhomogeneity reduces in this case only to a displacement of the real part of the oscillation frequency.

For $\eta \neq 0$ we find from (3.64) that the growth rate is proportional to

$$
\gamma \sim - \left[\operatorname{Re}\langle \lambda W \rangle \left(1 - \frac{\pi^{1/2}}{4}\frac{\eta}{1+\eta}\operatorname{Im}\langle \lambda^2 x W \rangle \right) \right.
$$
$$
\left. + \frac{\eta}{1+\eta}\frac{\pi^{1/2}}{4}\operatorname{Im}\langle \lambda^2 W \rangle \operatorname{Re}\langle \lambda^2 x W \rangle \right] \tag{3.67}
$$

where $x = -\omega_{Di}\lambda/ \mid k_z \mid v_{Ti}$. Let us show that $\gamma < 0$ for all $\eta > 0$. The second term of the expression in the square brackets is positive if $\eta > 0$ so that the sign of this expression could be negative only due to a negative value of the expression in the round brackets. However, for $\eta > 0$

$$\frac{\eta}{1+\eta}\operatorname{Im}\langle \lambda^2 x W \rangle < \langle \lambda^2 \rangle \max \operatorname{Im}(xW) = 2\max \operatorname{Im}(xW). \tag{3.68}$$

By means of tables for W, we ensure that

$$\max \operatorname{Im}(xW) < 2/\pi^{1/2}. \tag{3.69}$$

Hence, it follows that γ is negative.

Thus, it is seen that the ion temperature-gradient drift instability disappears in a plasma with $\beta \gg 1$ and $\eta > 0$.

To obtain a more detailed picture of the boundary of this instability for large β let us consider equation (3.64) in the limiting case $k_z v_{Ti} \gg \omega_{Di}$. We then find (cf. (3.65))

$$\omega = \omega_{ni}\left(1 + \frac{\eta}{2}\right) - \frac{2i \mid k_z \mid v_{Ti}}{\pi^{1/2}\beta}\left(1 - \frac{3\pi}{2}\frac{\eta}{1+\eta}\frac{\omega_{Di}^2}{k_z^2 v_{Ti}^2}\right). \tag{3.70}$$

The additional term in the imaginary part of ω is small compared with the leading one considered in (3.65). But due to this additional term, we can reveal a tendency for a temperature gradient influence on the perturbations discussed. Then it can be seen that the temperature gradient is a destabilizing factor if

$$\eta(1 + \eta) > 0. \tag{3.71}$$

Therefore, at the limit of validity of (3.70), i.e., when $\omega_{\mathrm{Di}}/k_z v_{Ti} \simeq 1$ and $\beta \simeq 1$, we should expect an instability if $\mid \eta \mid \gtrsim 1$. This result is in qualitative agreement with the extrapolation of the results of section 3.4.1 to the region of the mentioned values of $\omega_{\mathrm{Di}}/k_z v_{Ti}$ and β.

3.5   Systematization of the results

The above analysis show that, for $\nabla T = 0$ and finite β, the inertialess perturbations are not excited for all k_z excluding $k_z \lesssim \omega^*/v_{Te}$. However, the growth rate of the perturbations with $k_z \lesssim \omega^*/v_{Te}$ is very small, $\gamma \simeq (M_e/M_i)\omega^*$.

In the case of a positive temperature gradient, $\eta > 0$, generally speaking, a threshold of type $\eta \gtrsim 1$ is necessary for an instability. The perturbations with $k_z \simeq \omega^*/v_{Te}$ excited for a small positive η with the growth rate less than $\eta^2\omega^*$ are an exception to this rule.

The broadest class of the growing perturbations is that which occurs in a plasma with $\eta < 0$. The perturbations both with $k_z \lesssim \omega^*/v_{Te}$ and with large k_z, up to $k_z \simeq \omega^*/v_{Ti}$, belong to this class.

Using the results of this chapter, one can obtain a qualitatively complete picture of the inertialess instabilities in a collisionless plasma with an adopted η. Such a picture is presented in table 3.1. In this table the results concerning a plasma with $\eta \gtrsim 0$ are collected. By analogy, one can systematize instabilities also for $\eta < 0$.

Table 3.1. Inertialess instabilities of finite-β plasmas with $\eta > 0$.

Relative temperature gradient	Longitudinal wavelength	Oscillation frequency magnitude	Growth rate magnitude	Typical β	Temperatures	Equations
$\eta \approx 0$	$k_z < \omega^*/v_{Te}$	ω^*	$\dfrac{M_e}{M_i \beta} \exp\left(-\dfrac{2}{\beta}\right)\omega^*$	$\beta \lesssim 1$	$T_e \simeq T_i$	(3.19)
$0 < \eta \lesssim 1$	$k_z < \omega^*/v_{Te}$	ω^*	$\dfrac{\eta^2}{\beta}\exp\left(-\dfrac{2}{\beta}\right)\omega^*$	$\beta < 1$	$T_e \simeq T_i$	(3.26)
$\eta > 1$	$k_z \lesssim \omega^*/v_{Te}$	$\omega^*\beta$	$\omega^*\beta$	$\beta \lesssim 1$	$T_e \simeq T_i$	(3.19)
	$\omega^*/v_{Te} < k_z < \omega^*/v_{Ti}$	ω^*	ω^*/β^2	$\beta > 1$	$T_e < T_i$	(3.35), (3.37)
	$\omega^*/v_{Te} < k_z < \omega^*/v_{Ti}$	$\eta\beta\omega^*$	$\eta^3\omega^{*2}\beta^2/|k_z|v_{Te}$	$\beta < 1$	$T_e \simeq T_i$	(3.50), (3.51)
	$k_z \simeq \omega^*/v_{Ti}$	ω^*	ω^*	$\beta \lesssim 1$	$T_e \simeq T_i$	(3.57), (3.58), (3.62)

4 Alfven Magneto-drift Instabilities in a Finite-β Plasma

As explained in section 2.1, the low-frequency long-wavelength perturbations in an inhomogeneous finite-β plasma are subdivided into the *inertialess* and *Alfven oscillation branches*. Inertialess perturbations in a plasma with a Maxwellian velocity distribution of particles were studied in Chapter 3. We shall now consider the Alfven branches in such a plasma.

We begin with the theory of *Alfven branches* in the approximation $k_\perp \rho_i \to 0$. As noted in the introduction of Chapter 3, in this approximation the *resonant interaction* of particles with waves is not taken into account. In order to allow for this interaction we should augment the dispersion relation for the Alfven waves by terms of order $k_\perp^2 \rho_i^2$. This procedure was done in section 2.1.6. We shall now consider the specific processes of resonant interaction of Alfven waves with particles in a plasma with $\beta \simeq 1$ and a small temperature gradient compared with the density gradient, $| \eta | \ll 1$. The main emphasis is given to the analysis of the case where $\eta = 0$. This analysis is important because in the case $\eta = 0$, i.e., in the absence of a temperature gradient, the inertialess instabilities are not excited at all or their growth rate is very small. The results given below show that for $\eta = 0$ the Alfven instabilities are also not excited. Therefore, the theory as a whole predicts that the *confinement of a finite-β plasma in the homogeneous temperature regime is the most favourable from the viewpoint of the stability problem*.

When studying the *resonant interaction between Alfven waves (inertial perturbations) and particles*, we assume that $k_z = 0$ in section 4.2, $v_{Ti} \ll \omega/k_z \ll v_{Te}$ in section 4.3, and $\omega/k_z \simeq v_{Ti}$ in section 4.4.

The inertial perturbations with $k_z = 0$ in a finite-β plasma, in the approximation $k_\perp \rho_i \to 0$, were originally studied in [1.4]. The results of [1.4] were confirmed in [1.6]. In addition, in [1.6] the interaction between inertial perturbations with $k_z = 0$ and resonant particles was originally analysed.

The *inertial perturbations* with $k_z \neq 0$ in a finite-β plasma and the connection between them and the Alfven waves in a homogeneous plasma were originally studied in [1.8] in the approximation $k_\perp \rho_i \to 0$. (Neglecting the magneto-drift effects, i.e., assuming $\beta \ll 1$, this problem was considered in [2.10, 1.3].) The resonant effects related to these perturbations were initially discussed in [1.9] and later in [4.1].

The theory of Alfven perturbations in an inhomogeneous finite-β plasma was further developed in connection with the study of instabilities in a *two-energy component plasma*, discussed in Chapter 5, and also within the scope of the *hydrodynamic approach* considered in Chapters 9–11. The notion of these perturbations is significant also for the problem of instabilities in *force-free* and *almost force-free high-β plasma* (Chapter 12), and in the theory of *electromagnetic drift Kelvin–Helmholtz instabilities* (Chapter 13).

In addition, the Alfven perturbations in a straight magnetic field are of interest as a limiting case of such perturbations in *curvilinear magnetic fields*. The role of the field line curvature in the Alfven perturbations will be discussed in detail in Chapter 15. In particular, in Chapter 15 it will be shown that in a curvilinear magnetic field the stable Alfven perturbations may become hydromagnetically unstable; this corresponds to the *flute (interchange) instability*. As is known, this effect can also be demonstrated by a model of a plasma in the field of a *gravity force g* directed against the density gradient (see [2], Chapter 6, where the case of a low-β plasma was discussed). In section 4.5, following [2.4], this model problem is considered for the case of a finite-β plasma. The clarification of the correspondence between Alfven perturbations and the flute instability also clarifies the problem of the *finite ion Larmor radius stabilization* of such an instability at finite β. This problem was initially discussed in [4.2, 2.3, 2.4] and later in [4.3, 4.4] and some other papers.

It should be mentioned that the question of *coupling of the Alfven and slow magnetoacoustic waves* was also intensively discussed in the literature, see, in particular, [4.5–4.8]. In this connection, we wish to emphasize once again that in the approximation of straight magnetic field lines and long-wavelength perturbations, $k_\perp \rho_i \ll 1$, the coupling between the Alfven and magnetoacoustic waves is proportional to $(k_\perp \rho_i)^2$ (for $T_i \simeq T_e$ and $\beta \simeq 1$). Therefore, one should be wary of the results of papers where this coupling appears otherwise.

One of the main reasons for the intensive discussion of the coupling between Alfven and magnetoacoustic waves are the *magnetospheric observations* showing that Alfven perturbations also involve *compression and extension* of the magnetic field lines. In the theory of a homogeneous plasma, the compression and extension of the magnetic field lines are typical for magnetoacoustic perturbations but not for the Alfven ones. One should bear in mind, however, that the field-line compressibility is an inherent feature of the Alfven perturbations in an inhomogeneous finite-β plasma; this will be evident from Chapter 10. Consequently, from our point of view, the discussion concerning the coupling of the Alfven and magnetoacoustic waves is to some extent due to the lack of understanding of the above fact.

4.1   Alfven perturbation branches for finite β and $k_\perp \rho_i \to 0$

4.1.1   Perturbations with $k_z = 0$

Let us consider the Alfven perturbations with $k_z = 0$ in a plasma with a Maxwellian velocity distribution of particles. The dispersion relation for such perturbations is given by (2.13) with $\epsilon_{11}^{(0)}$ of the form of (2.9). It follows from the mentioned equations that

$$\omega^2 - \omega(\omega_{pi}^* + \omega_{Di}) + \omega_{pi}^* \omega_{Di} \sigma = 0 \qquad (4.1)$$

where σ is defined by (2.10).

Equation (4.1) has complex roots implying an instability if

$$-1 < \eta < -[1 + 2\beta(1 + \beta/2)^{-2}]^{-1}. \qquad (4.2)$$

For both low and high β, this corresponds to a narrow range of η near $\eta = -1$. This range attains a maximum width at $\beta = 2$:

$$-1 < \eta < -1/2. \qquad (4.3)$$

Let us now consider the solutions of (4.1) in the stability region. We shall designate the real roots of (4.1) by $\omega_\pm$, where

$$\omega_\pm = -\frac{\omega_{Di}}{2}\left\{\frac{2}{\beta} - 1 \pm \left[\left(\frac{2}{\beta} - 1\right)^2 + \frac{8\sigma}{\beta}\right]^{1/2}\right\}. \qquad (4.4)$$

The condition for these roots to be real, i.e., the condition opposite to (4.2), is

$$\sigma > -\beta(1 - 2/\beta)^2/8 \qquad (4.5)$$

or, in terms of η,

$$\eta > -[1 + 2\beta/(1 + \beta/2)^2]^{-1} \qquad \text{or} \qquad \eta < -1. \qquad (4.6)$$

For all β and $\sigma > 0$, the phase velocity of the perturbations with $\omega = \omega_+$ is directed towards the *electron magnetic drift*, $\omega_+/\omega_{De} > 0$, while that of the perturbations with $\omega = \omega_-$ is towards the *ion magnetic drift*, $\omega_-/\omega_{Di} > 0$.

Under the condition

$$-\beta(1 - 2/\beta)^2/8 < \sigma < 0 \qquad (4.7)$$

i.e.,

$$-[1 + 2\beta/(1 + \beta/2)^2]^{-1} < \eta < -1/2 \qquad (4.8)$$

and $\beta < 2$, both waves propagate towards the electron magnetic drift, $\omega_\pm/\omega_{De} > 0$. When $\beta > 2$, they propagate towards the ion magnetic drift, $\omega_\pm/\omega_{Di} > 0$. Waves with $\omega = \omega_-$ for $\beta < 2$ as well as waves with $\omega = \omega_+$ for $\beta > 2$ have a *negative energy*

$$\omega \partial \epsilon_{11}^{(0)}/\partial \omega < 0. \qquad (4.9)$$

The above results will be used in the analysis of the resonant interaction between Alfven waves and particles.

4.1.2   Perturbations with $k_z \neq 0$

For $k_z \neq 0$, it follows from (2.5), (2.7), and (2.9) that (4.1) is replaced by

$$\omega^2 - \omega(\omega_{pi}^* + \omega_{Di}) + \sigma\omega_{pi}^*\omega_{Di} - k_z^2 c_A^2 = 0. \qquad (4.10)$$

Neglecting the gradient terms, we find from (4.10) the dispersion relation for the Alfven waves in a homogeneous plasma

$$\omega^2 = k_z^2 c_A^2. \qquad (4.11)$$

It follows from (4.10) that, for $k_z \neq 0$, the instability of a plasma with η lying in the range of (4.2) is suppressed. Qualitatively, the stability condition means

$$\omega^* \lesssim k_z c_A. \qquad (4.12)$$

At the same time, condition (4.12) characterizes the order of magnitude of maximum k_z when the gradient effects are still important.

4.2 Interactions of drift-Alfven waves with resonant particles for $k_z = 0$

4.2.1 Resonant interaction of the waves with electrons
Using (2.34), we find that interaction between Alfven waves with $k_z = 0$ and electrons is characterized by the growth rate

$$\gamma = -\frac{1}{\partial \epsilon_{11}^{(0)}/\partial \omega} \frac{|\epsilon_{12}|^2}{|\epsilon_{22} - N^2|^2} \text{Im}\,\epsilon_{22}^{(e)} \qquad (4.13)$$

where, according to (2.3),

$$\text{Im}\,\epsilon_{22}^{(e)} = \frac{\pi\beta}{4} \frac{k_\perp^2 c^2 \omega}{|\omega_{\text{De}}|^3} \left(1 + \frac{2}{\beta}\frac{\eta}{1+\eta} - \frac{\omega_{ne}(1-\eta)}{\omega}\right) \exp\left(-\frac{\omega}{\omega_{\text{De}}}\right). \qquad (4.14)$$

(a) Waves with $\omega = \omega_+$
According to section 4.1.1, electrons interact with waves of type $\omega = \omega_+$ for all β if $\sigma > 0$, and for $\beta < 2$ if σ satisfies condition (4.7). Using (4.13) and (4.14), we then obtain the instability condition

$$1 + \frac{2}{\beta}\frac{\eta}{1+\eta} + \frac{4}{\beta}\frac{1-\eta}{1+\eta}\left\{\left[\left(\frac{2}{\beta}-1\right)^2 + \frac{8\sigma}{\beta}\right]^{1/2} + \frac{2}{\beta} - 1\right\}^{-1} < 0. \quad (4.15)$$

This condition is not satisfied for any $\eta \geqslant 0$, i.e., the perturbations discussed cannot be excited in plasma with $\eta \geqslant 0$. However, an instability is possible for $\eta < 0$. In particular, when $\beta \ll 1$, condition (4.15) means

$$-(1 - 2\beta) < \eta < -\beta. \qquad (4.16)$$

This range narrows with increasing β, and the instability disappears when $\beta = 2$.

(b) Waves with $\omega = \omega_-$
As stated in section 4.1.1, electrons interact also with waves of type $\omega = \omega_-$ under condition (4.7) and when $\beta < 2$. Allowing for (4.4), we find from (4.13) and (4.14) the following instability condition for these waves

$$1 + \frac{2}{\beta}\frac{\eta}{1+\eta} + \frac{4}{\beta}\frac{1-\eta}{1+\eta}\left\{-\left[\left(\frac{2}{\beta}-1\right)^2 + \frac{8\sigma}{\beta}\right]^{1/2} + \frac{2}{\beta} - 1\right\}^{-1} > 0. \quad (4.17)$$

The equation for the *instability boundary* for low β following from (4.17) is of the form

$$3\eta^2 + \eta - 1 = 0. \qquad (4.18)$$

The condition $\sigma < 0$ for low β means $-1 < \eta < -0.5$. Hence we find that a *resonant instability* occurs for low β if

$$-0.77 < \eta < -0.5. \tag{4.19}$$

As β increases, the range of the parameters η, corresponding to the instability, narrows. When $\beta = 2$, this range *disappears*, as well as that discussed in section 4.2.1.(*a*).

4.2.2   *Resonant interaction of waves with ions*

We start with the general relation for the growth rate of the Alfven perturbations with $k_z = 0$ following from (2.34):

$$\gamma = -\frac{1}{\partial \epsilon_{11}^{(0)}/\partial \omega}\mathrm{Im}\left(\epsilon_{11} - \frac{\alpha_{12}^2}{\epsilon_{22} - N^2}\right). \tag{4.20}$$

Here $\alpha_{12} = -i\epsilon_{12}$; the superscript (1) in $\mathrm{Im}\,\epsilon_{11}$ is omitted for simplicity. It follows from (2.3) and (2.28) that

$$\begin{aligned}
\mathrm{Im}\,\epsilon_{11} &= X/4 \\
\mathrm{Im}\,\alpha_{12} &= \Omega_i X/\omega \\
\mathrm{Im}\,\epsilon_{22} &= 4\Omega_i^2 X/\omega^2
\end{aligned} \tag{4.21}$$

where

$$X = \frac{\pi}{4}\frac{c^2}{c_A^2}z_i\frac{\omega^3}{|\omega_{\mathrm{Di}}|^3}\left(1 - \frac{\omega_{ni}(1-\eta)}{\omega} + \frac{2}{\beta}\frac{\eta}{1+\eta}\right)\exp\left(-\frac{\omega}{\omega_{\mathrm{Di}}}\right). \tag{4.22}$$

Allowing for (4.21) and (4.22), we reduce (4.20) to the form

$$\gamma = -\frac{X}{\partial \epsilon_{11}^{(0)}/\partial \omega}\frac{[(1/2)(\mathrm{Re}\,\epsilon_{22} - N^2) - 2(\Omega_i/\omega)(\mathrm{Re}\,\alpha_{12})]^2}{|\epsilon_{22} - N^2|^2}. \tag{4.23}$$

Obviously, using this equation, we should allow for the frequencies $\omega = \omega_\pm$ to be real, i.e., η should satisfy inequalities (4.6) or (4.8).

(*a*) *Waves with* $\omega = \omega_-$

In the case of waves where $\omega = \omega_-$, equations (4.22) and (4.23) yield the following instability condition (cf. equation (4.17))

$$1 + \frac{2}{\beta}\frac{\eta}{1+\eta} + \frac{4}{\beta}\frac{1-\eta}{1+\eta}\left\{\left[\left(\frac{2}{\beta}-1\right)^2 + \frac{8\sigma}{\beta}\right]^{1/2} - \frac{2}{\beta} + 1\right\}^{-1} < 0. \tag{4.24}$$

In particular, it follows from (4.24) that there is *no instability* for low β and permissible values of η. In addition, assuming $\beta \gg 1$, we find that in this case there is also no instability for permissible η. Therefore, we conclude that, for all values of the parameters β and η, the interaction between the wave of type ω_- and ions does not lead to instability.

(b) Waves with $\omega = \omega_+$

The ions interact with these waves for $\beta > 2$ and when σ satisfies condition (4.7). Allowing for (4.4), (4.22) and (4.23), we find that in this case the instability condition has the form

$$1 + \frac{2}{\beta}\frac{\eta}{1+\eta} + \frac{4}{\beta}\frac{1-\eta}{1+\eta}\left\{-\left[\left(\frac{2}{\beta}-1\right)^2 + \frac{8\sigma}{\beta}\right]^{1/2} - \frac{2}{\beta} + 1\right\}^{-1} > 0. \quad (4.25)$$

When $\beta \gg 1$, this condition together with (4.8) means

$$-(1 - 8/\beta) < \eta < -1/2. \qquad (4.26)$$

Thus, the interaction between waves of type ω_+ and the ions in a plasma with $\beta > 2$ results in an instability in the range of *negative* values of the parameter η determined (for $\beta \gg 1$) by condition (4.26). As β decreases to $\beta = 2$, this range narrows and, when $\beta = 2$, the instability disappears.

4.3 Role of resonant particles in Alfven perturbations with $k_z v_{Ti} \ll \omega \ll k_z v_{Te}$

4.3.1 The zeroth approximation in the parameters $\omega/k_z v_{Te}$ and $k_z v_{Ti}/\omega$

The Alfven perturbations with $k_z v_{Ti} \ll \omega \ll k_z v_{Te}$ in a plasma with $\beta \simeq 1$ in the zeroth approximation for parameters z_i and $(k_z v_{Ti}/\omega)^2$, as well as for $k_z = 0$, are described by the dispersion relation (4.1). The frequencies of these perturbations were given in section 4.1.1. For finite k_z, resonant interaction between the waves and the particles is modified in comparison with section 4.2. Under the condition $\omega \ll k_z v_{Te}$ the resonant interaction between waves and electrons is as small as $\omega/k_z v_{Te}$. Therefore, in the zeroth approximation in $\omega/k_z v_{Te}$ this interaction can be ignored; this case is dealt with in this section. However, such an interaction will be taken into account in section 4.3.2.

The condition of resonant interaction between waves with finite k_z and the ions means $\omega = k_z v_z + \omega_{di}$. When the ratio $k_z v_{Ti}/\omega$

is small, the term with k_z in this condition is essential only for an exponentially small number of particles. In this section we neglect such an effect and consider it later in section 4.3.3.

Under these assumptions, the general problem of resonant interaction between the Alfven perturbations and the particles reduces to the analysis of the resonance of type $\omega = \omega_{\mathrm{di}}$. The growth rate of the Alfven perturbations due to this resonance is given by the expression following from (2.33):

$$\gamma = -\frac{1}{\partial \epsilon_{11}^{(0)}/\partial \omega} \mathrm{Im}\left(\epsilon_{11} - \frac{\alpha_{12}^2}{\epsilon_{22} - N^2 - \alpha_{23}^2/\epsilon_{33}}\right) \qquad (4.27)$$

where $\alpha_{23} = -i\epsilon_{23}$.

Since the value $\alpha_{12} \equiv -i\epsilon_{12}$ depends solely on ions (see (2.3)), its form turns out to be the same as in the case where $k_z = 0$. On the other hand, the values of α_{23} and ϵ_{33} under the above assumption $(k_z v_{Ti} \ll \omega \ll k_z v_{Te})$ are determined by the electrons. These quantities are real when the terms of order $\omega/k_z v_{Te}$ are neglected. As for component ϵ_{22}, the contribution of the electrons to it in the zeroth approximation in $\omega/k_z v_{Te}$ is vanishingly small, and that of the ions is the same as in the case where $k_z = 0$.

Then, using relations (4.21), we reduce (4.27) to the form similar to (4.23)

$$\gamma = -\frac{X}{\partial \epsilon_{11}^{(0)}/\partial \omega} \frac{[(1/2)(\mathrm{Re}\,\epsilon_{22} - \alpha_{23}^2/\epsilon_{33} - N^2) - 2(\Omega_i/\omega)\mathrm{Re}\,\alpha_{12}]^2}{|\,\epsilon_{22} - \alpha_{23}^2/\epsilon_{33} - N^2\,|^2}.$$

$$(4.28)$$

Hence it can be seen that the problem of determining the sign of the growth rate of resonant interaction reduces to that discussed in section 4.2.2. Therefore, all the conclusions of section 4.2.2 concerning the resonant instability conditions holds in the case discussed. It is also evident that the absolute value of the growth rate for $k_z \gg \omega/v_{Te}$ is different from that in the case where $k_z = 0$. However, we shall not dwell upon the question of the absolute value of the growth rate since, for the most interesting values of the parameters η and β, it is negative.

4.3.2   Interaction of waves with resonant electrons

Using (2.3), we evaluate the quantities D_2, C_1 and C_2 determined by equations (2.6), (2.30) and (2.31) and conclude that, first, the contribution of C_1 into the general expression for the growth rate (2.33) can be neglected, and, second, the terms $\epsilon_{11}^{(0)}$ and ϵ_{13} in the expression for C_2 can be omitted. A small parameter which allows one such an approximation is $(k_z v_{Ti}/\omega)^2$. Then, we take into account the fact that *electron resonance* is important only for perturbations propagating against the ion magnetic drift, $\omega/\omega_{Di} < 0$, since otherwise the electrons would produce only a small correction to the ion growth rate. For such perturbations the quantity $\alpha_{12} = -i\epsilon_{12}$ as well as the ion contribution to ϵ_{22} are real. The resultant expression for the electron growth rate is

$$\gamma = -\frac{\alpha_{12}^2}{\partial\epsilon_{11}^{(0)}/\partial\omega}\frac{\mathrm{Im}\,(\epsilon_{22}^{(0)} - \alpha_{23}^2/\epsilon_{33})}{|\,\epsilon_{22} - N^2 - \alpha_{23}^2/\epsilon_{33}\,|^2}. \tag{4.29}$$

As in equation (4.27), the quantities α_{23} and ϵ_{33} used here include only the electron contribution. But now, in contrast to (4.27), imaginary terms are also allowed for in these quantities. In addition, the imaginary contribution of electrons to ϵ_{22} should be taken into account in deriving (4.29).

In this adopted approximation, the quantities $\epsilon_{22}^{(e)}, \alpha_{23}$, and ϵ_{33} are determined by equations (2.62)–(2.65). Using these equations, we find

$$\mathrm{Im}\left(\epsilon_{22}^{(e)} - \frac{\alpha_{23}^2}{\epsilon_{33}}\right) = \frac{\pi^{1/2}}{2}\beta_e N^2 \frac{\omega}{|\,k_z\,|\,v_{Te}}\frac{g_1}{g_2^2} \tag{4.30}$$

where

$$g_1 = \left(1 - \frac{\omega_{ne}}{\omega}\right)^3 - \frac{3}{2}\eta\frac{\omega_{ne}}{\omega}\left(1 - \frac{\omega_{ne}}{\omega}\right)^2$$
$$- \eta^2\frac{\omega_{ne}^2}{\omega^2}\left(1 - \frac{\omega_{ne}}{\omega}\right) + \frac{\eta^3}{2}\frac{\omega_{ne}^3}{\omega^3} \tag{4.31}$$

$$g_2 = 1 - \frac{\omega_{ne}}{\omega} - \frac{\omega_{De}}{\omega}\left(1 - \frac{\omega_{ne}}{\omega}(1+\eta)\right). \tag{4.32}$$

The instability condition following from equations (4.29) and (4.30) has the form

$$g_1\partial\epsilon_{11}^{(0)}/\partial\omega < 0. \tag{4.33}$$

This condition is not satisfied if $\eta \geqslant 0$, even for large positive η. However, there is an instability for large negative η; in this case the plasma turns out to be unstable for both low and high β.

4.3.3 Ion resonance of the type $\omega = k_z v_z + \omega_{di}$ for $\omega/\omega_{di} < 0$
In order to find out how the general expression for the growth rate (2.34) is simplified in the case of ion resonance of type $\omega = k_z v_z + \omega_{di}$

for $\omega/\omega_{\mathrm{di}} < 0$, we shall consider the following integrals over velocities in the anti-Hermitian parts of $\epsilon_{\alpha\beta}$

$$I_{ns} = \int F\lambda^n (v_z/v_T)^s \delta(\omega - k_z v_z - \omega_D\lambda)\mathrm{d}\mathcal{E}_\perp \mathrm{d}v_z \qquad (4.34)$$

where $\lambda = M\mathcal{E}_\perp/T$; $n = 0,1,2,3,4$; $s = 0,1,2$; and F is the Maxwellian function of (1.31). The indices are omitted for simplicity. After integrating over v_z, we obtain

$$\begin{aligned} I_{ns} = \frac{\pi^{1/2}n_0}{v_T |k_z|} &\exp\left(-\frac{\omega^2}{k_z^2 v_T^2}\right) \int_0^\infty \lambda^n \left(\frac{\omega - \omega_D\lambda}{k_z v_T}\right)^s \\ &\times \exp\left[-\lambda\left(1 - \frac{2\omega_D\omega}{k_z^2 v_T^2}\right) - \frac{\lambda^2 \omega_D^2}{k_z^2 v_T^2}\right]\mathrm{d}\lambda. \end{aligned} \qquad (4.35)$$

In this calculation we assumed ω and ω_D to be values of the same order, but opposite in sign, $\omega/\omega_D < 0$, and that $(|\omega_D|, |\omega|) \gg k_z v_T$. Consequently, a substantial contribution to the integrals of (4.35) is due only to sufficiently small values of λ

$$\lambda \lesssim k_z^2 v_T^2/|\omega_D\omega| \ll 1. \qquad (4.36)$$

Therefore, the integrals I_{ns} reduce to

$$\begin{aligned} I_{ns} = \frac{\pi^{1/2}n_0}{v_T |k_z|} &\left(\frac{\omega}{k_z v_T}\right)^s \exp\left(-\frac{\omega^2}{k_z^2 v_T^2}\right) \\ &\times \int_0^\infty \lambda^n \exp\left(-\frac{2\lambda|\omega_D\omega|}{k_z^2 v_T^2}\right)\mathrm{d}\lambda. \end{aligned} \qquad (4.37)$$

It is also clear that the integrals with the subscripts $n > 0$ contain additional small factors of order $(k_z^2 v_T^2/|\omega_D\omega|)^n$ whilst the greatest among the integrals with subscripts $s = 0,1,2$ is the integral with $s = 2$. Therefore, it is sufficient to take into account the contribution to the growth rate due only to the component $\epsilon_{\alpha\beta}$ containing I_{ns} with the smallest value of n and the greatest value of s, i.e., to the component $\epsilon_{33}(n=0, s=2)$. Then, using equations (2.3),(2.34) and (4.37), we obtain

$$\gamma = -\frac{1}{\partial\epsilon_{11}^{(0)}/\partial\omega} \frac{\alpha_{12}^2(\mathrm{Re}\,\alpha_{23})^2\mathrm{Im}\,\epsilon_{33}^{(i)}}{[(\epsilon_{22}^{(i)} - N^2)\mathrm{Re}\,\epsilon_{33} - (\mathrm{Re}\,\alpha_{23})^2]^2} \qquad (4.38)$$

where

$$\mathrm{Im}\,\epsilon_{33}^{(i)} = -\frac{\pi^{1/2}\omega_{\mathrm{pi}}^2\exp(-\omega^2/k_z^2 v_{Ti}^2)}{|k_z|\,v_{Ti}\omega_{Di}}g_3 \qquad (4.39)$$

$$g_3 = 1 - \frac{\omega_{ni}}{\omega}\left(1 - \frac{3}{2}\eta\right) - \frac{\omega\omega_{Ti}}{k_z^2 v_{Ti}^2}. \qquad (4.40)$$

The quantities $\mathrm{Re}\,\alpha_{23}$ and $\mathrm{Re}\,\epsilon_{33}$ are determined by (2.62) and (2.63), while the real quantities α_{12} and $\epsilon_{22}^{(i)}$ are calculated by means of (2.3) for $k_z = 0$.

From (4.38) we obtain the instability condition

$$-\omega_{\mathrm{Di}}g_3\partial\epsilon_{11}^{(0)}/\partial\omega < 0. \qquad (4.41)$$

In particular, it follows hence that the instability is absent when $\eta = 0$, but that it does occur at arbitrary small positive η

$$\eta > 0. \qquad (4.42)$$

When the assumption $\omega/k_z v_{T\mathrm{i}} \gg 1$ is used, the growth rate is exponentially small, but it becomes significant at the validity limits of this assumption.

Note that, when $\eta > 0$, perturbations with $\omega/\omega_{\mathrm{Di}} < 0$ correspond to the branch with $\omega = \omega_+$ determined by equation (4.4).

4.4   Role of resonant particles in Alfven perturbations with $\omega \simeq k_z v_{T\mathrm{i}}$ for $\nabla T = 0$

Since expression (2.33) for the growth rate of Alfven perturbations with $\omega \simeq k_z v_{T\mathrm{i}}$ is very complicated, we restrict ourselves to analysing the resonant excitation of such perturbations in a plasma with $\nabla T = 0$. Even with this simplifying assumption, the problem still remains too complicated. Therefore, we also assume that the electron temperature is lower than the ion temperature, $T_\mathrm{e} \ll T_\mathrm{i}$. Such an assumption substantially simplifies equation (2.33) since ϵ_{33} is large at low electron temperatures, i.e., the quantity $1/\epsilon_{33}$ is a small parameter. In the zeroth approximation in this parameter, equation (2.33) reduces to

$$\gamma = -\frac{A}{\partial D_1^{(0)}/\partial\omega} \qquad (4.43)$$

where

$$A = \mathrm{Im}\left(\epsilon_{11} + \frac{\epsilon_{12}^2}{\epsilon_{22} - N^2}\right). \qquad (4.44)$$

Note that the case $1/\epsilon_{33} \to 0$ corresponds to the approximation of *ideal plasma conductivity* along the magnetic field lines (see also Chapter 5).

We transform equation (4.44) into the form

$$
A = \mathrm{Im}\,\epsilon_{11}\left[\left(\mathrm{Re}\,\epsilon_{22} - N^2 - \mathrm{Re}\,\alpha_{12}\frac{\mathrm{Im}\,\alpha_{12}}{\mathrm{Im}\,\epsilon_{11}}\right)^2 \right.
$$
$$
\left. + \left(\frac{(\mathrm{Re}\,\alpha_{12})^2}{(\mathrm{Im}\,\epsilon_{11})^2} + \frac{\mathrm{Im}\,\epsilon_{22}}{\mathrm{Im}\,\epsilon_{11}}\right)\left(\mathrm{Im}\,\epsilon_{22}\mathrm{Im}\,\epsilon_{11} - (\mathrm{Im}\,\alpha_{12})^2\right)\right].
$$

(4.45)

The imaginary parts of $\epsilon_{11}, \alpha_{12}$ and ϵ_{22} are determined by the resonant interaction between the waves and the ions (we neglect the electron resonance as being small due to $(M_e/M_i)^{1/2}$). Taking into account equations (2.28) and (2.3) and the definition for α_{12} $(\alpha_{12} = -i\epsilon_{12})$, we find that when $\nabla T_i = 0$

$$
\mathrm{Im}\,\epsilon_{11} = \frac{\pi}{16}\frac{\omega_{pi}^2 z_i}{\omega\Omega_i^2}\left(1 - \frac{\omega_{ni}}{\omega}\right)\langle\lambda^2\omega_{di}^2\delta(t)\rangle
$$

$$
\mathrm{Im}\,\alpha_{12} = \frac{\pi}{4}\frac{\omega_{pi}^2 z_i}{\omega\Omega_i}\left(1 - \frac{\omega_{ni}}{\omega}\right)\langle\lambda^2\omega_{di}\delta(t)\rangle
$$

(4.46)

$$
\mathrm{Im}\,\epsilon_{22} = \pi\frac{\omega_{pi}^2 z_i}{\omega}\left(1 - \frac{\omega_{ni}}{\omega}\right)\langle\lambda^2\delta(t)\rangle.
$$

Here $t = \omega - \omega_{di} - k_z v_z$, $\lambda = M_i v_\perp^2/2T_i$, and the angle brackets $\langle\ldots\rangle$, in contrast to sections 3.4 and 3.5, means averaging over both the transverse and longitudinal velocity, i.e., they are determined by (2.12).

In the case of perturbations with $\omega/\omega_{ni} < 0$ the signs of the right-hand sides of (4.46) are the same as in the approximation of a homogeneous plasma. Therefore, such waves can be definitely considered to be stable. (This will be evident also from the subsequent analysis.) Consequently, we should study only the case where $\omega/\omega_{ni} > 0$.

It follows from (4.4) that in the case discussed of a plasma with $\nabla T = 0$, the frequency of an oscillation branch with $\omega/\omega_{ni} > 0$ is greater than ω_{ni} so that

$$
1 - \omega_{ni}/\omega > 0. \tag{4.47}
$$

Consequently, $\mathrm{Im}\,\epsilon_{11} > 0$. (We assume that $\omega > 0$ for simplicity.) Let us show that the expression in the square brackets in the right-hand side of (4.45) is also positive. Since, as follows from (4.46), $\mathrm{Im}\,\epsilon_{22} > 0$, it is sufficient for this purpose to prove the inequality

$$
\mathrm{Im}\,\epsilon_{22}\mathrm{Im}\,\epsilon_{11} \geqslant (\mathrm{Im}\,\alpha_{12})^2. \tag{4.48}
$$

Allowing for (4.46), we note that this inequality is equivalent to

$$
\langle\lambda^2\omega_{di}^2\delta(t)\rangle\langle\lambda^2\delta(t)\rangle \geqslant [\langle\lambda^2\omega_{di}\delta(t)\rangle]^2. \tag{4.49}
$$

To prove (4.49) we use the Schwartz inequality

$$\int f^2 \mathrm{d}x \int g^2 \mathrm{d}x \geqslant \left(\int fg \mathrm{d}x \right)^2 \tag{4.50}$$

where f, g are arbitrary functions of x. Assuming that

$$\begin{aligned} f &= \lambda \omega_{\mathrm{di}} [\delta(t)]^{1/2} \\ g &= \lambda [\delta(t)]^{1/2} \end{aligned} \tag{4.51}$$

we see that (4.49) is true. Thereby, we have proved that $A > 0$. Allowing for the denominator in (4.43) to be positive (see section 4.4.1), we find that $\gamma < 0$.

The same reasoning can be applied also to the branch with $\omega/\omega_{\mathrm{ni}} < 0$. The only difference is the fact that the inequality (4.47) for this branch is self-evident, i.e., it is satisfied without any relationship to the dispersion law (4.4).

Thus, we have shown that a plasma with $\nabla T = 0$ is stable for the perturbations considered.

4.5 Correspondence between Alfven perturbations and the flute instability in finite-β plasma

If a plasma is in the field of a *gravity force* $\boldsymbol{g} \parallel \boldsymbol{x}$, according to [2], section 6.1, the left-hand side of the dispersion relation (4.1) should be augmented by the term $-g\kappa_n k_y^2 / k_\perp^2$. Then, assuming $\nabla T = 0$, equation (4.1) is replaced by

$$(\omega - \omega_{\mathrm{ni}})(\omega - \omega_{\mathrm{Di}}) - g\kappa_n k_y^2 / k_\perp^2 = 0. \tag{4.52}$$

When $g\kappa_n < 0$ and the terms $\omega_{\mathrm{ni}}, \omega_{\mathrm{Di}}$ are neglected, this equation describes the *flute instability*. When these terms are allowed for, the instability is absent if

$$\omega_{\mathrm{ni}}^2 (1 + \beta/2)^2 > 4 \, | \, g\kappa_n \, | \, . \tag{4.53}$$

This is the condition for *flute instability suppression* by the effect of finite ion Larmor radius. It is evident that at finite β this stabilization effect is somewhat stronger than that at $\beta \to 0$. At the same time, inequality (4.53) characterizes the transition from the flute instability to the stable Alfven perturbations.

Equation (4.52) concerns the *small-scale perturbations* described by the *WKB approximation*. *Large-scale perturbations* of a *cylindrically symmetric plasma*, as was shown in [2.4], are described by the following differential equation

$$\frac{1}{r}\frac{\partial}{\partial r}\left(r^3 Q n_0 \frac{\partial \Phi}{\partial r}\right) + \left((1-m^2)Q n_0 + (m^2 g + r\omega^2)\frac{\partial n_0}{\partial r}\right)\Phi = 0. \quad (4.54)$$

Here

$$Q = (\omega - \omega_{ni})(\omega - \omega_{Di}) \qquad \Phi = E_\varphi / B_0. \quad (4.55)$$

The dependence of the perturbations on the coordinates is taken in the form $f(r)\exp(im\varphi)$ where r and φ are the cylindrical coordinates, E_φ is the φ-th component of the perturbed electric field; ω_{ni} and ω_{Di} are determined as above but with the substitution $k_y \to m/r$. The gravity force g is assumed to be directed along the cylinder radius.

As in the case of a plasma with $\beta \to 0$ (see [2], section 6.2), the most essential difference between the results following from (4.52) and (4.54) is for $m = 1$. Similarly to the case where $\beta \to 0$, the perturbations with $m = 1$ in a *plasma with a free boundary* turn out to be unstable even in the case of the strong inequality (4.53).

5 Low-frequency Instabilities in a Two-energy Component Plasma

In Chapters 3 and 4 we considered instabilities in a plasma where each charge species (ions and electrons) was characterized by the Maxwellian velocity distribution function. In addition, situations are possible in *laboratory* and *space conditions* when either charge species consists of more than a single energy group. The simplest case of such a plasma is that with *cold* and *hot particles* of the same charge species. In laboratory conditions this can be due to *thermonuclear α-particles, fast-neutral injection, high-frequency heating, runaway electrons*, special addition of a *cold plasma component*, etc. The presence of *cold* and *hot plasma components* is even more typical for various plasma regions in the *magnetosphere* of the Earth. Therefore, the study of the *gradient instabilities* in a *two-energy component plasma* is of interest. This chapter is devoted to this problem.

If the density of cold particles is sufficiently high, one can consider a plasma to be *ideally conducting* along the magnetic field lines when studying *long-wavelength low-frequency perturbations*. Formally, this fact means that the *permittivity tensor component* ϵ_{33} is infinite. Hence the instabilities excited by *hot particles* in a plasma also containing *cold particles* can be studied with the help of the following dispersion relation which is simpler than equation (1.1):

$$\begin{vmatrix} \epsilon_{11} - N^2 \cos^2\theta & \epsilon_{12} \\ \epsilon_{21} & \epsilon_{22} - N^2 \end{vmatrix} = 0. \qquad (5.1)$$

It is also clear that, if one keeps in mind the *long-wavelength magnetoacoustic instabilities*, their dispersion relation, in the presence of a *cold plasma*, is of the form

$$\epsilon_{22} - N^2 = 0. \qquad (5.2)$$

This equation can be derived from both (5.1) (allowing for ϵ_{12} to be small for small $k_\perp$, see section 2.1.1) and from (2.8) and (2.6) (allowing for $\epsilon_{33} \to \infty$).

If we are interested in the *resonant interaction of high-energy particles* with Alfven perturbations, the rather complicated expression (2.33) is replaced by the following equation for the *growth rate* in the presence of a *cold plasma* (cf.(4.43))

$$\gamma = -\frac{1}{\partial D_1^{(0)}/\partial\omega}\mathrm{Im}\left(\epsilon_{11} + \frac{\epsilon_{12}^2}{\epsilon_{22} - N^2}\right). \qquad (5.3)$$

Within the scope of (5.2) we shall consider the *gradient excitation* of *fast magnetoacoustic waves* by *high-energy particles* (section 5.1) and the so-called *drift-mirror instability* (section 5.2). We shall use (5.3) in the analysis of *Alfven instabilities driven by high-energy particles* with *finite β* (section 5.3).

Note that equations (5.2) and (5.3) formally coincide with equations (2.14) and (2.34). The difference is that equations (2.14) and (2.34) deal with the case $k_z = 0$, while equations (5.2) and (5.3) are valid for both $k_z = 0$ and $k_z \neq 0$. The same remark could be made about equation (4.43).

The study of low-frequency gradient-driven instabilities in a *two-energy component plasma* was initiated in [5.1, 5.2] (see also the review in [5.3] and the book of [5.4]). There the dispersion relation of [1.8] for the magnetoacoustic waves (i.e., equation (2.8)) has been generalized by taking into account a *cold plasma component*. As a result, dispersion relation (5.2) has been obtained. In [5.1] the first investigation of the *drift-mirror instability* was given. In [5.2] the *gradient excitation of the fast magnetoacoustic waves* was considered. Allowing for the analysis of fast magnetoacoustic waves by [5.2], the authors of [5.5, 1.10] paid attention to the importance of a *cold plasma component* in the study of the *Alfven perturbations*. They have obtained equation (5.1) and performed the initial analysis of equation (5.3). The paper [1.9] and the earlier paper [1.6], where similar problems were discussed for the case of a *single-energy component plasma*, were also the basis for [5.5, 1.10].

The gradient excitation of fast magnetoacoustic waves was initially discussed in [1.5] for the case of a single-energy component plasma. Qualitatively, the result of [1.5] is that the excitation can be due to the particles in the *tail of the Maxwellian distribution*, precisely, to the particles whose *magnetic-drift velocity* is greater than the *Alfven velocity*. When $\beta \simeq 1$, the number of those particles is exponentially small, the exponent power being of the order of the characteristic plasma size divided by the average ion Larmor radius.

It is clear that such an instability is rather exotic under the above conditions. However, the situation changes considerably when, due to the addition of a cold plasma, the *Alfven velocity decreases* and becomes comparable with the averaged *magnetic-drift velocity* of the ions. This effect was just considered in [5.2].

At the same time, it is clear that the gradient excitation of the fast magnetoacoustic waves can be of interest only for a rather *strongly inhomogeneous plasma*. The proper limitations on the plasma transverse size were discussed in [5.6]. There it was noted that this instability can be of interest for *laboratory confinement systems* containing *thermonuclear α-particles* and, in particular, for *tokamak reactors*. Gradient excitation of the fast magnetoacoustic waves was also discussed in [5.7] where the influence of a *high-frequency turbulence* on this instability was considered.

The *drift-mirror instability* is a *hybrid* caused by the effects of *anisotropy and inhomogeneity*. In [5.1] where this instability was initially considered, it was represented as the *ordinary mirror instability* (see [2.1], section 6.8) with the *shift* of the real part of the frequency to the drift frequency. In [5.1] it was also noted that the *boundary* of the drift-mirror instability is the same as that of the ordinary mirror instability, i.e., the gradient effects do not affect the instability boundary. The authors of [5.8] have confirmed the conclusion of [5.1] about the coincidence of the boundaries of the mirror and drift-mirror instabilities. However, it should be noted that this conclusion of [5.1, 5.8] is not universal: it is valid in the zeroth approximation in the parameter $\omega/k_z v_T$, but it does not hold at finite $\omega/k_z v_T$. This fact is shown in section 5.2.

We also note that the development of the drift-mirror instability theory was initially stimulated by [5.9] where the observations of *slow compressional waves* in the *dayside magnetosphere* during *substorms* were reported. The authors of [5.1, 5.10] related this phenomenon with the excitation of the drift-mirror instability. For further magnetospheric applications of the theory of this instability see also in particular [5.11, 5.12]. A critical survey of these applications has been made in [5.13, 5.14].

In a number of *laboratory experiments*, in particular, in the *theta-pinch* with *hot electrons*, an *anisotropic finite-β plasma* is realized. Then low-frequency instabilities of magnetoacoustic waves with an azimuthal phase velocity of the order of the drift velocity are observed [5.15, 5.16]. The *drift-mirror instability* was applied in [5.16] to interpret these observations.

The gradient excitation of the *Alfven waves* by high-energy particles was investigated in [5.5, 1.10]. This was stimulated by the problem of the interpretation of the low-frequency *geomagnetic pul-*

sations Pc3–5 (note, Pc stands for 'pulsations, continuous', see also [5.17]). The papers [5.5, 1.10] were a starting point for a wider analysis of the *Alfven instabilities* in the magnetospheric plasma allowing for the *curvature* and *longitudinal inhomogeneity* of the magnetic field (see [5.18] and Chapters 15 and 16 of this book).

Previously the importance of studying the *generation of Alfven waves* by *resonant particles* in the magnetospheric plasma has been repeatedly emphasized since many types of geomagnetic pulsations have a structure typical for these waves. The *generation mechanisms* caused by ion *cyclotron resonance* have been initially investigated (see the references in [5.5]). The condition for such a resonance has the form $\omega = k_z v_z + \Omega_i$. For sufficiently fast particles, when $v \gg c_A$, this condition means $k_z \simeq \Omega_i/v$ (since $\omega = k_z c_A \ll k_z v$). Therefore, the characteristic frequency of the waves excited by cyclotron resonance is of order $\omega \simeq \Omega_i c_A/v$. The so-called Pc1 pulsations lie in this frequency range.

The above-mentioned Pc3–5 *pulsations* have *essentially lower frequencies*. Therefore, the notion of a *cyclotron mechanism* of generation is *invalid* for such pulsations. On the other hand, the frequencies of these pulsations are found to be *comparable* with the drift frequencies of magnetospheric plasma calculated for reasonable values of the transverse wave numbers. In connection with this, the *drift mechanisms* for the generation of *Alfven waves* were discussed long before paper [5.5]. In some papers the notion of the *kinetic drift-Alfven instability* of the type discussed in [2.10] was used (these papers will be discussed in Chapter 14). However, it should be noted that, for a *finite* β value typical in some regions of the magnetosphere, this instability has a rather low growth rate and is suppressed due to *ion Landau damping* for not too small β. In addition, it becomes essentially weaker in the presence of a cold plasma because it strongly depends on the perturbed electric field E_z. This field becomes smaller the greater the *longitudinal conductivity* of the plasma. In contrast with this, the growth rate of the drift-Alfven instability of type [5.5] increases with increasing β, and its nature is not related to the E_z field. Therefore, such an instability and its various *modifications* caused by the *curvature* and *inhomogeneity* of the magnetic field are more promising prospects for applications to the problems of magnetospheric pulsation generation.

Resonant interaction of high-energy particles with Alfven waves can lead to *quasilinear diffusion* in the coordinate and velocity spaces. Such a diffusion was studied in [5.19].

Instabilities in a two-energy component magnetospheric plasma with finite β were also considered, in particular, in [5.20–5.27]. However, in the majority of these papers, mistakes were made (see [5.27]).

Therefore, one should consider these papers warily.

5.1 Gradient excitation of the fast magnetoacoustic waves by high-energy particles

According to (2.3), in the presence of cold and hot particles we can write

$$\epsilon_{22} = c^2/c_A^2 + \epsilon_{22}^{(h)} \tag{5.4}$$

where

$$\epsilon_{22}^{(h)} = \sum_h \frac{4\pi e^2 k_\perp^2}{M\omega\Omega^2} \langle \frac{v_\perp^4}{4} G\zeta_0 \rangle. \tag{5.5}$$

The superscript h means the contribution of *high-energy* (or *hot*) particles.

Neglecting $\epsilon_{22}^{(h)}$ as a small correction, we find that the real part of the frequency of the waves described by equation (5.2) is determined by

$$\omega^2 = k_\perp^2 c_A^2. \tag{5.6}$$

This dispersion law corresponds to *fast magnetoacoustic waves*. The phase velocity of these waves in the *"drift"* direction y is greater than or equal to the Alfven velocity, $\omega/k_y \gtrsim c_A$. On the other hand, it is necessary for the *gradient effects* of *high-energy particles* to be important such that $\omega/k_y \lesssim V^* \simeq v\rho/a$ where v and ρ are the characteristic velocity and Larmor radius of the particles and a is the characteristic size of their inhomogeneity. It follows from the above inequalities that an allowance for the drift effects in the problem of fast magnetoacoustic waves is important only under the condition

$$\rho/a \gtrsim c_A/v \tag{5.7}$$

i.e., for a *sufficiently steep inhomogeneity* of the high-energy component.

Assuming condition (5.7) to be valid, we shall calculate the growth rate of the fast magnetoacoustic waves due to their interaction with high-energy particles. For simplicity we consider the pressure of these particles to be small, $\beta_h \ll 1$, and neglect the *magneto-drift effects* related to a finite ω_d value. In addition, we restrict ourselves to the case of a single component of high-energy particles and assume that their distribution is *Maxwellian* with *homogeneous temperature*, $\nabla T = 0$. Then, according to (5.5), we find that

$$\epsilon_{22}^{(h)} = -\beta_h \frac{k^2 c^2}{\omega} \left(1 - \frac{\omega_n}{\omega}\right) \int \frac{f_\parallel(v_z) dv_z}{\omega - k_z v_z}. \tag{5.8}$$

Here $f_\parallel(v_z)$ is the *one-dimensional Maxwellian function* determined by (1.35) with $U = 0$ and $n_0 \to 1$. Exact definitions of β_h and ω_n are

$$\beta_h = 8\pi n_h T_h / B_0^2$$
$$\omega_n = ck_y \kappa_n T_h e_h / B_0 \tag{5.9}$$

where n_h, T_h, and e_h are the density, temperature and charge of high-energy (hot) particles, and $\kappa_n = \partial \ln n_h / \partial x$.

Using (5.2), (5.4), and (5.8), we find the following expression for the growth rate

$$\gamma = -\frac{\pi^{1/2}}{2} \frac{\beta_h}{|k_z| v_{Th}} \left(1 - \frac{\omega_n}{k_\perp c_A}\right) c_A^2 k_\perp^2 \exp\left[-\left(\frac{k_\perp c_A}{k_z v_{Th}}\right)^2\right] \tag{5.10}$$

where $v_{Th} = (2T_h/M_h)^{1/2}$, and M_h is the mass of the hot particles.

The instability condition is qualitatively the same as (5.7). The growth rate attains a maximum for $k_z/k_\perp \simeq c_A/v_{Th}$, so that

$$\gamma_{\max} \simeq \beta_h \operatorname{Re}\omega. \tag{5.11}$$

It can be seen that when parameter β_h increases to the value $\beta_h \simeq 1$, the growth rate also increases and becomes of order $\operatorname{Re}\omega$. In this case a more exact analysis is necessary, see [5.2].

It follows from (5.6) and (5.11) that the growth rate increases with increasing $k_\perp$. However, (5.11) becomes invalid for $k_\perp \rho \gtrsim 1$. We then find the estimate

$$\gamma_{\max} \lesssim \beta_h \Omega_h c_A / v_{Th} \tag{5.12}$$

where Ω_h is the cyclotron frequency of high-energy particles.

5.2   Drift-mirror instability

In contrast to section 5.1, we now assume $\omega/k_\perp \ll c_A$ and the distribution function of the high-energy particles to be *anisotropic*. As in section 5.1, we shall study the perturbations described by dispersion relation (5.2).

In the case of an anisotropic plasma it is more convenient to use the equilibrium distribution function of the form of (1.32), but not (1.29). The recipe for modifying ϵ_{22} under the replacement of (1.29) by (1.32) was elucidated in section 1.4: this modification reduces to the substitution of G_1, defined by (1.59), into equation (2.3) instead of G. Taking the above into account, we present the start value of ϵ_{22} in the form

$$\epsilon_{22} = \sum \frac{4\pi e^2 k_\perp^2}{M\omega\Omega^2} \int \frac{v_\perp^4}{4} \left[\frac{\partial F_1}{\partial \mathcal{E}_\perp} + \frac{k_z}{\omega} \left(\frac{\partial F_1}{\partial v_z} - v_z \frac{\partial F_1}{\partial \mathcal{E}_\perp} \right) \right.$$
$$\left. + \frac{k_y}{\omega\Omega} \frac{\partial F_1}{\partial x} \right] (\omega - \omega_d - k_z v_z)^{-1} v_\perp dv_\perp dv_z \qquad (5.13)$$

where F_1 is the distribution function of high-energy particles. For simplicity we assume the *distribution function F_1 to be bi-Maxwellian*:

$$F_1 = \frac{n_0 M}{T_\perp} \left(\frac{M}{2\pi T_\parallel} \right)^{1/2} \exp\left(-\frac{Mv_\perp^2}{2T_\perp} - \frac{v_z^2}{v_{T\parallel}^2} \right). \qquad (5.14)$$

Here $v_{T\parallel} = (2T_\parallel/M)^{1/2}$; $n_0, T_\perp$ and $T_\parallel$ are functions of the coordinate x.

Integrating (5.13) over the longitudinal velocity v_z, we find

$$\epsilon_{22} = \frac{k_\perp^2 c^2}{\omega^2} \sum \frac{\beta_\perp}{2} \int_0^\infty \lambda^2 \exp(-\lambda)d\lambda$$
$$\times \left[\alpha_1 + \alpha_2 \frac{\omega}{\omega - \omega_d} Z\left(\frac{\omega - \omega_d}{k_z v_{T\parallel}} \right) \right]. \qquad (5.15)$$

Here $\lambda = Mv_\perp^2/2T_\perp$ and

$$\alpha_1 = \frac{T_\perp}{T_\parallel} - 1 - \frac{\omega_{T\perp}}{k_z^2 v_{T\parallel}^2}(\omega - \omega_d)$$

$$\alpha_2 = \frac{T_\perp}{T_\parallel} + \frac{\omega_d}{\omega}\left(1 - \frac{T_\perp}{T_\parallel} \right) - \frac{\omega_{n\perp}}{\omega}\left[1 + \eta\left(\lambda - \frac{3}{2} + \frac{(\omega - \omega_d)^2}{k_z^2 v_{T\parallel}^2} \right) \right]$$
$$\qquad (5.16)$$

where $\omega_{n\perp} = ck_y T_\perp \kappa_n/eB_0$, $\omega_{T\perp} = \eta\omega_{n\perp}$, and $\eta = \partial \ln T_\perp/\partial \ln n_0$ (we assume $\partial \ln T_\parallel/\partial \ln T_\perp = 1$). Below we restrict ourselves to the case of one species of high-energy particles.

Because of the complexity of expression (5.15) we consider the perturbations with $(\omega, \omega_d) \ll k_z v_{T\parallel}$ keeping quadratically small terms. We then obtain from (5.2) that the frequency of perturbations at the instability boundary is equal to

$$\omega = \frac{T_\parallel}{T_\perp}\left(\omega_{n\perp} + \frac{3}{2}\omega_{T\perp} \right) + 3\omega_D\left(1 - \frac{T_\parallel}{T_\perp} \right) \qquad (5.17)$$

while the instability boundary is determined by the equation

$$\beta_\perp(T_\perp/T_\| - 1) = 1 + \delta. \tag{5.18}$$

Here

$$\delta = \frac{\beta_\perp \omega_{n\perp}^2 h}{k_z^2 v_{T\|}^2} \frac{T_\|}{T_\perp}\left(1 + \frac{3}{2}\beta_\perp + \frac{3}{2}\eta(1 + \beta_\perp)\right). \tag{5.19}$$

The plasma is unstable if the temperature anisotropy is greater than that defined by equation (5.18), i.e.,

$$\frac{T_\perp}{T_\|} - 1 > \frac{1 + \delta}{\beta_\perp}. \tag{5.20}$$

This is the condition for *drift-mirror instability*, while for $\delta = 0$ equation (5.20) represents the condition for *mirror instability*.

It can be seen that $\delta > 0$ for $\eta > 0$ and $\eta < \eta_0$, where

$$\eta_0 = -\frac{\beta_\perp + 2/3}{\beta_\perp + 1}. \tag{5.21}$$

Hence we find that, for the given η, the condition for mirror instability is less rigid than that for the drift-mirror instabilty. This fact is in accordance with the discussion presented in the introduction of this chapter. Therefore, the linear statement of the problem of drift-mirror instability is justified only when

$$\eta_0 < \eta < 0. \tag{5.22}$$

For $\beta_\perp \ll 1$ it follows from (5.21) that

$$\eta_0 = -2/3. \tag{5.23}$$

For $\beta_\perp \gg 1$ we have, instead of (5.23),

$$\eta_0 \approx -1. \tag{5.24}$$

Note also that the case $\beta_\perp \gg 1$ and $\eta_0 \approx -1$ corresponds to the *almost force-free high-β plasma*. The drift-mirror instability in such a plasma requires a special consideration and therefore the relation for the instability boundary differs from (5.18).

5.3   Gradient excitation of Alfven waves by high-energy particles

One of the essential points in the problem of the *Alfven waves* in a *two-energy component plasma* is the *modification* of the dispersion law of these waves.

In the presence of cold and hot particles we find from (2.3) that (2.9) is replaced by

$$\epsilon_{11}^{(0)} = \epsilon_{11}^{(0)c} + \epsilon_{11}^{(0)h} \tag{5.25}$$

where

$$\epsilon_{11}^{(0)c} = c^2/c_A^2 \tag{5.26}$$

$$\epsilon_{11}^{(0)h} = \frac{c^2}{c_A^2} \frac{n_h}{n_c} \left(1 - \frac{\omega_n}{\omega}\right)\left(1 - \frac{\omega_D}{\omega}\right). \tag{5.27}$$

Here n_h and n_c are the densities of hot (high-energy) and cold particles (we assume $n_h \ll n_c$); the Alfven velocity c_A is determined by the cold plasma density, so that the ratio n_h/n_0 appears in (5.27). For simplicity we assume that the *temperature gradient* of the hot particles is *vanishing*; ω_n and ω_D are the density-gradient drift and magnetic-drift frequencies of the hot particles. The pressure of hot particles is assumed to be essentially larger than that of cold ones, so that the parameter β is determined by hot particles only. It follows from the equilibrium equation (1.42) that

$$\omega_D = -\omega_n \beta/2. \tag{5.28}$$

Recall that in a single-energy component plasma the allowance for gradient effects in the problem of Alfven waves leads to *oscillation branches* with frequencies of the order of drift or magnetic-drift frequencies (see, e.g., (4.1)). In contrast to this, in a two-energy component plasma with $n_c \gg n_h$ the frequencies of these waves turn out to be *essentially less* than drift or magnetic-drift frequencies determined by the hot particles. Considering, for example, the Alfven perturbations with $k_z = 0$, taking the right-hand side of (5.25) to vanish (in accordance with (2.13)) and using (5.26) and (5.27), we find

$$\omega^2 = -\omega_n \omega_D n_h/n_c. \tag{5.29}$$

Hence, $\omega \ll (\omega_n, \omega_D)$ if β is neither extremely low or high, so that

$$n_h/n_c \ll \beta \ll n_c/n_h. \tag{5.30}$$

The allowance for $k_z \neq 0$ leads to the known *Alfven addition* in the right-hand side of (5.29), so that in this case we have, instead of (5.29),

$$\omega^2 = k_z^2 c_A^2 + 2\omega_D^2 n_h/n_c \beta. \tag{5.31}$$

Here relation (5.28) was taken into account.

The importance of the modification considered in the Alfven dispersion law will be more obvious if we remember the analysis of the *resonant interaction* between waves and particles given in Chapter 4, and especially in section 4.4. In section 4.4, when considering the resonant interaction of type $\omega = k_z v_z + \omega_d$ with the branch $\omega/\omega_{ni} > 0$, we arrived at the conclusion that the excitation of this branch is impossible since the frequency of this branch satisfies the inequality (4.47) but not the opposite one. We now have (5.29) or (5.31) instead of (4.4), so that $\mid \omega \mid \ll \omega_n$ (if k_z is not too large). Hence inequality (4.47) is replaced by the opposite one. Therefore, it is clear without additional analysis that, in contrast to section 4.4, in the case of a *two-energy component plasma*, excitation but not damping of Alfven waves takes place.

Hence we should make a quantitative calculation, in order to prove the above qualitative consideration, and to obtain an estimate of the growth rate. We shall study not only the excitation of the *long-wavelength perturbations* described by equation (5.3) and treated below in section 5.3.1, but also the excitation of *short-wavelength perturbations* (section 5.3.2). The results will be summarized in section 5.3.3.

We restrict ourselves to the case of *Maxwellian high-energy particles* only. The case of a *loss-cone distribution* was discussed in [5.5].

5.3.1   *Instability of long-wavelength Alfven perturbations*

We start with the general expression for the growth rate, equation (5.3). The thermal velocity of the *hot particles*, v_T, is assumed to be greater than the Alfven velocity, $v_T \gg c_A$. Then the contribution of ω in the expression for ζ_0 in the permittivity tensor components $\epsilon_{\alpha\beta}$ (see equations (2.3), (2.28), and (1.56)) is unimportant, so that we have approximately

$$\zeta_0 \equiv (\omega - k_z v_z - \omega_d)^{-1} \approx -\mathcal{P}(k_z v_z + \omega_d)^{-1} - i\pi\delta(k_z v_z + \omega_d) \quad (5.32)$$

where $\mathcal{P}$ denotes the principal value of the integral. In addition, since $\omega \ll \omega^*$, we can neglect the term $\partial F/\partial\mathcal{E}$ compared with $\partial F/\partial x$ in $\epsilon_{\alpha\beta}$. As a result, expressions (2.28) and (2.3) for Im $\epsilon_{11}, \epsilon_{21}$, and ϵ_{22} reduce to the form

$$\text{Im}\,\epsilon_{11} = \frac{\pi}{16}\frac{k_\perp^2 c^2}{\omega^2 \Omega^2}\langle\lambda\delta(k_z v_z + \omega_d)\omega_d^3\rangle$$

$$\epsilon_{21} = -\frac{i}{4}\frac{k_\perp^2 c^2}{\omega^2 \Omega}\langle\lambda(3 + \zeta_0\omega_d)\omega_d\rangle \quad\quad (5.33)$$

$$\epsilon_{22} - \frac{k_\perp^2 c^2}{\omega^2} = \frac{k_\perp^2 c^2}{\omega^2}\langle\lambda\zeta_0 k_z v_z\rangle.$$

Substituting (5.33) into (5.3) we find

$$\gamma = -\frac{1}{32} \frac{c_A^2 k_\perp^2}{\Omega^2 \text{Re}\,\omega} \Lambda \tag{5.34}$$

where

$$\Lambda = \pi \langle \lambda \delta(k_z v_z + \omega_d) \omega_d^3 \rangle + \text{Im} \frac{\langle \lambda (3 + \omega_d \zeta_o) \omega_d \rangle^2}{\langle \lambda \zeta_0 k_z v_z \rangle} \tag{5.35}$$

and $\text{Re}\,\omega$ is determined by (5.31), i.e., in the explicit form

$$\text{Re}\,\omega = \pm k_z c_A (1 + 2\omega_D^2/k_z^2 v_T^2)^{1/2}. \tag{5.36}$$

Since Λ does not depend on ω (see (5.32)), it can be seen from (5.34) that one of the oscillation branches is unstable, $\gamma > 0$. An additional analysis similar to that in section 4.4 shows that the *unstable branch* is that with $\text{Re}\,\omega/\omega_n > 0$. This fact is natural since otherwise the imaginary parts of $\epsilon_{\alpha\beta}$ have the same signs as in the case of a homogeneous plasma, while the energy of the waves considered is positive.

In order of magnitude

$$\gamma \simeq \beta(k_\perp \rho)^3 c_A/a \tag{5.37}$$

where ρ is the characteristic Larmor radius of high-energy particles and a is the characteristic length of their inhomogeneity.

Note that the analysis of the same problem in [1.10, 5.5] is not precise. In the mentioned papers only the term $\text{Im}\,\epsilon_{11}$ was taken into account, while the term $\text{Im}\,[\epsilon_{12}^2/(\epsilon_{22} - N^2)]$ in (5.3) was neglected assuming it to be small, of order $k_\perp^2 \rho^2$, in comparison with $\text{Im}\,\epsilon_{11}$. In reality, both terms have the same order of magnitude. It is clear that this inaccuracy is quantitative but not qualitative. Therefore, the qualitative results of [1.10, 5.5] remain valid and are in accordance with the ones reached above.

5.3.2 Instability of the short-wavelength Alfven perturbations

According to (5.37), the growth rate of *Alfven waves* increases with increasing $k_\perp \rho$. However, for $k_\perp \rho \simeq 1$ expression (5.34) and the estimate (5.37) become invalid. In order to study the excitation of Alfven waves for $k_\perp \rho \gg 1$ we shall use the general dispersion relation (1.1) with $\epsilon_{\alpha\beta}$ of the form of (2.1) and (2.2). As has been considered throughout this chapter, we assume ϵ_{33} to be sufficiently large, so that equation (1.1) reduces to (5.1) again. Using the *asymptotics* of the Bessel functions for *large arguments*, we then find that the components ϵ_{12} and ϵ_{21} are unimportant. As a result, we arrive at the following dispersion relation similar to (2.7):

$$\epsilon_{11} - k_z^2 c^2 / \omega^2 = 0. \tag{5.38}$$

We then obtain the same expression for the real part of the frequency as in the case of a homogeneous plasma

$$\mathrm{Re}\,\omega = \pm \mid k_z \mid c_{\mathrm{A}}. \tag{5.39}$$

The growth rate is equal to

$$\gamma = \frac{T}{M n_{\mathrm{c}}} \left\langle \left(\frac{\partial F}{\partial \mathcal{E}} + \frac{k_y}{\omega \Omega} \frac{\partial F}{\partial x} \right) \frac{\omega_{\mathrm{d}}^2 \lambda}{\mid \xi \mid^3} \delta(\omega - k_z v_z - \omega_{\mathrm{d}}) \right\rangle. \tag{5.40}$$

Here the notations are the same as those adopted in (2.2).

The result of (5.40) holds for an arbitrary ratio v_T / c_{A}. If, by analogy with section 5.3.1, we adopt $v_T \gg c_{\mathrm{A}}$, it follows from (5.40) that

$$\gamma = -\frac{2}{\beta n_{\mathrm{c}}} \left\langle \frac{F}{\mid \xi \mid^3} \frac{\mid v_z \mid}{c_{\mathrm{A}}} \omega_{\mathrm{d}}^2 \delta(k_z v_z + \omega_{\mathrm{d}}) \mathrm{sgn}\left(\frac{\omega_{\mathrm{d}}}{\omega} \right) \right\rangle. \tag{5.41}$$

In order of magnitude

$$\gamma \simeq \beta c_{\mathrm{A}} / a (k_\perp \rho)^2 \tag{5.42}$$

so that the growth rate decreases as $k_\perp^{-2}$ with increasing $k_\perp$.

It is clear from (5.41) that the branch with $\mathrm{Re}\,\omega/\omega_n > 0$ is unstable (see section 5.3.1).

5.3.3   Summary and discussion of the results

At the validity limits of the qualitative relations (5.37) and (5.42), i.e., at $k_\perp \rho \simeq 1$, these relations reduce to the same estimate:

$$\gamma \simeq \gamma_{\max} \simeq \beta c_{\mathrm{A}} / a. \tag{5.43}$$

The longitudinal wave number of such perturbations is of order

$$k_z \simeq v_T \beta / c_{\mathrm{A}} a. \tag{5.44}$$

It should be kept in mind that for $k_\perp \rho \simeq 1$ the wave frequency becomes of the order of the growth rate

$$\mathrm{Re}\,\omega \simeq \gamma \simeq \beta c_{\mathrm{A}} / a. \tag{5.45}$$

Hence the *method of expanding in a series* in the small parameter $\gamma/\mathrm{Re}\,\omega$, used in the above, proves to be at the validity limits for such $k_\perp \rho$.

Thus, it follows from our analysis that in a plasma with a group of *high-energy ions with an inhomogeneous density*, the Alfven waves can be excited. The characteristic transverse wavelength is of the order of the Larmor radius of these ions. The growth rate and the frequency of the unstable perturbations have the same order of magnitude determined by (5.45). The longitudinal wavelength is given by (5.44).

The Alfven growth rate (5.45) is *small* compared with the magnetoacoustic growth rate (5.12). However, the Alfven instability does not require the *threshold* of (5.7). Therefore, under condition (5.7) one can restrict oneself only to the allowance for the magnetoacoustic instability, while under the opposite condition to (5.7) only the Alfven instability is possible.

6 Electromagnetic Instabilities in a Strongly Inhomogeneous Plasma

For a *finite* ratio of the *Larmor radius* of particles to the characteristic length of a plasma inhomogeneity, it is necessary to allow for the *cyclotron* and *high-frequency instabilities* which are the subject of this chapter. Under the condition where the ion temperature is large or at least not too small compared with the electron temperature, such instabilities are associated with *short-wavelength perturbations* (with respect to the ion Larmor radius). In this connection, we start by analysing the short-wavelength perturbations of a *Maxwellian plasma* neglecting the cyclotron and high-frequency resonances and assuming $k_z = 0$, and $E \perp B_0$. This is the goal of section 6.1. Then we shall examine finite-β effects in the *drift-cyclotron instability* (section 6.2), the *lower-hybrid-drift instability* (section 6.3), the *drift loss-cone instability* (section 6.4), and the *ion cyclotron excitation* of the *transverse drift waves* (section 6.5). As in section 6.1, we assume $k_z = 0$ in sections 6.2–6.5.

The study of instabilities with $\omega/k_z \lesssim v_{Te}$ is also of interest. This is the goal of sections 6.6 and 6.7 where the instabilities associated with *short-wavelength drift-Alfven oscillation branches* are considered. It is assumed in section 6.6 that $\omega \ll \Omega_i$, i.e., *low-frequency perturbations* are studied. In contrast to this, in section 6.7 we consider that $\omega \simeq \Omega_i$ and allow for *ion-cyclotron effects*. In addition, the excitation of the above-mentioned oscillation branches is studied in Chapter 8, where it is assumed that the reason for such an excitation is an *additional transverse current* (see sections 8.5 and 8.6). Section 14.5 also concerns the topic considered, where *low-frequency short-wavelength drift-Alfven perturbations* in a *low-β plasma* are discussed.

80

The short-wavelength perturbations of finite-β plasma were initially studied in [2.8]. In the main we follow this paper in section 6.1. One of the principal results of this section is the fact that in the *approximation of zeroth electron Larmor radius*, the dispersion relation of the *short-wavelength perturbations with $k_z = 0$ coincides*, to an accuracy of the replacement of indices of charge species, *with that of the long-wavelength perturbations with $v_{Ti} \ll \omega/k_z \ll v_{Te}$.* This facilitates the study of short-wavelength perturbations and, in particular, allows one to observe without additional analysis the *existence* of the *short-wavelength ion-drift oscillation branch* similar to the *long-wavelength electron drift oscillation branch* discussed in section 3.3.1.

This ion-drift oscillation branch and its modification for the case of *non-Maxwellian ions* are of interest since they are related to instabilities *excited by ions* (such instabilities are discussed in sections 6.2–6.4). Therefore, in section 6.1 substantial attention is devoted to the calculation and qualitative estimations of the *decay rate of these perturbations due to electrons.* In particular, it is shown in section 6.1 that, when $\nabla T = 0$ and $k_\perp \rho_e$ is small, such a decay rate is proportional to $(k_\perp \rho_e)^4$. In section 6.1 the dispersion law of the *ion-drift oscillation branch* in a plasma with *cold electrons, $T_e \to 0$, and finite β* is obtained; this is a generalization of the corresponding electrostatic results necessary for section 6.3.

The *drift-cyclotron instability* was originally studied in the *electrostatic approximation* in [6.1]. In this approximation the detailed theory of this instability was presented in [6.2], and the main results of this approximation were summarized in [2], sections 2.2 and 2.3. *Experimental confirmation* of the prediction of [6.1] was initially reported in [6.3], where the authors dealt with plasma with $\beta \simeq 10^{-3}$ which was produced in a *toroidal multipole*. For further experimental studies of the *drift-cyclotron instability* see [6.4].

The role of the magnetic drift of particles due to finite β in the problem of cyclotron waves with $k_z = 0$ and $E \perp B_0$ was initially discussed in [6.5]. However, as was shown in [6.6], the results of [6.5] are inaccurate. The systematic analysis of the *drift-cyclotron instability in finite-β plasma* was initiated in [2.8]. The corresponding results of [2.8] are presented in section 6.2.

It follows from section 6.2 that a finite-β plasma is *more stable* with respect to drift-cyclotron perturbations than a plasma with $\beta \to 0$. For finite β *a few stabilizing effects* are revealed. One of them is associated with the *broadening of the cyclotron resonance* due to the *velocity dependence of the ion magnetic drift.* This effect does not cause full stabilization but only a *decrease* of the growth rate. It is revealed when $\beta \gtrsim (M_e/M_i)^{1/4}$. If β is not too small compared with unity ($\beta \geqslant 0.3$), the *electron magnetic drift* becomes substantial. The interaction of perturbations with *resonant electrons* in a plasma with $\beta \simeq 1$ causes *full stabilization* of the drift-cyclotron instability if the ion Larmor radius *is not too large*, $\rho_i/a \simeq (M_e/M_i)^{1/2}$, and $T_e \simeq T_i$.

If a plasma is *very strongly inhomogeneous*, $\rho_i/a \gg (M_e/M_i)^{1/2}$, and $\beta \simeq 1$, the instability is suppressed for the *highest harmonics* but *holds* for harmonics with small numbers. However, in the latter case the growth rate of the instability is a few times smaller than that when $\beta \ll 1$. *One more stabilizing effect* is revealed for $\beta \geqslant 2$. For such values of β the phase velocity of the ion-drift oscillations becomes larger than the ion density-gradient drift velocity; therefore, the interaction of *ions* with such oscillations leads to their *damping*.

The theory of a drift-cyclotron instability in a finite-β plasma was further developed in [6.7–6.10]. These papers *confirm* the above-mentioned qualitative notions and yield a more complete quantitative illustration of the role of finite β in this instability.

Magnetospheric applications of the drift-cyclotron instability were discussed, in particular, in [6.11, 6.12]. The authors of these papers paid attention to the fact that this instability may be substantial at the *boundary of the polar cusp*.

The authors of [6.7] have compared the above-mentioned experiment of [6.3] on plasma with $\beta \ll 1$ and an experiment of [6.13] which dealt with *high-β plasma* in the *theta-pinch*. In contrast to [6.3], they did *not observe* a drift-cyclotron instability. According to [6.3], this fact may be considered as *experimental proof of the stabilizing role of finite β*. The theory of drift-cyclotron instability in a finite-β plasma is of interest also for the so-called *Tandem Mirror Experiment* [6.14, 6.15]. This was pointed out in [6.9].

The *lower-hybrid-drift instability*, also called the *high-frequency drift instability*, in the case of *strongly inhomogeneous plasma with hot ions* was initially predicted in [2], section 2.2. In the cited reference the plasma pressure is assumed to be small, $\beta \to 0$, and the perturbations are considered to be electrostatic. The electrostatic expression for the growth rate obtained there ([2], equation (2.19)) has been reproduced in [6.16]. The authors of [6.16] have also introduced the term *"lower-hybrid-drift instability"*. The allowance for finite β in the problem of this instability was made in [6.17–6.20]. The statement of the problem in these papers somewhat *differs* from that presented in section 6.3 since an additional *transverse electric current* is allowed for there. We shall separately consider such a situation in Chapters 7 and 8. However, the mentioned difference in the statement of the problem is only *of slight importance* in the case where $\omega \ll k_\perp v_{Ti}$ discussed in section 6.3. Therefore the results of [6.17–6.20] are consistent with those of section 6.3. Both studies indicate that, as in the case of the drift-cyclotron instability, an *increase in β* causes a *decrease of the growth rate*, although the instability is *not totally suppressed when* $T_e \ll T_i$.

In [6.19] the importance of the lower-hybrid-drift instability in the problem *of anomalous transport* in the *theta-pinches* has been emphasized. According to [6.20], the *satellite observations* [6.21] of *electrostatic and magnetic noise* in the *distant magnetotail* may be attributed to the development of this instability. In addition, according to [6.22], the same instability may be *the source of the field fluctuations* observed in [6.23] at the *magnetopause*.

The drift loss-cone instability ([2], section 2.3) was initially studied in [6.2] and then in more detail in [6.24]. In these papers it was assumed to be *electrostatic*. The role of *electromagnetic effects* in this instability was initially analysed in [1.1, 6.25] and further in [6.26–6.29]. In [1.1, 6.25] the electron temperature was assumed to be negligibly small compared with the average ion energy, so that only the role of finite ρ_i was studied. We follow such a statement of the problem in section 6.4.1. Then it is revealed that the *critical value* of ρ_i/a required for instability *increases with* β_i, while the characteristic value of the *growth rate decreases*. However the instability is *not totally suppressed* even when $\beta_i \gtrsim 1$.

An examination of the role of *increasing the electron temperature*, i.e. β_e, was made in [2.9]. The results of this paper are presented in section 6.4.2. It emerges that when *the electrons are heated* to the extent that $\beta_e \simeq 1$, the *drift loss-cone instability* caused by ions with $\beta_i \gtrsim 1$ may be *totally suppressed*.

Evidence of the *drift loss-cone instability* in experiments on the confinement of *hot-ion plasma* was reported in [6.30–6.36] and others.

One of the main and experimentally verified methods of *suppression* of the drift loss-cone instability is the *injection* of a *plasma* of a relatively *low density* with *slightly warm ions* [6.30–6.33]. Such a method is called *"stabilization by warm plasma"*. In [1.1] and then in [6.26] the problem of the *stabilizing effect of increasing β* in combination with the *injection of warm plasma* on this instability was analysed. This problem is discussed in section 6.3.3. In particular, it turns out that *the larger* the value of β_i the *smaller* the *warm-plasma temperature* necessary for stabilization (see inequality (6.64)).

A possibility of the existence of the *drift loss-cone instability* in the *tokamak divertor* was considered in [6.37].

The role of the *drift loss-cone instability* in *magnetospheric plasma* was discussed in [6.29]. According to [6.29], certain varieties of this instability can play an important role in the dynamics of the *ring current* and the *inner edge of the plasma sheet region* of the magnetosphere, and would give rise to *precipitation of protons* on the auroral field lines, which may contribute to the excitation of the *diffuse aurora*. In addition, the authors of [6.29] consider that these instabilities may be relevant to the observation of the *ion-cyclotron harmonics* by [6.38].

In contrast to the instabilities presented in sections 6.2–6.4 relating to *hot-ion plasma*, the instabilities studied in section 6.5 can be excited in a strongly inhomogeneous *hot-electron plasma*. The theory of these instabilities was developed in [2.5] and further in [3.8, 3.9, 3.21]. *Space applications* of this theory were given in [3.19–3.21]. As mentioned in Chapter 3, these applications concern the problem of *electromagnetic noise* in the *hydromagnetic discontinuities* of the *solar wind*.

The *short-wavelength drift-Alfven (SDA) waves* discussed in sections 6.6 and 6.7 were initially studied in connection with the problem of *anomalous plasma transport* [2.10, 6.39]. It was then assumed that their frequencies were *small* compared with the ion cyclotron frequency. Subsequently the authors of [6.40–6.42] have noted the importance of SDA waves in the problem of *high-frequency heating* of plasma and extended the notions about these waves into the region of frequencies *higher than the ion cyclotron frequency*, $\omega \gg \Omega_i$. According to [6.40–6.42], SDA waves with $\omega \gg \Omega_i$ can be excited by a *transverse current* arising from the influence on plasma of an *alternating electromagnetic field* (see in detail Chapters 7 and 8). As a result, such an instability can lead to *fast heating* by this current. Recently the notions about SDA waves were used to explain *anomalous electron heat conductivity in tokamaks* [6.43, 6.44]. This is associated with the fact that SDA waves result in *bending* of the magnetic field lines, so that the electrons moving along such field lines can be *displaced across the equilibrium magnetic surfaces*. Further development of the *non-linear theory of SDA waves* can be found in [2.11, 6.45–6.47].

6.1   Short-wavelength perturbations in a finite-β Maxwellian plasma

Using the results of section 2.2.1, we shall consider *short-wavelength* $(k_\perp \rho_i \gg 1)$ *perturbations with* $k_z = 0$ *and* $\boldsymbol{E}_\perp \boldsymbol{B}_0$ *in a finite-β Maxwellian plasma*.

In section 6.1.1 the perturbations are assumed to be of *long wavelength*, with respect to the electron Larmor radius $(k_\perp \rho_e \to 0)$, while in section 6.1.2 the finiteness of $k_\perp \rho_e$ is taken into account. In both cases we shall restrict ourselves to the analysis of a plasma with $T_e = T_i \equiv T$. However, in section 6.1.3 we shall assume $T_e = 0$.

6.1.1 The $k_\perp \rho_e \to 0$ approximation

Under the above-mentioned assumptions and with $k_\perp \rho_e \to 0$, the perturbations are described by dispersion relation (2.35) with $\epsilon_{11} = \epsilon_{11}^{(0)}$ and $\epsilon_{12} = \epsilon_{12}^{(0)}$, where $\epsilon_{11}^{(0)}$ and $\epsilon_{12}^{(0)}$ are defined by equations (2.42), while ϵ_{22} is of the form of (2.43). Substituting the above-mentioned expressions for $\epsilon_{\alpha\beta}$ into (2.35), we arrive at the dispersion relation

$$\hat{\imath}_i \left(\frac{\omega_{Di}}{\omega} \right) \hat{\imath}_i + \hat{\imath}_e \left(\left\langle \frac{\omega_{de}}{\omega - \omega_{de}} \right\rangle \right) \hat{\imath}_i \left(1 - \frac{\omega_{Di}}{\omega} \right) = 0. \qquad (6.1)$$

Note that within the accuracy of an interchange of the species subscripts, equation (6.1) coincides with equation (3.28) taken for $T_e = T_i$. Therefore, to find physical results from (6.1) we can use the analysis given in section 3.3.

Then we conclude that equation (6.1) describes *two oscillation branches*: the branch of *ion-drift waves (ion-drift oscillation branch)* and the *magneto-drift branch*.

In the absence of a temperature gradient, $\nabla T = 0$, the frequency of the ion-drift oscillation branch for arbitrary β is

$$\omega = \omega_{ni} \qquad (6.2)$$

so that in this case (i.e., for $\nabla T = 0$) the oscillation frequency is real. For small $\nabla T (|\eta| \ll 1)$ the solution of equation (6.1) for this branch is (cf. (3.39))

$$\omega = \omega_{ni} \left(1 + \eta \frac{\beta/2}{1 + \beta/2} - \frac{\eta^2}{4} \frac{\beta}{(1 + \beta/2)^3 \langle \zeta_{0e} \lambda \rangle} \right) \qquad (6.3)$$

where $\lambda = M_e v_\perp^2 / 2T$, $\zeta_{0e} = (\omega - \omega_{de})^{-1}$. It can be seen that under these conditions the *oscillations are damped* with a decay rate proportional to η^2 due to their interaction with resonant electrons.

The *short-wavelength magneto-drift oscillation branch* for $0 < \eta^2 - 1 \ll 1$ is characterized by the frequency (cf. (3.42))

$$\omega = -\omega_{De} \exp \left[\frac{1 + \eta}{1 - \eta} \left(\beta + \frac{1}{(1 + \eta)^2} \right) \right]. \qquad (6.4)$$

Resonant interaction of such perturbations *with electrons* is absent, while that with ions is negligibly small due to the adopted approximation $k_\perp \rho_i \to \infty$. In a plasma with $\beta \ll 1$ and $1 \ll |\eta| \ll 2/\beta$, equation (6.4) is replaced by (cf. (3.48))

$$\omega = -\eta \omega_{De} / 2. \qquad (6.5)$$

For $\eta < 0$ these waves can interact with *resonant electrons*, but, since $|\omega| \gg \omega_{\mathrm{De}}$, the growth rate of such an interaction is *exponentially small*.

6.1.2   *Electron damping of the ion-drift oscillation branch*

According to the above, when $\eta = 0$ and $k_\perp \rho_e \to 0$, in a plasma there is an *ion-drift oscillation branch* (6.2) with zero growth rate. The phase velocity of such oscillations, ω/k_y, has the same direction as the *electron magnetic drift*. Therefore, if small terms proportional to high powers of $k_\perp \rho_e$ are taken into account, as well as for $\eta \neq 0$ we must obtain an imaginary addition to the frequency due to the interaction of the waves with resonant electrons.

Since the operator $\hat{l}_e$ for $\eta = 0$ is equal to $1 - \omega_{ne}/\omega$, the sign of $\hat{l}_e$ in the case of $\omega \approx \omega_{ni} = -\omega_{ne}$ is *positive*. Hence as in the case of a homogeneous plasma, resonant electrons *remove energy* from the waves, $\omega \mathrm{Im}\, \epsilon_{22}^{(e)} > 0$, and the *waves must be damped*, $\gamma < 0$. Let us verify this by a straightforward calculation of the growth rate (i.e., in this case, of the decay rate).

We take into account that, in the zeroth approximation in $k_\perp \rho_e$, the solution of equation (2.35) has the form of (6.2). Then, according to (2.42), $\epsilon_{12}^{(0)} = 0$ (we allow for $\nabla T = 0$). Therefore, to an accuracy of the terms of order $(k_\perp \rho_e)^2$ inclusively, equation (2.35) reduces to the form

$$\epsilon_{11}^{(0)} + \epsilon_{11}^{(1)} = 0 \tag{6.6}$$

where $\epsilon_{11}^{(0)}$ is defined by the first equation of (2.42), and $\epsilon_{11}^{(1)}$ is defined by the first equation of (2.41). It follows from the mentioned relationships that

$$\omega = \omega_{ni}[1 - 2z_e(1 - \beta/2)(1 + \beta/2)^{-1}]. \tag{6.7}$$

It can be seen that the addition to the oscillation frequency of order z_e is real. Hence, to find the growth rate (or decay rate) of the oscillations it is necessary to take into account the terms of order $(k_\perp \rho_e)^4$.

Starting with equation (3.6) and allowing for the above-mentioned smallness of $\epsilon_{12}^{(0)}$ for $\nabla T = 0$, we find a general expression for the growth rate (cf. (2.34) and (4.20)):

$$\gamma = -\frac{1}{\partial \epsilon_{11}^{(0)}/\partial \omega} \mathrm{Im}\left(\epsilon_{11}^{(2)} + \frac{(\epsilon_{12}^{(0)})^2 + 2\epsilon_{12}^{(0)}\epsilon_{12}^{(1)}}{\epsilon_{22} - N^2} \right). \tag{6.8}$$

We use the fact that (cf. (4.21))

$$\begin{aligned}
\operatorname{Im} \epsilon_{11}^{(2)} &= X/4 \\
\operatorname{Im} (-\mathrm{i}\epsilon_{12}) &= \Omega_e X/\omega \\
\operatorname{Im} \epsilon_{22} &= 4X(\Omega_e/\omega)^2
\end{aligned} \qquad (6.9)$$

where in the case discussed

$$X = \frac{\pi}{2} \frac{z_e^2}{k^2 \lambda_{De}^2} \frac{\omega^3}{|\omega_{De}|^3} \exp\left(-\frac{\omega}{\omega_{De}}\right). \qquad (6.10)$$

With the help of (6.9) we transform (6.8) into

$$\gamma = -\frac{1}{\partial \epsilon_{11}^{(0)}/\partial \omega} \frac{\bar{\alpha}_{12}^2}{|\epsilon_{22} - N^2|^2} \operatorname{Im} \epsilon_{22} \qquad (6.11)$$

where

$$\begin{aligned}
\bar{\alpha}_{12} &= -\mathrm{i}(\epsilon_{12}^{(0)} + \epsilon_{12}^{(1)}) - (\epsilon_{22} - N^2)z_e \omega \kappa_B / 4\omega_{De} k \\
&= -(k\kappa_B \lambda_{De}^2)^{-1} z_e \beta (1 - \beta/2).
\end{aligned} \qquad (6.12)$$

Since $\omega \partial \epsilon_{11}^{(0)}/\partial \omega \equiv 2$, $\omega \operatorname{Im} \epsilon_{22} > 0$, it can be seen from (6.11) that, in accordance with the above, $\gamma < 0$, i.e., the *oscillations are damped.* When $\beta \simeq 1$, it follows from (6.11) that the decay rate is of order

$$\gamma \simeq - |\operatorname{Re} \omega| z_e^2. \qquad (6.13)$$

In the case where $\beta \ll 1$ the decay rate is exponentially small, so that for $z_e \simeq 1$ we have

$$\gamma \simeq - |\omega_{ni}| \exp(-2/\beta)/\beta. \qquad (6.14)$$

For $z_e \ll 1$ the factor z_e^2 should be added to the right-hand side of (6.14), i.e.,

$$\gamma \simeq -z_e^2 |\omega_{ni}| \exp(-2/\beta)/\beta. \qquad (6.15)$$

Thus, we have shown that for $\eta = 0$ the *short-wavelength pertur-bations* $(k_\perp \rho_i \gg 1)$ *with* $k_z = 0$ *associated with the ion-drift branch* (6.2) *are damped in finite-β plasma.* For $z_e \ll 1$ the decay rate is of order z_e^2 compared with the frequency. Extrapolation of this result to the region $z_e \simeq 1$ leads to the conclusion that for such z_e the decay rate is of the order of the frequency, $\gamma \simeq \operatorname{Re} \omega$ if $\beta \simeq 1$, and is exponentially small compared with the frequency, $\gamma/\operatorname{Re} \omega \simeq \exp(-2/\beta)$ if $\beta \ll 1$.

6.1.3   Plasma with cold electrons

In contrast to sections 6.1.1 and 6.1.2 we now take $T_e = 0$ (*the cold-electron plasma approximation*). Then

$$\epsilon_{11}^{(2)} = \epsilon_{12}^{(1)} = \epsilon_{22} = 0 \qquad (6.16)$$

$$\epsilon_{11}^{(1)} = \omega_{pe}^2/\Omega_e^2. \qquad (6.17)$$

Using these relations and expressions (2.42) for $\epsilon_{11}^{(0)}$ and $\epsilon_{22}^{(0)}$, we find from (2.35) that the frequency of the ion-drift oscillation branch is given by

$$\omega = \frac{\omega_{ni}[1 + \beta(1 + \eta)/2]}{1 + \beta/2 + k^2\rho_{ei}^2} \qquad (6.18)$$

where $\rho_{ei}^2 = T_i/M_e\Omega_e^2$ is the square of the electron Larmor radius calculated at the ion temperature.

Equation (6.18) is a *generalization* of a similar equation for the *electrostatic approximation* (see [2], equation (2.8)) to the case of a *plasma with arbitrary β*.

It follows from (6.18) that the maximum of ω as a function of the wave number depends on β as $(1 + \beta/2)^{1/2}$.

6.2   Drift-cyclotron instability in a finite-β plasma

Let us augment the analysis of section 6.1 by including the ion cyclotron effects. According to section 2.2.1, for this aim, the term $\delta\epsilon_{11}$ should be allowed for in the expression for ϵ_{11}. In the case of *Maxwellian ions* which is of interest to us, this term is given by the second equation of (2.42). Then in the case where $\beta \to 0$ it follows from (2.35) that

$$\epsilon_{11}^{(0)} + \epsilon_{11}^{(1)} + \delta\epsilon_{11} = 0. \qquad (6.19)$$

It is assumed that the terms with the magnetic-drift frequency in the expressions for $\epsilon_{11}^{(1)}$ and $\delta\epsilon_{11}$ are omitted. Therefore, according to (2.48), equation (6.19) can be written in the form

$$k^2\lambda_{Di}^2\epsilon_{11} \equiv \hat{l}_i + \frac{T_e}{T_i}\hat{l}_e[1 - I_0(z_e)\exp(-z_e)] - \hat{l}_i(2\pi z_i)^{-1/2}\sum_n \frac{\omega}{\omega - n\Omega_i} = 0.$$

$$(6.20)$$

This dispersion relation is the *starting point* in the theory of the *drift-cyclotron instability* (also called the *gradient-cyclotron instability*) *in a zero-β plasma* (see [2], section 2.1). It follows from (6.20) that the mentioned instability takes place when the ion Larmor radius ρ_i is not too small compared with the characteristic length of the density inhomogeneity a. In the particular case where $T_e = 0$ the *instability condition* means

$$\rho_i/a > 2(M_e/M_i)^{1/2}. \tag{6.21}$$

If inequality (6.21) is not strong, *a small number* of ion cyclotron harmonics with *growth rate* of order

$$\gamma \simeq (M_e/M_i)^{1/4}\Omega_i \tag{6.22}$$

is excited. The *wave number* of unstable waves is of order

$$k \simeq \rho_i^{-1}(M_i/M_e)^{1/2} \tag{6.23}$$

following from the approximate relationship

$$\omega \simeq \omega_{ni} \simeq \Omega_i \tag{6.24}$$

and allowing for $\rho_i/a \simeq (M_e/M_i)^{1/2}$.

If inequality (6.21) is strong, the number of the highest excited harmonic is of order

$$n_{max} \simeq (M_i/M_e)^{1/2}\rho_i/a. \tag{6.25}$$

In this case the wave number of the highest harmonic is of order (6.23), while the growth rate differs from (6.22) by a factor n_{max}, so that

$$\gamma(n_{max}) \simeq \frac{\rho_i}{a}\left(\frac{M_i}{M_e}\right)^{1/4}\Omega_i. \tag{6.26}$$

If the inhomogeneity is so large that

$$\rho_i/a > (M_e/M_i)^{1/4} \tag{6.27}$$

the notion of a *discrete spectrum of cyclotron harmonics* at $n \simeq n_{max}$ becomes invalid since under these conditions it would follow from (6.26) that $\gamma > \Omega_i$, but in this case we must allow for *many harmonics* at the same time. Therefore, when inequality (6.27) holds, the *drift-cyclotron instability* of the *highest harmonics changes* into the so-called *lower-hybrid-drift instability*, i.e., into the *high-frequency drift instability* (see [2] section 2.1.3, and [6.16]).

In the present section we shall assume the *plasma inhomogeneity* to be *not too large*, so that the instability of each harmonic can be examined *separately*. In section 6.3 we shall consider the *lower-hybrid-drift instability*.

Hereinafter we assume $T_e = T_i \equiv T$.

In transforming (2.35) to (6.19), we have neglected the terms of order $\beta, \exp(-2/\beta), \omega_{Di}/(\omega - n\Omega_i)$ and $\exp[-(\omega - n\Omega_i)/\omega_{Di}]$. If ω is close to $n\Omega_i$, the terms with $\omega_{Di}/(\omega - n\Omega_i)$ and $\exp[-(\omega - n\Omega_i)/\omega_{Di}]$ are more important than the terms of order β and $\exp(-2/\beta)$. In other words, when β is fairly small, it is necessary, first of all, to take into account the *ion magnetic drift*, and only for large β should the *electron magnetic drift* and the *electromagnetic character of the perturbations* be allowed for.

Thus, assuming $\beta \ll 1$ we should primarily study the following modification of equation (6.20):

$$k^2 \lambda_D^2 \epsilon_{11} \equiv 2 - \hat{l}_e I_0(z_e) \exp(-z_e) - \hat{l}_i \langle \omega J_n^2(\xi_i)/(\omega - n\Omega_i - \omega_{di}) \rangle = 0. \tag{6.28}$$

Let us estimate for which values of β it is necessary to change from (6.20) to (6.28). Since the difference $\omega - n\Omega_i$ is of the order of the right-hand side of (6.22) for small β, the term of order $\omega_{di}/(\omega - n\Omega_i)$ becomes important when

$$\beta \gtrsim \beta_{\min} \simeq (M_e/M_i)^{1/4}. \tag{6.29}$$

In section 6.2.1 a plasma with $\beta_{\min} < \beta \ll 1$ and in section 6.2.2 that with $\beta \simeq 1$ will be considered.

6.2.1   *Electrostatic magneto-drift drift-cyclotron instability*

Let us assume $(M_e/M_i)^{1/4} < \beta \ll 1$. According to the above-mentioned, for this case the perturbations are described by (6.28).

In contrast to the case where $\beta \to 0$ (see [2], section 2.1), even when $\omega - n\Omega$ is small, in the presence of a term containing the magnetic drift in equation (6.28), the cyclotron contribution in this equation is small for $\beta > \beta_{\min}$. As a result, for $\beta > \beta_{\min}$ the *instability of the hydrodynamic type* (see [2], section 2.1) which causes the growth rate of the form of (6.22) and (6.26), no longer occurs. However, in this condition the instability is not completely suppressed. It becomes *essentially kinetic*, i.e., the growth rate is no longer determined by all the ions, but only by those which are *in resonance* with the wave (resonance of type $\omega_{di} = \omega - n\Omega_i$).

Allowing for the imaginary contribution of the *cyclotron term* in (6.28), we find the oscillation frequency and the growth rate:

$$\mathrm{Re}\,\omega = \frac{\omega_{ni} I_0(z_e) \exp(-z_e)}{2 - I_0(z_e) \exp(-z_e)} \tag{6.30}$$

$$\gamma = \frac{-\pi \omega^2 (1 - \omega_{ni}/\omega)}{2 - I_0(z_e) \exp(-z_e)} \langle \delta(\omega - n\Omega_i - \omega_{di}) J_n^2(\xi_i) \rangle. \tag{6.31}$$

When $\xi_i \gg 1$, the approximate formula $J_n^2(\xi_i) \approx (\pi \xi_i)^{-1}$ is true (see also section 2.2.1). Using this formula, we calculate the average in (6.31) and arrive at the following expression for the growth rate

$$\gamma = -\frac{\omega^2(1 - \omega_{ni}/\omega)}{|\omega_{Di}|[2 - I_0(z_e)\exp(-z_e)]} \frac{\exp[-(\omega - n\Omega_i)/\omega_{Di}]}{[2z_i(\omega - n\Omega_i)/\omega_{Di}]^{1/2}}. \quad (6.32)$$

The growth rate should satisfy the above condition $\gamma \lesssim \omega - n\Omega_i$. This yields a lower limit for the difference $\omega - n\Omega_i$ entering in (6.32). Hence an estimation of the maximum growth rate can be obtained by assuming $\gamma \simeq \omega - n\Omega_i$. In this case we have $\omega - n\Omega_i \ll \omega_{Di}$, so that the exponential in (6.32) may be replaced by unity. We then obtain

$$\gamma_{max} \simeq \frac{\omega}{z_i^{1/3}}\left(\frac{\omega}{\omega_{Di}}\right)^{1/3}\left(1 - \frac{\omega_{ni}}{\omega}\right)^{2/3}. \quad (6.33)$$

When $\rho_i/a \simeq (M_e/M_i)^{1/2}$, only perturbations with $\omega \simeq \Omega_i(n = 1)$ can be excited. In this case it follows from (6.33) that

$$\gamma_{max} \simeq \Omega_i(M_e/M_i\beta)^{1/3}. \quad (6.34)$$

At the validity limits of inequality (6.29), i.e., when $\beta \simeq (M_e/M_i)^{1/4}$, equation (6.34) is replaced by equation (6.22).

When $\rho_i/a \gg (M_e/M_i)^{1/2}$, it follows from (6.33) that for *the highest harmonics*

$$\gamma_{max} \simeq n\Omega_i(M_e/M_i\beta)^{1/3} \quad (6.35)$$

while for *the first harmonic*

$$\gamma_{max} \simeq \Omega_i(M_e a/\beta^{1/2}M_i\rho_i)^{2/3}. \quad (6.36)$$

Similarly to (6.34), when $\beta \simeq \beta_{min}$, expressions (6.35) and (6.36) are transformed to the respective relationships of the approximation $\beta \to 0$ (see [2], section 3.1).

The instability considered here can be called the *electrostatic magneto-drift drift-cyclotron instability*.

6.2.2 *Electromagnetic drift-cyclotron instability*

With increasing β, the *electromagnetic part* of the perturbations and the *electron magnetic drift* become important. These effects are taken into account by equation (2.35) with the components $\epsilon_{\alpha\beta}$ given in section 2.2.1. Now we proceed to the analysis of this equation. The corresponding instability can be called the *electromagnetic drift-cyclotron instability*.

In section 6.1 we analysed equation (2.35) neglecting the ion resonance. It was shown that the *perturbations were damped* due to their interaction with the *resonant electrons*. The decay rate was expressed by the approximate relationships (6.13)–(6.15).

Besides the electron damping, there is one more effect associated with the electrons and revealed at finite β; this effect is described by equation (6.7) and is the *change of sign of the value* $1 - \omega_{ni}/\omega$. This sign becomes positive when

$$\beta > 2. \tag{6.37}$$

Then $\mathrm{sgn}(\omega \mathrm{Im}\, \epsilon_{11}) > 0$, so that the ions do not result in an excitation, but in a damping of the perturbations.

Hereinafter we shall assume $\beta < 2$ and examine under which conditions *electron damping overcomes ion excitation*.

(a) Stabilization of perturbations for $\rho_i/a > (M_e/M_i)^{1/2}$
In this case $z_e \simeq 1$, so that the electron decay rate is of order (6.14) while the ion growth rate is of order (6.34). Comparing these two expressions, we conclude that stabilization takes place even for small β

$$\beta \gtrsim 3[\ln(M_i/M_e)]^{-1} \approx 0.3. \tag{6.38}$$

(b) Stabilization of the highest harmonics at $\rho_i/a \gg (M_e/M_i)^{1/2}$
As in section 6.2.2.(a), here we also have $z_e \simeq 1$, and (6.14) still holds, while the ion growth rate is now determined by (6.35). As a result, criterion (6.38) still holds.

(c) Perturbations with $\omega \simeq \Omega_i$ for $\rho_i/a \gg (M_e/M_i)^{1/2}$
Now

$$z_e \simeq (a/\rho_i)^2 M_e/M_i \ll 1 \tag{6.39}$$

so that we have estimates (6.13) and (6.15) for the electron decay rate. For $\beta \ll 1$ the ion growth rate is of the order of (6.36). When $\beta \simeq 1$, it is determined by the dispersion relation of the form of (6.19) where $\epsilon_{11}^{(0)}, \epsilon_{11}^{(1)}$, and $\delta\epsilon_{11}$ are given by (2.41) and (2.42). Solving (6.19), we find that (6.36) holds up to $\beta \simeq 1$ (but, in reality, it is considerably modified if $\beta > 2$, see condition (6.37)). Due to the smallness of the electron decay rate (6.15), for small β stabilization does not take place, so that the instability occurs up to $\beta \approx 2$.

6.3 Electromagnetic lower-hybrid-drift (high-frequency drift) instability in a finite-β plasma

As mentioned in section 6.2, under condition (6.27), the drift-cyclotron instability in a zero-β plasma turns into the *lower-hybrid-drift instability*. Then the growth rate of separate harmonic becomes of the order of the cyclotron frequency and, as a result, a *substantial broadening of the cyclotron resonance* occurs.

The broadening of the cyclotron resonance is also caused by *ion magnetic drift*. For a sufficiently large ion magnetic drift, i.e., for sufficiently *high* β, the frequency range of such a broadening can also be compared with the ion cyclotron frequency, i.e.,

$$\omega_{di} \gtrsim \Omega_i. \tag{6.40}$$

As mentioned in section 2.2.1, in this case it is necessary to use *the approximation of a continuous spectrum of cyclotron harmonics*, i.e., the so-called *high-frequency approximation*. Let us consider under what β this approximation is valid.

We allow for *magneto-drift broadening* of the cyclotron resonance to be due to particles with thermal velocities, i.e., with $v_\perp^2 \simeq T_i/M_i$, in contrast to the situation considered in section 6.2.1 when $v_\perp^2 \ll T_i/M_i$. Then, taking into account (6.23), condition (6.40) means

$$\beta \gtrsim (M_e/M_i)^{1/2} a/\rho_i. \tag{6.41}$$

Comparing this inequality with (6.28) and keeping in mind (6.27), we conclude that the condition $\beta \gtrsim (M_e/M_i)^{1/4}$, treated in section 6.2 as a condition of importance for the ion magnetic drift in the drift-cyclotron instability, can also be treated as a condition that the broadening due to the ion magnetic drift prevails upon the broadening due to the growth rate. In other words, under condition (6.29), the transition to the high-frequency approximation is determined not by condition (6.27), which can be invalid, but by condition (6.41).

If β becomes of order unity, it can be seen from (6.41) that condition (6.27) is replaced by a condition of type (6.21). In other words, when $\beta \simeq 1$, the instability of a plasma with finite ion Larmor radius caused by density inhomogeneity resembles a high-frequency instability, but not a cyclotron instability. Of course, some features of a cyclotron type of instability should be revealed. Furthermore, in the case of a strongly inhomogeneous plasma our estimates correspond to *the highest frequency range of the spectrum* associated with rather short-wavelength perturbations. A longer wavelength and lower frequency range of the spectrum must remain substantially cyclotronic even though condition (6.41) will be valid.

Based upon the above, we shall assume condition (6.41) to be valid and examine the *lower-hybrid-drift instability* at first in *the approximation of cold electrons*, $T_e = 0$ (section 6.3.1), and then by allowing for *finite* T_e (section 6.3.2).

6.3.1   *Cold-electron plasma*

For $T_e = 0$ (*cold-electron plasma*) the problem of *lower-hybrid-drift instability* is reduced to the allowance for *ion resonant interaction* described by $\delta\epsilon_{00}$ of the form of (2.52) with *the ion-drift oscillation branch* (6.18). Then we obtain that

$$\gamma = \frac{\pi^{1/2}\omega^2 k\rho_{ei}^2}{v_{Ti}(1 + \beta/2)(1 + \beta/2 + k^2\rho_{ei}^2)}. \qquad (6.42)$$

Remembering the statement at the end of section 6.1.3, we conclude that the maximum of γ as a function of the wave number depends on β as $(1 + \beta/2)^{1/2}$.

6.3.2   *Plasma with finite electron temperature*

When $T_e \simeq T_i$, it follows from comparing (6.42) with (6.14) that the *lower-hybrid-drift instability*, considered in section 6.3.1 in the range of maximum frequencies, is possible only for sufficiently small β, namely for (cf. (6.38))

$$\beta \lesssim 2[\ln(a/\rho_i)]^{-1}. \qquad (6.43)$$

In this connection, an instability of *perturbations with longer wavelength* seems to be more substantial since in that case the electron decay rate contains an additional smallness z_e^2 (see (6.13) and (6.15)). We then conclude that perturbations with $\omega \simeq \Omega_i$ can be excited for $T_e \simeq T_i$ even in a plasma with $\beta \simeq 1$ if

$$\rho_i/a > (M_e/M_i)^{1/3}. \qquad (6.44)$$

Numerical calculations of the growth rate of the lower-hybrid-drift instability at various T_e/T_i and β were performed, in particular, in [6.20].

6.4   Drift loss-cone instability in a finite-β plasma

For a *loss-cone ion distribution* (see section 2.2.1), $\beta \to 0$ and $T_e \to 0$, equation (2.35) reduces to the following

$$1 + \frac{\kappa_n k_y \Omega_e}{k^2\omega} + \frac{M_i}{\pi M_e} \frac{1}{(k\rho_i)^3} \sum_n \frac{\omega}{\omega - n\Omega_i} = 0. \qquad (6.45)$$

Here for simplicity we have assumed the ion distribution to be a δ-function (see (2.54)) and have omitted a term with ω_{ni}/ω (see (2.55)) which is insignificant for the subsequent analysis; $\rho_i = v_0/\Omega_i$ is the ion Larmor radius corresponding to the velocity distribution of the form of (2.54).

When

$$\rho_i/a > (M_e/M_i)^{2/3} \tag{6.46}$$

equation (6.45) describes the *electrostatic drift loss-cone instability* (see [2], section 2.3). When inequality (6.46) is not strong, the characteristic wave number of unstable perturbations is defined by the relation

$$k\rho_i \simeq (M_i/M_e)^{1/3} \tag{6.47}$$

while the characteristic frequency and the growth rate are of the order of the ion cyclotron frequency, $\mathrm{Re}\,\omega \simeq \gamma \simeq \Omega_i$.

Let us consider how these results will be modified for *finite β*. In section 6.4.1 we shall assume, as in (6.45), *the electrons to be cold*, $T_e \to 0$ (*cold-electron plasma*), while in section 6.4.2 we shall examine the *role of finite T_e*. In addition, in section 6.4.3 we shall deal with the problem of the *suppression* of the *drift loss-cone instability in a finite-β plasma* by a small *admixture* of *slightly warm ions*.

6.4.1 Cold-electron plasma

Using (2.35) and the corresponding expression for ϵ_{ik} given in section 2.2.1, we obtain the following dispersion relation generalizing (6.45) to the case of ions with $\beta \lesssim 1$:

$$1 + \frac{\omega_{pe}^2}{c^2 k^2} + \frac{\kappa_n k_y \Omega_e}{k^2 \omega} + \frac{M_i}{M_e} \frac{1}{(k\rho_i)^3} \sum_{n=-\infty}^{\infty} \frac{\omega}{\omega - n\Omega_i} = 0. \tag{6.48}$$

When $\beta \gtrsim 1$, in addition to considering ∇n_0, we should also take into account ∇B_0.

Equation (6.48) differs from (6.45) by the term $(\omega_{pe}/ck)^2$, which is associated with the *electromagnetic part of perturbations*. It follows from (6.48) that *electromagnetic effects* are important when

$$k \lesssim \omega_{pe}/c. \tag{6.49}$$

Using equations (6.49) and (6.47), we conclude that the *zero-β approximation* is invalid for

$$\beta \gtrsim (M_e/M_i)^{1/3}. \tag{6.50}$$

Let us assume this condition to be satisfied. Then (6.48) has complex solutions if the wave frequency

$$\omega^{(0)} \equiv -\frac{\kappa_n k_y \Omega_e}{k^2(1 + \omega_{pe}^2/c^2 k^2)} \tag{6.51}$$

is of the order of or greater than the ion cyclotron frequency Ω_i. This condition can be satisfied only when

$$\mid \kappa_n \mid \equiv \mid \partial \ln n_0/\partial x \mid \geqslant 2(M_e/M_i)\omega_{pe}/c. \tag{6.52}$$

This is the *boundary of the drift loss-cone instability* in a plasma with $\beta > (M_e/M_i)^{1/3}$.

Condition (6.52) can be written in a form similar to (6.46)

$$\rho_i/a \geqslant (M_e/M_i)^{1/2}\beta^{1/2}. \tag{6.53}$$

It can be seen that the critical value of the relative density gradient $1/a$, for the onset of plasma instability, increases with β, going as $\beta^{1/2}$. This situation can be regarded as being *favorable* for *stable plasma confinement*. When $\beta \simeq 1$, the density gradient must be such that (cf. (6.21))

$$\rho_i/a > (M_e/M_i)^{1/2}. \tag{6.54}$$

The growth rate is reduced due to the electromagnetic nature of the perturbations. When $\beta > (M_e/M_i)^{1/3}$ and ρ_i/a is not very much larger than the right-hand side of (6.53), we find

$$\gamma \simeq \frac{\Omega_i}{(2\pi)^{1/2}} \left(\frac{M_e}{M_i \beta^3} \right)^{1/4}. \tag{6.55}$$

In contrast to the case where $\beta \to 0$, the growth rate is small compared with the ion-cyclotron frequency.

6.4.2 Hot-electron plasma

Now we consider β_e to be finite (*hot-electron plasma*). For simplicity we assume $\beta_i > 1$. We describe the *resonant interaction of ions with waves* by means of (2.56). Let us study separately the cases where $\beta_e < 1$ and $\beta_e > 1$, using the approximate relationships from section 2.2.1. The electron temperature gradient will be neglected.

When $\beta_e < 1$ and $\beta_i > 1$, the real and imaginary parts of the wave frequency become the following (cf. (6.18) and (6.42)):

$$\mathrm{Re}\,\omega = \omega_{ni}(1 + 2/\beta) \tag{6.56}$$

$$\gamma = \frac{4\omega_{ni}^2}{kv_0\beta_i} \left(1 - \frac{\pi}{2} \frac{1}{\mid \kappa_B \mid \rho_i} \Phi(\beta_e) \right) \tag{6.57}$$

where

$$\Phi(\beta_e) = (2/\beta_e)^2 \exp(-2/\beta_e). \tag{6.58}$$

The first term in the large brackets of the right-hand side of (6.57) describes the *ion excitation*, while the second describes the *electron damping*.

Recall that our analysis concerns the case of a *weakly inhomogeneous magnetic field*, $|\kappa_B| \rho_i \ll 1$. Therefore, it follows from (6.57) that by *increasing β_e the instability can be suppressed*.

We now assume that $\beta_e > 1$ and, as above, $\beta_i > 1$. In this case we have

$$\operatorname{Re}\omega = \omega_{ni}\left[1 + \frac{2}{\beta}\left(1 + \frac{T_e}{T_i}\right)\right] \tag{6.59}$$

$$\gamma = \frac{2T_e}{T_i}\frac{\omega_{ni}^3}{\Omega_e k v_0}\left(1 + \frac{T_e}{T_i}\right)\left(1 + \frac{T_e}{T_i} - \frac{\pi}{2\rho_i |\kappa_B|}\right) \tag{6.60}$$

where $T_i \equiv M_i v_0^2 / 2$. It can be seen that, when $T_e \lesssim T_i$, *the perturbations are damped* for all values of $|\kappa_B| \rho_i$. When $T_e \gg T_i$, it is a necessary condition for stabilization that

$$|\kappa_B| \rho_i < T_i/T_e. \tag{6.61}$$

Thus, with the objective of the *suppression of drift loss-cone instability*, too high a *heating of electrons* $(T_e > T_i)$ for $\beta > 1$ is undesirable.

6.4.3 Stabilization by a warm plasma
In the presence of a *small admixture of slightly warm Maxwellian ions*, equation (6.48) is replaced by

$$D_0 + \delta D_0 = 0 \tag{6.62}$$

where D_0 is the left-hand side of (6.48), while δD_0 is of the form

$$\delta D_0 = \frac{\alpha M_i}{M_e(k\rho_1)^2}\left(1 - \sum_n \frac{\omega I_n(z_1)\exp(-z_1)}{\omega - n\Omega_i}\right). \tag{6.63}$$

Here $\alpha = n_1/n_0 \ll 1$, n_1 is the number density of slightly warm ions, $\rho_1^2 = T_1/M_i\Omega_i^2$ is their squared Larmor radius, T_1 is their temperature, and $z_1 = (k\rho_1)^2$.

Using (6.62) and (6.63), we can evaluate the *effect of the warm ions* on the *drift loss-cone instability* in the case where the density gradient is not very large, so that inequality (6.53) is not strong. In this case the characteristic value of k is of the order of ω_{pe}/c so that $z_1 \simeq \beta T_1/M_e v_0^2$. Let

$$T_1 \gtrsim M_e v_0^2/\beta. \tag{6.64}$$

In this case $z_1 \gg 1$ so that $I_n(z_1)\exp(-z_1) \approx (2\pi z_1)^{-1/2}$. Using these estimates, we find by means of (6.62) and (6.63) that the instability is impossible if

$$\alpha \gtrsim \pi^{1/2}(T_i/M_i v_0^2)^{3/2}. \tag{6.65}$$

When $\beta \simeq 0.1$, it follows from (6.64) and (6.65) that the *drift loss-cone instability is suppressed*, e.g., if $\alpha \simeq 10^{-2}$, $T_1/M_i v_0^2 \simeq 10^{-2}$.

In [6.26] a *numerical calculation* of the critical value of ρ_i/a as a function of α for various β_i and $T_1/M_i v_0^2$ has been performed. The plot given in that paper shows that the dependence of the critical value of ρ_i/a on β_i is unimportant when $\alpha \simeq 10^{-2}$.

6.5   Ion-cyclotron excitation of transverse drift waves

Consider *short-wavelength perturbations with $k_z = 0$ and $\boldsymbol{E} \parallel \boldsymbol{B}_0$*. According to section 2.2.2, such perturbations are described by dispersion relation (2.57) with ϵ_{33} in the form of (2.58)–(2.60). We restrict ourselves to the case of a *Maxwellian plasma* when the expressions for $\epsilon_{33}^{(0)}$ and $\delta\epsilon_{33}$ are of the form of (2.61) assuming also $\nabla T = 0$. In this case the $\hat{l}$-operators in (2.61) turn into differences $1 - \omega_n/\omega$ where ω_n is associated with the respective charge species. The plasma pressure is assumed to be low, $\beta \ll 1$, and the magnetic drift in the denominator of the right-hand side of equation (2.61) for $\epsilon_{33}^{(0)}$ is neglected. The term with $\delta\epsilon_{33}$ is considered to be correctional. If it is completely neglected, equation (2.57) yields

$$\omega = \frac{\omega_{ne}}{1 + c^2 k^2/\omega_{pe}^2}. \tag{6.66}$$

This result is a generalization of (3.16) for *transverse drift waves* in a plasma with $\nabla T = 0$ to the range of the wave number (cf. (6.49))

$$k \gtrsim \omega_{pe}/c. \tag{6.67}$$

The maximum for the frequency of (6.66) as a function of the wave number is attained when $k = k_0 \equiv \omega_{pe}/c$ and is given by

$$\omega_{\max} = \frac{\omega_{pe}}{c}\frac{T_e}{M_e a \Omega_e} = \frac{\beta_e^{1/2}}{2}\frac{v_{Te}}{a} \tag{6.68}$$

where, as usual, $a^{-1} = \partial \ln n_e / \partial x$. The frequency ω_{max} is equal to or larger than the ion cyclotron frequency when

$$\rho_e / a \gtrsim M_e / M_i \beta_e^{1/2}. \tag{6.69}$$

In this case the allowance for $\delta \epsilon_{33}$ describing the *interaction* of the transverse drift waves *with the ion cyclotron motion* becomes substantial. Such an interaction as in the case of waves with $E \perp B_0$, as examined in sections 6.2–6.4, *causes an instability*. We shall consider this instability at first in neglecting the ion magnetic drift (section 6.5.1) and then when allowing for such a drift (section 6.5.2). In both cases we restrict ourselves to allowing only for the ion cyclotron harmonic with $n = 1$. The *ions* will be assumed to be *colder* than the electrons, $T_i \ll T_e$.

6.5.1 *Ion-cyclotron instability of hydrodynamic type*
Neglecting the ion magnetic drift, we find from (2.57) and (2.61) under condition (6.69) that the oscillation branch (6.66) of frequency $\text{Re}\,\omega \simeq \Omega_i$ is excited with growth rate

$$\gamma \simeq \Omega_i (M_e / M_i)^{1/2} [I_1(z_i) \exp(-z_i)]^{1/2}. \tag{6.70}$$

Using (6.67), we conclude that the maximum of the growth rate (6.70) as a function of β_i is attained when

$$\beta_i \simeq M_e / M_i. \tag{6.71}$$

In this case

$$\gamma \simeq (M_e / M_i)^{1/2} \Omega_i. \tag{6.72}$$

When β_i is sufficiently large compared to that in (6.71), we have from equation (6.70) that (6.72) is replaced by

$$\gamma \simeq (M_e / M_i)^{3/4} \beta_i^{-1/4} \Omega_i. \tag{6.73}$$

The validity of this result is limited by the condition of *smallness of the ion magnetic drift* and will be discussed in section 6.5.2.

The above relations describe *the ion-cyclotron instability of hydrodynamic type* related to the *transverse drift oscillation branch*.

6.5.2 *Kinetic magneto-drift ion-cyclotron instability*
The role of the *ion magnetic drift* in the *cyclotron excitation* of *transverse drift waves* may be analysed similarly to section 6.2. Then, by analogy with (6.29), we find that this role is substantial when

$$\beta_i \gtrsim (M_e / M_i)^3 \beta^{-4} \tag{6.74}$$

where $\beta = \beta_e + \beta_i$. In particular, when $\beta_i \simeq \beta_e$, this inequality means

$$\beta > (M_e/M_i)^{3/5}. \tag{6.75}$$

Under condition (6.74), formula (6.73) for the growth rate of the instability of hydrodynamic type becomes invalid. Now the growth rate must be calculated *kinetically* from the analysis of the *resonant interaction of ions* with the wave. In this case one can find equations similar to (6.31) and (6.32) from which an estimate for the maximum growth rate of the *kinetic magneto-drift ion-cyclotron instability* follows (cf. (6.34)):

$$\gamma_{max} \simeq \Omega_i (\beta_i \beta)^{-1/3} M_e/M_i. \tag{6.76}$$

At the limits of its validity, i.e., under the conditions when both sides of relationship (6.74) are of the same order, equation (6.76) means the same as (6.73). When $\beta_i \simeq \beta_e \simeq 1$, we have

$$\gamma_{max} \simeq \Omega_i M_e/M_i. \tag{6.77}$$

Thus, in this case the ratio γ/ω becomes of the same order as in the case of *low-frequency magneto-drift resonance* (cf. (6.77) and (3.19)).

Other results of the theory of *ion-cyclotron instabilities* related to the *transverse drift waves* can be seen in [3.8, 3.9, 3.21].

6.6 Short-wavelength drift-Alfven (SDA) perturbations in a finite-β plasma

6.6.1 Oscillation branches

Let us take the dispersion relation (2.66) and neglect the correctional terms contained in $D^{(1)}$. For simplicity we assume $T_i = T_e$. The temperature gradient is neglected, i.e., $\nabla T = 0$. In this case we arrive at the dispersion relation

$$\omega^2 = \omega_{ne}^2 + \frac{2}{1 + \beta/2} k_\perp^2 \rho_i^2 k_z^2 c_A^2. \tag{6.78}$$

In deriving (2.66), we assumed $k_\perp \rho_i \gg 1$. Therefore, at the validity limits of (6.78), i.e., at $k_\perp \rho_i \simeq 1$, and for $\beta \lesssim 1$ we have $\omega \simeq \max(\omega_{ne}, k_z c_A)$. Consequently, the oscillation branches of (6.78) can be considered as an extension of the *long-wavelength drift-Alfven oscillation branches* (see Chapter 4) into the region $k_\perp \rho_i \gtrsim 1$. In this connection, waves of type (6.78) can be called the *short-wavelength drift-Alfven (SDA) perturbations.*

6.6.2   *Hydrodynamic treatment of the SDA perturbations*
The dispersion relation (6.78) can also be found by starting with rather simple hydrodynamic equations. Allowing for $k_\perp \rho_i \gg 1$ we take the ion density in *Boltzmann form*

$$n = n_0 \exp(-e_i \phi / T_i) \qquad (6.79)$$

where $n = n_0 + \tilde{n}$, and $\tilde{n}$ is the perturbed density. The electrons are described by the equations of continuity and motion. Using the fact that *the electron inertia is unimportant* in the waves discussed, we find that, according to the equation of transverse motion, the electron current density across the field B_0 is given by

$$\boldsymbol{j}_{\perp e} = \frac{e_e c n_0}{B_0} \boldsymbol{E} \times \boldsymbol{e}_z - \frac{c T_e}{B} \boldsymbol{e}_z \times \nabla n \qquad (6.80)$$

while, allowing for (6.79), the equation of longitudinal motion reduces to the form

$$\left(\frac{\partial}{\partial t} + V_{ne} \frac{\partial}{\partial y} \right) A_z + c \left(1 + \frac{T_e}{T_i} \right) \frac{\partial \phi}{\partial z} = 0. \qquad (6.81)$$

Here $B = B_0 + \tilde{B}_z$, where $\tilde{B}_z$ is the perturbation of the z-component of the magnetic field and $V_{ne} = c T_e \kappa_n / e_e B_0$ is the *electron density-gradient drift velocity*. Using equation (6.79) and Maxwell's equations, we find

$$\tilde{B}_z = -\frac{4\pi e_e n_0}{B_0} \left(1 + \frac{T_e}{T_i} \right) \phi. \qquad (6.82)$$

Taking into account (6.79), (6.80), and (6.82), we obtain from the electron continuity equation

$$\left(1 + \frac{\beta}{2} \right) \left(\frac{\partial}{\partial t} + V_{ni} \frac{\partial}{\partial y} \right) \phi - \frac{c T_i}{4\pi e^2 n_0} \frac{\partial}{\partial t} \nabla_\perp^2 A_z = 0 \qquad (6.83)$$

where $V_{ni} = c T_i \kappa_n / e_i B_0$ is the *ion density-gradient drift velocity*.

When $T_e = T_i$, dispersion relation (6.78) follows from (6.81) and (6.83).

6.6.3   *Interaction of the SDA perturbations with resonant electrons*
By means of dispersion relation (2.66), one can find that, *when $\nabla T = 0$, the SDA perturbations are damped due to interaction with resonant*

electrons. When $\nabla T \neq 0$ and $\beta \ll 1$, an instability takes place if $\eta < 0$ or $\eta > 4$ (see, in detail, section 14.5).

Let us consider in detail the resonant interaction in the case of a high-pressure plasma, $\beta \gg 1$, and with sufficiently small k_z. According to (2.67) and (2.68), in the approximation $k_z c_A k_\perp \rho_i / \omega^* \to 0$ and allowing for the above-mentioned assumptions, we have

$$D^{(0)} = \frac{\beta}{2} \left(1 - \frac{\omega_{pe}^*}{\omega} \right) \left(1 - \frac{\omega_{pi}^*}{\omega} \right) \tag{6.84}$$

$$D^{(1)} = \frac{i\pi^{1/2}}{2} \frac{\omega}{|k_z| v_{Te}} \beta_e^2 \left(1 + \frac{T_i}{T_e} \right) \left(1 - \frac{\omega_{pi}^*}{\omega} \right)^2 \chi_3. \tag{6.85}$$

In this case the oscillation branch $\omega = \omega_{pi}^*$ has zero growth rate, while the branch $\omega = \omega_{pe}^*$ is characterized by the growth rate

$$\gamma = \frac{\pi^{1/2}}{4} \frac{\beta_e \omega_{pe}^{*2}}{|k_z| v_{Te}} \left(1 + \frac{T_i}{T_e} \right)^2 \frac{\eta}{1+\eta}. \tag{6.86}$$

It can be seen that an instability takes place if

$$\eta > 0 \qquad \text{or} \qquad \eta < -1. \tag{6.87}$$

In order to elucidate the possibility of excitation of the branch $\omega = \omega_{pi}^*$ we shall consider the parameter $(k_z c_A k_\perp \rho_i / \omega^*)^2$ to be small, but finite. In this case it follows from equation (2.64) that, when $T_e = T_i$, the mentioned oscillation branch is excited if

$$-1 < \eta < -4/5 \tag{6.88}$$

i.e., only for a very narrow range of negative values of the parameter η.

6.7   Ion-cyclotron instability of the SDA perturbations

Starting with equation (2.69), we shall study the possibility of *ion-cyclotron excitation of the SDA perturbations*. For simplicity we assume that $T_e = T_i \equiv T$.

We consider perturbations with *sufficiently small* k_z (cf. section 6.6). As a result, we find that, when $\beta \ll 1$, there is an instability related to the oscillation branch $\omega \approx \omega_{ne}$ if

$$\eta > 3/4 \tag{6.89}$$

and an instability related to the branch $\omega \approx \omega_{ni}$ if

$$\eta < 0. \qquad (6.90)$$

In the case $\beta \gg 1$ the perturbations with $\omega \approx \omega_{pe}^{*}$ are excited when

$$\eta > 4 \qquad \text{or} \qquad \eta < -1 \qquad (6.91)$$

while those with $\omega \approx \omega_{pi}^{*}$ are excited when

$$-1 < \eta < 0. \qquad (6.92)$$

In obtaining the above instability conditions, we assumed that $(\omega - n\Omega_i)/\omega_{Di} \ll 1$.

7 Description of Perturbations in a Plasma with Transverse Current

It was assumed above that the transverse electric current was due to the plasma pressure gradient and determined by relation (1.39). We shall now consider situations where (1.39) is invalid. Physically, this means that there is an additional transverse current in the plasma. We shall study the effects due to such a current. The associated instabilities will be called *transverse-current-driven instabilities* (or simply *current-driven instabilities*).

The additional transverse current can arise in the presence of a *shock wave* or a *soliton* [7.1], in the presence of a *periodic magnetoacoustic wave* used for *high-frequency plasma heating* [7.2], and also in the case of a *plasma rotating in crossed electric and magnetic fields* [7.3]. The case of a *rotating plasma* can be considered as *strictly stationary* (but, of course, by neglecting the dissipative effects). In contrast to this, the transverse current due to a shock wave, a soliton or a magnetoacoustic wave is *not stationary*. However, this nonstationarity can be neglected in the case of instabilities whose *growth rate is large compared with the ion cyclotron frequency*. In this case one can consider that the electric field of the quasi-stationary wave leads to an *electron crossed-field drift* but has no effect on the ions. In this approach, the additional transverse current is due to such a drift. In more detail the problem of the equilibrium with an additional transverse current will be discussed in sections 7.1–7.4. There we discuss properties of the *electron and ion equilibria* (sections 7.1 and 7.2), find expressions for the *magnetic-field gradient* (section 7.3), and mention *equilibria which have been discussed in the literature* (section 7.4). Note that when speaking about a transverse current, we usually keep in mind the additional transverse current.

As before, when considering the current-driven instabilities we shall initially derive equations describing the perturbations and then examine the instabilities relevant to those perturbations. The first part of this programme will be realized in the present chapter, sections 7.5–7.8, while the second will be dealt with in Chapter 8.

In sections 7.5–7.7 we deal with the *small-scale perturbations* described by the local dispersion relation. In contrast to this, in section 7.8 our goal is to derive a differential equation for *large-scale perturbations*.

In section 7.5 we begin with the fact that the local dispersion relation can be represented in the form of (1.4) with some *modified* expressions for the *permittivity tensor* components $\epsilon_{ik}(i, k = 0, 2, 3)$. In the absence of a transverse current, the values ϵ_{ik} are defined by (1.64). A main goal of section 7.5 is to determine the effect of a transverse current on these values. In section 7.6 we consider how the dispersion relation of *electrostatic perturbations is modified* due to a transverse current. (Recall that such a dispersion relation means simply $\epsilon_{00} = 0$.)

In section 7.7 we are interested in small-scale perturbations of *low-frequency* with respect to the electrons. It is more convenient to describe such perturbations by the dispersion relation of the form of (1.1) with the permittivity tensor components $\epsilon_{\alpha\beta}(\alpha, \beta = 1, 2, 3)$. In section 7.7 we find a current modification of these components. There we also obtain a dispersion relation for the *electromagnetic perturbations in a cold plasma* describing the *fast magnetoacoustic waves* and the *whistlers modified by a transverse current*. Such perturbations are of sufficiently long wavelength, $k_\perp < \omega_{pe}/c$. In this connection, it is necessary to go *beyond the WKB approximation*; that is the goal of section 7.8.

7.1 Electron equilibrium

We assume that in a plasma there is an *equilibrium electric field* E_0 directed along the x-axis. Let us analyse how this field modifies the *electron equilibrium*. We shall make such an analysis following the approach of section 1.2.

Neglecting the coordinate dependence of E_0, we conclude that one of the constants of motion is the *particle energy* (per unit mass)

$$\mathcal{E} = \frac{v^2}{2} - \frac{e}{M} x E_0. \tag{7.1}$$

We omit the particle species subscripts.

Expression (1.25) is the starting point for the *inhomogeneous constant of motion* V_y which holds in the presence of the electric field.

However, in order to consider the value X to be equal to the x-coordinate of the *guiding centre* of a particle, one should adopt

$$V_y(X) = V_E \qquad (7.2)$$

where V_E is the *electron crossed-field drift velocity* along the y-axis under the effect of $\boldsymbol{E}_0$ and $\boldsymbol{B}_0$:

$$V_E = -cE_0/B_0. \qquad (7.3)$$

It is assumed that $V_E = V_E(X)$.

As in section 1.2.1, we calculate the integral in (1.25) by expanding in a series in the difference $x - X$. Then we find the following modification of the inhomogeneous constant of motion X (cf. (1.24))

$$X = x + (v_y - V_E)/\Omega. \qquad (7.4)$$

Let us restrict ourselves to the case of an isotropic electron velocity distribution. Then, as in section 1.2.2, certain $F(\mathcal{E}, X)$ could be taken as the equilibrium distribution function f_0. However, when $\boldsymbol{E}_0 \neq 0$, such a representation is not convenient since the contribution of the crossed-field velocity (7.3) is not extracted explicitly in the energy constant (7.1). Therefore, instead of $\mathcal{E}$, we introduce a *modified kinetic energy* $\mathcal{E}_0$ determined by the relation

$$\mathcal{E}_0 = (\boldsymbol{v} - \boldsymbol{V}_E)^2/2. \qquad (7.5)$$

Using (7.4), we find that

$$\mathcal{E} = \mathcal{E}_0 + V_E^2/2 + eXE_0/M. \qquad (7.6)$$

Hence it is clear that $\mathcal{E}_0$ is also a constant of the motion. Therefore, one can take (cf. (1.29))

$$f_0 = F(\mathcal{E}_0, X). \qquad (7.7)$$

By analogy with (1.31) we have for the case of *Maxwellian electrons*:

$$f_{0e} = n_0(X) \left(\frac{M}{2\pi T_e(X)} \right)^{3/2} \exp\left(-\frac{M\mathcal{E}_0}{T_e(X)} \right). \qquad (7.8)$$

Hence we find for a weakly inhomogeneous spatial distribution and for $V_E \ll v_{Te}$ that

$$f_{0e} = n_0 \left\{ 1 + \frac{v_y}{\Omega}\kappa_n \left[1 + \eta\left(\frac{M_e v^2}{2T_e} - \frac{3}{2} \right) \right] - \frac{M_e v_y V_E}{T_e} \right\}$$
$$\times \left(\frac{M_e}{2\pi T} \right)^{3/2} \exp\left(-\frac{M_e v^2}{2T_e} \right). \qquad (7.9)$$

Here $n_0 = n_0(x), T_e = T_e(x), \kappa_n = \partial \ln n_0/\partial x, \eta = \partial \ln T_e/\partial \ln n_0$.

Note that, historically, the transition from $\mathcal{E}$ to $\mathcal{E}_0$ was initially performed in [1.3], Appendix II, where the quantity $\mathcal{E}_0 - eXE_0/M$, i.e., a *modified total energy*, as a constant of motion was used. The constant of motion $\mathcal{E}_0$ was effectively used in [6.18] and in subsequent papers by the same authors (see also section 7.4 for more about those papers). In the literature, other representations of f_0 have also been used (cf. section 7.4).

We now proceed to a modification of *the equilibrium electron trajectories*. It is clear that the expressions for the trajectories in the presence of E_0 differ from those in the case where $E_0 = 0$ only by the crossed-field drift (recall that an inhomogeneity of E_0 is neglected). Therefore we now have instead of (1.48) (cf. [1.3], Appendix II)

$$
\begin{aligned}
v_x &= v_{\perp 0} \cos \alpha \\
v_y &= v_{\perp 0} \sin \alpha + V_d + V_E \\
x &= x_0 - (v_{\perp 0})(\sin \alpha - \sin \alpha_0) \\
y &= y_0 + (v_{\perp 0}/\Omega)(\cos \alpha - \cos \alpha_0) + (V_D + V_E)(t - t_0).
\end{aligned} \tag{7.10}
$$

Here

$$
\begin{aligned}
v_{\perp 0} &= [v_{x0}^2 + (v_{y0} - V_E)^2]^{1/2} \\
\alpha_0 &= \tan^{-1}[(v_{y0} - V_E)/v_{x0}] \\
v_{x0} &= v_x(t_0) \\
v_{y0} &= v_y(t_0).
\end{aligned} \tag{7.11}
$$

It is clear that $v_{\perp 0}$ represents the modulus of the cyclotron rotation velocity of the particle.

The form of writing the electron equilibrium equations given above is convenient due to the fact that, when one changes to the reference frame moving along the y-axis with the velocity V_E, the electric field E_0 cancels out throughout. Thus, the modifications due to E_0 are reduced in the adopted approximation to a *shift of the electron component* with velocity V_E

$$
v_y \to v_y + V_E. \tag{7.12}
$$

Therefore, one can operate in two reference frames : firstly, one in which the electron trajectories are of the form of (7.10), and second, when we have (1.48) instead of (7.10). We shall call the first *the reference frame with moving electrons*, and the second *the reference frame with electrons at rest*. However, one should keep in mind that, in the reference frame with electrons at rest, the particles move across the magnetic field with *magnetic-drift velocity* V_d (see (1.27)). At the same time, in this reference frame there is also a macroscopic velocity of electrons related to their spatial inhomogeneity, i.e., to the so-called *diamagnetic (Larmor) drift velocity* V_{pe}^* (cf. (1.38) and (1.39)).

7.2   Ion equilibrium

In the description of the *ion equilibrium* there is some arbitrariness. In order to remove such an arbitrariness, it is necessary to examine the *preceding plasma processes*. This concerns the *quasi-stationary plasma equilibrium*, e.g., the plasma in the field of a *shock wave or a periodic magnetoacoustic wave*. In contrast to this, the description of the ion equilibrium in the case of a *rotating plasma* is simpler since such an equilibrium is strictly stationary.

Let us consider some examples.

7.2.1   *Ion distribution just after applying an alternating electromagnetic field to a plasma*

Let us assume that, before applying an *alternating electromagnetic field* to a plasma, the ion equilibrium distribution function is of the form of (1.29). If the field is applied suddenly, i.e., for a time sufficiently small compared with the ion cyclotron period, then the ion distribution just after switching-on the field is unchanged. It is clear that in this case we can consider only instabilities whose growth rate is large compared with the ion cyclotron frequency.

The simplest example of such an equilibrium is that of *cold ions*. In the reference frame with moving electrons (see section 7.1) such ions are at rest. An alternative to this is the case of *hot ions* with the above-mentioned distribution function of the form of (1.29). When this distribution is spatially inhomogeneous, they have, in this reference frame, a *macroscopic velocity* non-vanishing for $\nabla p_{\perp i} \neq 0$ (cf. (1.38) and (1.39)).

7.2.2   *Distribution of partially relaxing ions*

If we wish to consider a *quasi-equilibrium* when the ions are not yet completely carried away by the crossed-field motion it is convenient to approximate the ion distribution function by that of form (1.29) with a shift

$$v_y \to v_y + \alpha V_E \qquad (7.13)$$

where $0 \leqslant \alpha \leqslant 1$. When $\alpha = 0$, the ions do not take part in the crossed-field motion, while in the case where $\alpha = 1$ they are completely carried away.

It is clear from comparing (7.13) with (7.12) that the case considered reduces to the case of section 7.2.1 with the substitution $V_\mathrm{E} \to V_{\mathrm{Eeff}}$, where

$$V_{\mathrm{Eeff}} = (1 - \alpha)V_\mathrm{E}. \qquad (7.14)$$

It is also clear that, changing to the reference frame with the electrons at rest, the equilibrium with $\alpha = 1$ reduces to that discussed in section 1.2, i.e., to the equilibrium with $V_\mathrm{E} = 0$.

Note also that all the equilibria discussed here turn into those studied in section 1.2 when $E_0 = 0$. Such equilibria can be called the *"formally evolutionary equilibria"*. Other ion quasi-equilibria discussed in the literature will be mentioned in section 7.4.

7.2.3   Ions in a rotating plasma

Formally, in describing a *rotating plasma*, we should use the cylindrical coordinates (cf. [2], section 7.1). However, keeping in mind *small-scale perturbations*, we can consider the axisymmetry direction to be the y-axis. Then, using the results of [2], section 7.1, we find that the ions move along the y-axis with the velocity

$$V_\mathrm{i} = V_\mathrm{E} + V_\mathrm{c} \qquad (7.15)$$

where V_c is the *centrifugal drift velocity* defined by the relation

$$V_\mathrm{c} = -V_\mathrm{E}^2/r\Omega_\mathrm{i}. \qquad (7.16)$$

Here r is the distance from the axis of symmetry.

It is clear that in changing to the reference frame with electrons at rest, the plasma equilibrium considered reduces to that with the ions moving with velocity V_c.

7.3   Magnetic-field gradient

According to Chapters 1 and 2, knowledge of the *magnetic-field gradient* is necessary, first, for calculating the *magnetic-drift velocity* of a particle, and, second, for using the *pressure-balance equation*, which is important when allowing for *electromagnetic effects*. In the standard situations discussed in section 1.2, when the transverse current is expressed by (1.39), the magnetic-field gradient is defined by relationship (1.42), i.e., it is due only to the plasma transverse-pressure gradient. We shall now consider how this relationship is modified in the presence of an additional transverse current.

Allowing for the discussion of section 7.1 and sections 7.2.1 and 7.2.2, we conclude that in the presence of a wave (*shock wave, soliton or periodic magnetoacoustic wave*)

$$\frac{\mathrm{d}B_0}{\mathrm{d}x} = -\frac{4\pi}{B_0}\frac{\mathrm{d}p_\perp}{\mathrm{d}x} - \frac{4\pi e_e n_0 V_E}{c}(1-\alpha). \tag{7.17}$$

Instead of this, in the case of a *rotating plasma*

$$\frac{\mathrm{d}B_0}{\mathrm{d}x} = -\frac{4\pi}{B_0}\frac{\mathrm{d}p_\perp}{\mathrm{d}x} - \frac{4\pi e_i n_0}{c}V_c. \tag{7.18}$$

Both results can be written in a unified form

$$\frac{\mathrm{d}B_0}{\mathrm{d}x} = -\frac{4\pi}{B_0}\frac{\mathrm{d}p_\perp}{\mathrm{d}x} - \frac{4\pi e_i n_0}{c}V_j \tag{7.19}$$

where V_j is an effective additional current velocity related to the ions and is defined by

$$V_j = \begin{cases} -(1-\alpha)V_E & \text{for the case of (7.14)} \\ V_c & \text{for the case of (7.16).} \end{cases} \tag{7.20}$$

In terms of V_j the transverse current is written in the form

$$j_{0y} = \frac{c}{B_0}\frac{\mathrm{d}p_\perp}{\mathrm{d}x} + e_i n_0 V_j. \tag{7.21}$$

When the current velocity V_j is large compared with the *diamagnetic drift velocities* of the plasma components, equation (7.19) is replaced by

$$\frac{\mathrm{d}B_0}{\mathrm{d}x} = -\frac{4\pi e_i n_0}{c}V_j. \tag{7.22}$$

Hence we find that the *averaged magnetic-drift velocity* of particles is of order

$$V_D \simeq \beta V_j \tag{7.23}$$

where β is related to a respective plasma component. Therefore, when $\beta \ll 1$, we have $V_D \ll V_j$.

7.4   Equilibria previously discussed in the literature

In [7.4] and in other papers by the same group of authors (see, in particular, [6.40–6.42, 7.5, 7.6]), it was assumed that the electrons are at rest (given the meaning explained in section 7.1), while the ions move across the magnetic field with a velocity u. (In these papers the y-axis was taken as the inhomogeneity direction, and the x-axis was the drift direction.) In fact, the equilibrium of [7.4] reduces to that mentioned in section 7.2.1 if the latter is considered in the reference frame with electrons at rest. The equilibrium of [7.4] seems to be natural for the problem of the *high-frequency heating* of a plasma at the stage of *non-relaxing ions* (cf. section 7.2.1). Its advantage is the fact that it is "formally evolutionary" when $E_0 \to 0$, given the meaning mentioned in section 7.2.2.

In [7.7] it has been assumed that the electron equilibrium distribution function has the form

$$f_{0e} = n_0 \left[1 - \left(\epsilon' + \frac{\alpha M_e v_1^2}{2T_e} \right) X \right] \left(\frac{M_e}{2\pi T_e} \right)^{3/2} \exp\left(-\frac{M_e \mathcal{E}}{T_e} \right) \quad (7.24)$$

where $\mathcal{E}$ and X are defined by equations(7.1) and (1.24), respectively, and n_0, T_e, ϵ', and α are constants. The representation in (7.24) is correct since the equilibrium distribution function is expressed in terms of the constants of motion. However, it does not seem particularly convenient since ϵ' has no simple meaning for $E_0 \neq 0$ and also for $E_0 = 0$ and $\nabla T \neq 0$. In this connection, the representation of (7.9) seems to be more convenient.

The ion distribution function is taken in [7.7] in the form

$$f_{0i} = n_0 (M_i/2\pi T_i)^{3/2} \exp(-M_i v^2/2T_i) \quad (7.25)$$

where n_0 and T_i are constants. A formal difficulty of such a distribution is related to the fact that it is non-evolutionary in the meaning pointed out in section 7.2.2. In other words, when $E_0 \to 0$, the function in (7.25) does not satisfy the condition of a strict equilibrium (1.29). However, the accuracy of (7.25) is sufficient in the case where V_E is greater than the ion diamagnetic drift velocity due to a pressure gradient.

In [6.16] distributions (7.24) and (7.25) were used. The authors of [6.16] then discussed how the values of V_E could be realized. They allowed for—in the case of *unmagnetized Maxwellian ions*—the following relationship to hold

$$E_0 = \frac{1}{n_0 e_i} \frac{\partial p_i}{\partial x} + \frac{M_i}{e_i} \frac{dV_{xi}}{dt} \quad (7.26)$$

where $d/dt = \partial/\partial t + V_i \cdot \nabla$, and that, as a result

$$V_E = -V_{pi}^* - \Omega_i^{-1} dV_{xi}/dt \quad (7.27)$$

where V_i is the ion macroscopic velocity and V_{pi}^* is the ion diamagnetic drift velocity. On the basis of (7.27), the authors of [6.16] have noted that two cases can be distinguished. In the first case, $dV_{xi}/dt = 0$, so that

$$V_E = -V_{pi}^*. \qquad (7.28)$$

In the second case, $dV_{xi}/dt \neq 0$ and, as a result

$$V_E \neq -V_{pi}^*. \qquad (7.29)$$

However, it can be seen that in the case when (7.28) holds and when $V_E \ll v_{Ti}$, changing to the other reference frame, we obtain, in fact, the trivial equilibrium considered in section 1.2, i.e., an equilibrium *without* additional transverse current. When $V_E \gtrsim v_{Ti}$, the assumption of a Maxwellian distribution of the form of (7.25) does not seem to be sufficiently validated.

Note also that the implication of (7.29) is equivalent to that of condition (7.13).

The electron equilibrium model (7.24) was then improved in [6.18]. As mentioned in section 7.1, the electron distribution function in [6.18] was of the form of (7.8) and (7.9). The authors of [6.18] took the *shifted Maxwellian distribution* to be the ion distribution function

$$F_{i0} = n_0 \left(\frac{M}{2\pi T_i} \right)^{3/2} \exp\left(-\frac{v_\perp^2 + v_z^2}{v_{Ti}^2} \right) \qquad (7.30)$$

where $v_\perp^2 = v_x^2 + (v_y - V_{yi})^2$ and

$$V_{yi} = V_E + V_{pi}^*. \qquad (7.31)$$

It was assumed that $\nabla T = 0$.

Note that when V_{yi} is of the form of (7.31) and making the transition to the reference frame with electrons at rest, the electric field E_0 is excluded from the problem. In other words, the equilibrium considered in [6.18], in fact, does not allow for the presence of an additional transverse current.

7.5 Modification of the permittivity tensor

7.5.1 General remarks

As mentioned in section 7.1, we are interested in a plasma whose components have, besides the diamagnetic drift velocities, some unequal additional *macroscopic transverse velocities*. We shall consider the perturbations in a reference frame moving relative to the laboratory reference frame with a velocity equal to the electron additional velocity. In this reference frame the electron contribution into the *permittivity tensor* $\epsilon_{ik}(i, k = 0, 2, 3)$ will be the same as that into the right-hand sides of equations (1.64). As for the ions, by analogy with Chapter 6, we take into account only their contribution into the component ϵ_{00}, i.e., we deal only with the electrostatic part of the *ion permittivity*. Then the general problem of modifying the permittivity tensor $\epsilon_{ik}(i, k = 0, 2, 3)$ for a plasma with transverse current is reduced to modifying the ion contribution into ϵ_{00}, i.e., into the value $\epsilon_{00}^{(i)}$.

We shall modify $\epsilon_{00}^{(i)}$ in the *high-frequency approximation* in section 7.5.2 and in the *discrete cyclotron harmonic spectrum* in section 8.7. In accordance with the above-mentioned, in the second case we shall keep in mind only the problem of a *rotating plasma*.

In section 7.6 we shall discuss the dispersion relation of the *electrostatic approximation*. In this case the main focus of attention is paid to the various representations of $\epsilon_{00}^{(e)}$ (the electron contribution into ϵ_{00}) used in the problems of instabilities in a plasma with a transverse current. This analysis seems all the more necessary in that in the literature erroneous representations of $\epsilon_{00}^{(e)}$ are found.

7.5.2 The high-frequency approximation for $\epsilon_{00}^{(i)}$

An expression for $\epsilon_{00}^{(i)}$ in the *high-frequency approximation* can be obtained by means of the well-known relationship (see [1], equation (2.10))

$$\epsilon_{00}^{(i)} = \frac{4\pi e^2}{M_i k^2} \int \frac{\boldsymbol{k} \cdot \partial f_0 / \partial \boldsymbol{v}}{\omega - \boldsymbol{k} \cdot \boldsymbol{v}} \mathrm{d}\boldsymbol{v} \qquad (7.32)$$

where for f_0 expression (1.37) should be taken with the substitution (see (7.20))

$$v_y \to v_y + V_j. \qquad (7.33)$$

By replacing the integration variable $v_y \to v_y - V_j$, equation (7.32) reduces to

$$\epsilon_{00}^{(i)} = \frac{4\pi e^2}{M_i k^2} \int \frac{\boldsymbol{k} \cdot \partial f_0 / \partial \boldsymbol{v}}{\omega - \omega_j - \boldsymbol{k} \cdot \boldsymbol{v}} \mathrm{d}\boldsymbol{v} \qquad (7.34)$$

where $\omega_j = k_y V_j$ and f_0 is of the form of (1.37). Substituting (1.37) here, we find

$$\epsilon_{00}^{(i)} = \frac{4\pi e^2}{M_i k^2} \int \frac{\mathrm{d}\boldsymbol{v}}{\omega - \omega_j - \boldsymbol{k} \cdot \boldsymbol{v}} \left[\boldsymbol{k} \cdot \boldsymbol{v} \left(\frac{\partial F}{\partial \mathcal{E}} + \frac{v_y}{\Omega_i} \frac{\partial^2 F}{\partial x \partial \mathcal{E}} \right) + \frac{k_y}{\Omega_i} \frac{\partial F}{\partial x} \right].$$
$$(7.35)$$

The term with $\partial^2 F/\partial x \partial \mathcal{E}$ is important here for $|\omega - \omega_j| \gg kv_{Ti}$. In this case, it is *completely compensated by* the contribution of the last term in the square brackets. Then $\epsilon_{00}^{(i)}$ takes the same form as in the case of a homogeneous plasma:

$$\epsilon_{00}^{(i)} = -\frac{\omega_{pi}^2}{(\omega - \omega_j)^2}. \tag{7.36}$$

Compensating for the gradient terms mentioned here is the effect mentioned at the end of section 1.4.

It follows from (7.35) for the *Maxwellian ion distribution* and $|\omega - \omega_j| \ll kv_{Ti}$ that

$$\epsilon_{00}^{(i)} = \frac{1}{(k\lambda_{Di})^2}\left\{1 + \frac{i\pi^{1/2}}{kv_{Ti}}\left[\omega - \omega_j - \omega_{ni}\left(1 - \frac{\eta}{2}\right)\right]\right\}. \tag{7.37}$$

It can be seen that the role of a current in this approximation is similar to that of a temperature gradient, Then the case $V_j/V_{ni} > 0$ corresponds to that of $\eta < 0$.

If the usual *drift effects* are small compared with the current ones, i.e., $V_j \gg V_i^*$, we have for the Maxwellian distribution

$$\epsilon_{00}^{(i)} = (k\lambda_{Di})^{-2}\{1 + Z[(\omega - \omega_j)/kv_{Ti}]\} \tag{7.38}$$

where Z is defined by (2.25).

7.6   Dispersion relation for electrostatic perturbations

The *electrostatic approximation* can be justified for a sufficiently small β and also for the case when the transverse wavelength of perturbation is small compared with the electron Larmor radius, $k_\perp \rho_e \gg 1$. (When $k_\perp \rho_e \gg 1$, the components ϵ_{02} and ϵ_{03} are small due to the smallness of the Bessel functions.)

We represent the dispersion relation of the electrostatic approximation in the form (cf. (2.80))

$$1 + \epsilon_{00}^{(i)} + \epsilon_{00}^{(e)} = 0. \tag{7.39}$$

For $\epsilon_{00}^{(i)}$ one should substitute here the expressions given in section 7.5.2. As for $\epsilon_{00}^{(e)}$, in the case of *Maxwellian electrons*, it can be presented in a similar form to [2], equation (1.106):

$$\epsilon_{00}^{(e)} = \frac{1}{k^2\lambda_{De}^2}\left[1 - \omega\sum_n \left\langle \left(\hat{l}_e + (\hat{l}_e - 1)\frac{n\omega}{z_e\Omega_e}\right)\frac{J_n^2(\xi_e)}{\omega - n\Omega_e - \omega_{de} - k_z v_z}\right\rangle\right]. \tag{7.40}$$

Here the notation of section 2.1.2 is used. Note also that we operate in the reference frame with electrons at rest.

Note that the terms with $\hat{l}_e - 1$ in (7.40) are important in the limit of *high-frequency perturbations* and *cold electrons*, i.e., for $\omega \gg (\Omega_e, k_\perp v_{Te})$. They are similar to the terms with $\partial^2 F/\partial x \partial \mathcal{E}$ in (7.35).

Let us consider some particular cases of (7.40).

7.6.1   The low-frequency approximation, with respect to electrons, for $\beta \ll 1$

In the *low-frequency approximation*, only the term with $n = 0$ is important in the sum of (7.40). If $\beta \ll 1, V_{de}/V_j$ is small, of order β. Therefore, the term with ω_{de} in (7.40) must be omitted. This term was retained, in particular, in [7.7, 7.8, 6.16]. Therefore, the corresponding results of these papers are subject to inaccuracies.

For small $k_\perp \rho_e$ and $k_z v_{Te}/\omega$ the above-mentioned small term of order β is in the sum with formally small terms of order $k_\perp^2 \rho_e^2$. Therefore, the term with ω_{de} must be taken into account when $k_\perp^2 \rho_e^2 \lesssim \beta$. However, the last inequality means $k_\perp \lesssim \omega_{pe}/c$, and this is simply the well-known *condition of violation of the electrostatic approximation*. A description of perturbations with $k_\perp \lesssim \omega_{pe}/c$ will be given in section 7.7.

When $k_z = 0$, an exponentially small effect of resonant interaction of electrons with perturbations is also associated with the term ω_{de}. Such an effect was considered in [7.9]. However, for a quantitative study of this effect, it is necessary to go beyond the electrostatic approximation.

Thus, the correct *low-frequency limit* of (7.40) for $\beta \ll 1$ is given by

$$\epsilon_{00}^{(e)} = (k\lambda_{De})^{-2}[1 + \hat{l}_e Z(\omega/ \mid k_z \mid v_{Te}) I_0(z_e) \exp(-z_e)]. \qquad (7.41)$$

Let us now discuss under which conditions the gradient effects are important in this expression. When $z_e \gtrsim 1$ or $\omega/ \mid k_z \mid v_{Te} \lesssim 1$, the contribution of the *gradient effects* (they are formally associated with the departure of $\hat{l}_e$ from unity) into (7.41) is of order ω^*/ω, i.e., allowing for $\omega \simeq \omega_j$, the contribution is of order V^*/V_j. Since $\beta \ll 1$, we conclude (allowing for (7.23)) that the last ratio is not small compared with unity only if $\kappa_B \ll (\kappa_n, \kappa_T)$. Therefore, in situations where the characteristic lengths of the magnetic field and density gradients are of the same order of magnitude (e.g., in the case of a *shock wave or a soliton*), the gradient contribution into (7.41) can be neglected. Then

$$\epsilon_{00}^{(e)} = (k\lambda_{De})^{-2}[1 + Z(\omega/ \mid k_z \mid v_{Te}) I_0(z_e) \exp(-z_e)]. \qquad (7.42)$$

Dispersion relation (7.39) with $\epsilon_{00}^{(i)}$ and $\epsilon_{00}^{(e)}$ of the form of (7.38) and (7.42) was analysed in [1], Chapter 13, and in the papers mentioned there.

When $z_e \ll 1$ and $\omega \gg k_z v_{Te}$, the leading terms in (7.41) are *mutually compensated*. As a result, the contribution of $\hat{l}_e - 1$ becomes more substantial. It then follows from (7.41) that

$$\epsilon_{00}^{(e)} = \frac{\omega_{pe}^2}{\Omega_e^2} + \frac{\omega_{pe}^2 k_y \kappa_n}{k^2 \omega \Omega_e} - \frac{\omega_{pe}^2}{\omega^2} \frac{k_z^2}{k^2}. \tag{7.43}$$

Dispersion relation (7.39) with $\epsilon_{00}^{(i)}$ and $\epsilon_{00}^{(e)}$ defined by (7.36) and (7.43) is simply equation (2.60) of [2]. It was studied in section 2.5 of [2].

7.6.2   *Short-wavelength perturbations, with respect to electrons, near the electron–cyclotron harmonics*

The region of very large wave numbers, $k_\perp \rho_e \gg 1$, is of interest due to the fact that excitation of perturbations with frequencies $\omega \gtrsim \Omega_e$ can be related to such $k_\perp$ when $V_j \ll v_{Te}$. The possibility of such an excitation is clear if one recalls the general condition of excitation $\omega_j \gtrsim \omega$ and if one allows for the above inequalities.

When $k_\perp \rho_e \gg 1$, the components ϵ_{02} and ϵ_{03} are small for arbitrary β (see (1.64)). Consequently, in this case the dispersion relation of the *electrostatic approximation* (7.39) has *no restrictions* with respect to β. For the given $k_\perp \rho_e$, expression (7.40) for $\epsilon_{00}^{(e)}$ is simplified. In the present section we consider simplification of $\epsilon_{00}^{(e)}$ assuming $\beta \ll 1$ and $k_z v_{Te} \ll \Omega$. These assumptions correspond to the *approximation of a discrete spectrum of electron–cyclotron harmonics*. Allowing for only one specific harmonic, we find from (7.40)

$$\epsilon_{00}^{(e)} = \frac{1}{k^2 \lambda_{De}^2} \left(1 - \frac{\omega}{\pi} \hat{l}_e \left\langle \frac{\xi_e^{-1}}{\omega - n\Omega_e - \omega_{de} - k_z v_z} \right\rangle \right). \tag{7.44}$$

The departure of $\hat{l}_e$ from unity is significant here only when $(\kappa_n, \kappa_T) \gg \kappa_B$. As for ω_{de} in the denominator of (7.44), an allowance for this quantity is important if $\omega - n\Omega_e$ and $k_z v_z$ are sufficiently small.

Expression (7.44) will be used in section 8.6.

7.6.3   *The high-frequency approximation for hot electrons*

For $k_\perp \rho_e \gg 1$ in (7.40) one can perform a summation over cyclotron harmonics if ω_{de} or $k_z v_z$ are comparable to or larger than the electron cyclotron frequency Ω_e (cf. section 2.2.1). We then obtain (cf. (2.52) and (7.37))

$$\epsilon_{00}^{(e)} = (k\lambda_{\mathrm{De}})^{-2}[1 + i\omega\pi^{1/2}\hat{l}_e(kv_{Te})^{-1}].\qquad(7.45)$$

In this case the electrons behave in perturbations as *unmagnetized particles*. The dependence of the right-hand side of (7.45) on the magnetic field (through the operator $\hat{l}_e$) is due to the fact that this field is in the expression for the equilibrium distribution function.

We shall use equation (7.45) in section 8.6.

7.6.4   The high-frequency approximation for cold electrons
In the case of *cold electrons* and for $\omega \gg \Omega_e$ it follows from (7.40) that (cf. (7.36))

$$\epsilon_{00}^{(e)} = -\omega_{pe}^2/\omega^2.\qquad(7.46)$$

Note that in this approximation the gradient term cancels out from the expression for $\epsilon_{00}^{(e)}$. Such a cancelling out is formally due to the compensating additional term with $\hat{l}-1$ in (7.40).

The *high-frequency approximation* for *cold electrons* is important, in particular, in the problem of *Buneman's instability*, see section 1.6 of [1]. The authors of [7.10] discovered the role of the *density gradient* in this problem. They started with the expression for ϵ_{00} of the form of (7.40) but *without* the compensating term with $\hat{l}_e - 1$. As a result, they have arrived at the conclusion that the right-hand side of (7.46) contains the same gradient term as (7.43). Hence, it is clear that the analysis of the role of the density gradient in the Buneman instability in [7.10] is erroneous.

7.7   Description of electromagnetic small-scale perturbations of low-frequency, with respect to electrons

In studying the *electromagnetic perturbations* with frequencies smaller than the electron cyclotron frequency, it is convenient to use $\epsilon_{\alpha\beta}(\alpha,\beta = 1,2,3)$ as defined in section 1.1.4, instead of $\epsilon_{ik}(i,k = 0,2,3)$. The starting dispersion relation is then written in the form of (1.1).

7.7.1   General expressions for $\epsilon_{\alpha\beta}(\alpha,\beta = 1,2,3)$
The general approach for obtaining $\epsilon_{\alpha\beta}(\alpha,\beta = 1,2,3)$ from ϵ_{ik} was given in section 1.1.4. Following this approach, we find that in the

case of a plasma with a transverse current

$$\epsilon_{11} = \epsilon_{11}^{(0)} + \epsilon_{11}^{(1)} + \epsilon_{11}^{(2)}$$
$$\epsilon_{12} = \epsilon_{12}^{(0)} + \epsilon_{12}^{(1)} \tag{7.47}$$
$$\epsilon_{13} = \epsilon_{13}^{(1)}.$$

Here

$$\epsilon_{11}^{(0)} = 1 + \epsilon_{00}^{(i)} + \frac{\omega_{pe}^2 k_y}{k^2 \omega \Omega_e}\left[\kappa_n + \kappa_B\left(1 - \frac{\omega_{pi}^* + \omega_j}{\omega}\right)\right]$$
$$\epsilon_{12}^{(0)} = \frac{i}{\kappa_B k_y \lambda_{De}^2}\frac{\omega_{de}}{\omega}\left(1 - \frac{\omega_{pi}^* + \omega_j}{\omega}\right). \tag{7.48}$$

When $k_\perp \rho_e \ll 1$, the quantities $\epsilon_{\alpha\beta}$ with superscript (1) are as small as $(k_\perp \rho_e)^2$. For arbitrary $k_\perp \rho_e$, they are defined by the electron contribution into the sums (2.1) provided that the part of $\epsilon_{11}^{(2)}$ containing the function ζ_0 is separated from ϵ_{11}. The quantity $\epsilon_{11}^{(2)}$ is defined thereby.

As for expressions for $\epsilon_{22}, \epsilon_{23}, \epsilon_{33}$, they can be found by means of (2.2) whilst keeping only the electron contribution.

Note that, in obtaining (7.48), the *equilibrium equation* (7.19) has been taken into account. As a result the terms with $(\omega_{pi}^* + \omega_j)/\omega$ appear in the right-hand side of (7.48). Thereby, we have extracted in explicit form *both modifying factors* associated with the transverse current: first, a shift of the ion component relative to the electron one which manifests as a modification of $\epsilon_{00}^{(i)}$ (see section 7.5.2), and, second, an additional term with ω_j/ω which is a consequence of the new equilibrium condition (7.19).

7.7.2   Description of perturbations with $k_z = 0$

When $k_z = 0$, the dispersion relation reduces to (2.35) with ϵ_{11} and ϵ_{12} of the form of (7.47) and ϵ_{22} of the form of (2.40). In this case the expressions for $\epsilon_{11}^{(1}$, $\epsilon_{11}^{(2)}$ and $\epsilon_{12}^{(1)}$ are given by equations (2.37) and (2.39). For $\epsilon_{00}^{(i)}$ one of the expressions (7.36)–(7.38) should be substituted in (7.47).

In particular, when $\omega \ll k v_{Ti}$, in the *zeroth approximation in* $k_\perp \rho_e$ and when neglecting small imaginary terms of order $\omega/k v_{Ti}$, the following dispersion relation is obtained (cf. (6.1))

$$\frac{\omega_{Di}}{\omega}\left(1 - \frac{\omega_{ni}}{\omega}\right)\left(1 - \frac{\omega_{pi}^*}{\omega} - \frac{\omega_j}{\omega}\right)$$
$$+ \frac{T_i}{T_e}\left[\hat{i}_i\left(1 - \frac{\omega_{Di}}{\omega}\right) + \frac{\omega_{Di}\omega_j}{\omega^2}\right]\hat{i}_e\left\langle\frac{\omega_{de}}{\omega - \omega_{de}}\right\rangle = 0. \tag{7.49}$$

We shall use equation (7.49) in section 8.3.

7.7.3 Description of perturbations in a plasma with cold electrons

When $T_e \to 0$, in neglecting the *transverse electron inertia*, only the values with superscript zero remain in the right-hand sides of equations (7.47). In the limit $T_e \to 0$ one should use the identity $\omega_{De}/\lambda_{De}^2 = -\omega_{Di}/\lambda_{Di}^2$ in the second equality of (7.47). The component ϵ_{33} is defined by

$$\epsilon_{33} = -\omega_{pe}^2/\omega^2 \tag{7.50}$$

while the components ϵ_{23} and ϵ_{22} as well as ϵ_{13} vanish. In the limit $\epsilon_{33} \to \infty$ the dispersion relation reduces to

$$\tilde{\epsilon}_{11} - N^2 \cos^2\theta = 0. \tag{7.51}$$

Here

$$\tilde{\epsilon}_{11} \equiv \epsilon_{11}^{(0)} - \epsilon_{12}^{(0)2}/N^2. \tag{7.52}$$

Substituting $\epsilon_{11}^{(0)}$ and $\epsilon_{12}^{(0)}$ from (7.48) into (7.52), we find

$$\tilde{\epsilon}_{11} = \epsilon_{00}^{(i)} + \frac{\omega_{pe}^2 \kappa_n k_y}{k^2 \omega \Omega_e} + \frac{\omega_{pe}^4}{\Omega_e^2 c^2 k^2}\left(1 - \frac{\omega_{pi}^* + \omega_j}{\omega}\right). \tag{7.53}$$

In the case of *cold ions*, i.e., when $\omega_{pi}^* \to 0$ and $\epsilon_{33}^{(i)}$ is of the form of (7.36), using (7.52) and (7.53) we arrive at the dispersion relation

$$\omega\left[\omega - \omega_j\left(1 - \frac{\kappa_n}{\kappa_B}\right)\right] - \frac{k^2 k_z^2 c^4 \Omega_e^2}{\omega_{pe}^4} - \frac{k^2 c_A^2 \omega^2}{(\omega - \omega_j)^2} = 0. \tag{7.54}$$

Initially, this result has been obtained in [7.11] and has then been reproduced in [7.4, 7.12].

Note that in the absence of a transverse current, i.e., when $V_j = 0$, it follows from (7.50) that

$$\omega^2 = k^2 c_A^2 + k^2 k_z^2 c^4 \Omega_e^2/\omega_{pe}^4. \tag{7.55}$$

For $k_z = 0$ this corresponds to *fast magnetoacoustic waves* with the dispersion law

$$\omega^2 = k^2 c_A^2 \tag{7.56}$$

while if k_z is sufficiently large, equation (7.55) reduces to the dispersion relation for *whistlers*,

$$\omega^2 = k^2 k_z^2 c^4 \Omega_e^2/\omega_{pe}^4 \equiv \omega_w^2. \tag{7.57}$$

Allowing for the *transverse electron inertia* and for terms of order $1/\epsilon_{33}$, we find, instead of (7.51),

$$(\tilde{\epsilon}_{11} + \delta\epsilon_{11})(1 - N^2/\epsilon_{33}) - N^2 \cos^2\theta = 0 \qquad (7.58)$$

where

$$\delta\epsilon_{11} = \omega_{pe}^2/\Omega_e^2. \qquad (7.59)$$

For the case of cold ions, it follows from (7.58) and (7.59) that

$$(1+q)\left\{1 + q\left[1 - \frac{\omega_j}{\omega}\left(1 - \frac{\kappa_n}{\kappa_B}\right)\right] - \frac{\omega_{LH}^2}{(\omega - \omega_j)^2}\right\} - \frac{\Omega_e^2}{\omega^2}\frac{k_z^2}{k^2} = 0.$$
$$(7.60)$$

Here q is the parameter characterizing the *electromagnetic character of perturbations* and is defined by

$$q = \omega_{pe}^2/c^2 k^2 \qquad (7.61)$$

and $\omega_{LH}^2 = \Omega_e^2 M_e/M_i$ is the square of the *lower-hybrid frequency*.

Initially, equation (7.60) has been obtained in [7.4] (see also [7.12]).

Evidently, when $q \gg 1$, equation (7.60) reduces to (7.54). On the other hand, when $q \ll 1$ (the *electrostatic approximation*), it yields

$$1 - \frac{k_y\kappa_n}{k^2}\frac{\Omega_e}{\omega} - \frac{\Omega_e^2 k_z^2}{\omega^2 k^2} - \frac{\omega_{LH}^2}{(\omega - \omega_j)^2} = 0. \qquad (7.62)$$

The authors of [7.13] have obtained an equation of the type of (7.60) but without the terms with $(\omega_j/\omega)(1 - \kappa_n/\kappa_B)$. Therefore, the corresponding results of [7.13] are wrong.

7.7.4 Dispersion relation for perturbations with $k_\perp\rho_e \ll 1$ and $\omega \ll k_z v_{Te}$ in a plasma with finite electron temperature

Let us assume that $k_\perp\rho_e \ll 1$ and $\omega \ll k_z v_{Te}$. Then neglecting small anti-Hermitian terms of order $\omega/k_z v_{Te}$, by means of (2.3) we find

$$\begin{aligned}
\epsilon_{33} &= \bar{\epsilon}_{33} + \hat{\epsilon}_{33} \\
\epsilon_{22} &= 0 \\
\epsilon_{23} &= i\frac{\omega_{pe}^2}{\omega\Omega_e}\frac{k_\perp}{k_z}(1 - \omega_{pe}^*/\omega)
\end{aligned} \qquad (7.63)$$

where

$$\begin{aligned}
\bar{\epsilon}_{33} &= (k_z\lambda_{De})^{-2}(1 - \omega_{ne}/\omega) \\
\hat{\epsilon}_{33} &= -\frac{k_y\kappa_B}{k_z^2}\frac{\omega_{pe}^2}{\omega\Omega_e}(1 - \omega_{pe}^*/\omega).
\end{aligned} \qquad (7.64)$$

We represent expression (7.48) for ϵ_{11} in the form

$$\epsilon_{11} = \bar{\epsilon}_{11} + \hat{\epsilon}_{11} \tag{7.65}$$

where

$$
\begin{aligned}
\bar{\epsilon}_{11} &= \epsilon_{00}^{(i)} + \omega_{pe}^2 k_y \kappa_n / k^2 \omega \Omega_e \\
\hat{\epsilon}_{11} &= \frac{k_y \kappa_B}{k^2} \frac{\omega_{pe}^2}{\omega \Omega_e} \left(1 - \frac{\omega_{pi}^* + \omega_j}{\omega} \right).
\end{aligned} \tag{7.66}
$$

Taking into account equation (7.48) for ϵ_{12} and (7.63)–(7.66), the dispersion relation (1.1) reduces to

$$\bar{\epsilon}_{11}\bar{\epsilon}_{33} - N^2(\bar{\epsilon}_{11} + \bar{\epsilon}_{33} \cos^2 \theta) \left[1 - (\omega - \omega_{pe}^*)(\omega - \omega_{pi}^* - \omega_j)/\omega_w^2 \right] = 0 \tag{7.67}$$

where ω_w is defined by the second equality of (7.57).

We shall analyse (7.67) in section 8.5.

7.8 Starting equations for large-scale perturbations

Whilst discussing the perturbations of a *cold plasma* with a transverse current in section 7.7.3, we mentioned that they should be considered as *electromagnetic perturbations* if $q \gg 1$, where q is defined by (7.61). This inequality means that the perturbations have a transverse wavelength $\lambda_\perp \simeq 1/k_\perp$ larger than c/ω_{pe}. At the same time, in section 7.7.3 we assumed such perturbations to be of small scale compared with the characteristic size of the transverse gradient δ. Together, these two assumptions mean $c/\omega_{pe} < \lambda_\perp < \delta$. Now we shall find equations for perturbations with $\lambda_\perp \gtrsim \delta$, i.e., for *large-scale perturbations*. Obviously, in this case the WKB approximation is invalid. Therefore, our goal is to obtain a differential equation for the perturbation amplitude.

Studying the small-scale perturbations, we assumed above the current velocity to be independent of the transverse coordinate x. Therefore, we can use the reference frame with electrons at rest. Now we allow for a coordinate dependence of the current velocity. This excludes the possibility to choose the mentioned reference frame. Therefore we shall use below a reference frame with the electrons moving and the ions at rest.

Considering a plasma with an *inhomogeneous current–velocity profile*, we have the possibility of studying the instabilities caused by just this inhomogeneity, but not a relative motion of electrons and ions underlying the mechanism of small-scale instabilities. Therefore, below in this section we shall allow for only the perturbed motion of the electron component assuming the ions to be infinitely heavy.

As in obtaining (7.54), we use the *approximation of infinite longitudinal conductivity*. In terms of the total fields $\boldsymbol{E}$ and $\boldsymbol{B}$ this means

$$\boldsymbol{E} \cdot \boldsymbol{B} = 0. \tag{7.68}$$

Linearizing this equation and allowing for an equilibrium electric field $\boldsymbol{E}_0 \parallel \boldsymbol{e}_x$, we obtain

$$\tilde{E}_z = -\tilde{B}_x E_0 / B_0 \tag{7.69}$$

where the tilde means the perturbation. Using (7.69) and the Maxwell equation $\nabla \times \tilde{\boldsymbol{E}} = \mathrm{i}(\omega/c)\tilde{\boldsymbol{B}}$, we find

$$
\begin{aligned}
\tilde{E}_x &= \frac{\omega}{ck_z}\tilde{B}_y + \frac{\mathrm{i}}{ck_z}\frac{\partial}{\partial x}(V_{\mathrm{E}}\tilde{B}_x) \\
\tilde{E}_y &= -\frac{\omega - k_y V_{\mathrm{E}}}{ck_z}\tilde{B}_x
\end{aligned}
\tag{7.70}
$$

where V_{E} is defined by (7.3). In addition, it follows from div $\tilde{\boldsymbol{B}} = 0$ that

$$k_z \tilde{B}_z + k_y \tilde{B}_y = \mathrm{i}\partial \tilde{B}_x / \partial x. \tag{7.71}$$

A relationship between perturbed quantities can be found by using the electron motion equation

$$M_e \mathrm{d}\boldsymbol{V}_e / \mathrm{d}t = e_e \boldsymbol{E} + (e_e/c)\boldsymbol{V}_e \times \boldsymbol{B}. \tag{7.72}$$

Since the ion motion is ignored, we have

$$\boldsymbol{V}_e = \boldsymbol{j}/e_e n_e \equiv c\nabla \times \boldsymbol{B}/4\pi e_e n_e. \tag{7.73}$$

Using (7.73), we express $\boldsymbol{V}_e$ in (7.72) in terms of $\tilde{\boldsymbol{B}}$. Then we linearize the result whilst allowing for the quasi-neutrality condition which means $n_e = 0$ for the infinitely heavy ions. Subsequently, we express all the perturbed quantities in terms of $\tilde{B}_x$ using the equilibrium condition (7.17) which means in the case considered

$$\frac{B_0'}{B_0} = -\frac{\omega_{pe}^2}{c^2}\frac{V_{\mathrm{E}}}{\Omega_e}. \tag{7.74}$$

Here the prime represents the derivative with respect to the x-coordinate.

As a result, we obtain a second order differential equation of the form

$$\tilde{B}_x'' + [-k^2 + U(x,\omega)]\tilde{B}_x = 0 \tag{7.75}$$

where $k^2 = k_y^2 + k_z^2$. *In neglecting the transverse electron inertia*, i.e., the left-hand side of (7.72), the function $U(x,\omega)$ has the form

$$U(x,\omega) = \frac{\omega_{pe}^2}{c^2}\frac{V_E'}{\Omega_e} + \frac{\omega_{pe}^4}{\Omega_e^2 c^4 k_z^2}(\omega - k_y V_E)\left(\omega - \frac{\kappa_n}{\kappa_B}k_y V_E\right). \tag{7.76}$$

Allowing for the transverse electron inertia we have, instead of (7.76) :

$$\begin{aligned}
U(x,\omega) = {}&\frac{\omega_{pe}^2}{c^2}\frac{V_E'}{\Omega_e} + \frac{\omega_{pe}^4(\omega - k_y V_E)}{\Omega_e^2 c^4 k_z^2(1 + V_E'/\Omega_e)} \\
&\times\left[\omega - \frac{k_y V_E}{\kappa_B}\left(\kappa_n + \frac{(V_E'/\Omega_e)'}{(1 + V_E'/\Omega_e)^2}\right)\right].
\end{aligned} \tag{7.77}$$

Equation (7.75) with U of the form of (7.76) has been obtained in [7.14], while that of form (7.77) has been shown in [7.15].

Note that in the WKB approximation equations (7.75) and (7.76) yield the local dispersion relation (7.54) with $c_A^2 \to 0$ when one substitutes $\omega_j \to -k_y V_E$, $\omega - \omega_j \to \omega$ in equation (7.54). The last substitution is related to using the other reference frame.

It follows from the comparison of (7.77) and (7.76) and using equation (7.74) that allowing for the transverse electron inertia is necessary when $\delta \lesssim c/\omega_{pe}$, i.e., in the case of *sufficiently abrupt profiles* of the quasi-stationary wave.

We shall analyse (7.75)–(7.77) in section 8.2.

8 Transverse-current-driven Instabilities

Certain current-driven instabilities can be considered to be electrostatic and can be studied in the approximation of a *homogeneous plasma* and a *homogeneous magnetic field*. A series of such instabilities have been discussed in Chapter 13 of [1]. They are:

(1) *modified two-stream instability*, [1], section 13.1;
(2) *electron-acoustic instability*, [1], section 13.2;
(3) *ion-acoustic instability in the magnetized electron approximation*, [1], section 13.3.

One should augment this list by:

(4) *Buneman's instability*, [1], section 1.6;
(5) *ion-acoustic instability in the unmagnetized electron approximation*, [1], section 3.4.

Both these instabilities are formally independent of the magnetic field.

One more *electrostatic current-driven instability* which can be studied in the approximation of a *homogeneous plasma* and *homogeneous magnetic field* is

(6) *the electron-cyclotron instability*.

We shall discuss it below.

In addition, in the case of a *low-β plasma* and of *not too large a current velocity*, there are electrostatic instabilities in an *inhomogeneous plasma* which can be studied in the approximation of a *homogeneous magnetic field*. Some of them were considered in [2]. They are :

(7) *lower-hybrid-drift instability in a cold plasma*, [2], section 2.5;
(8) *ion-cyclotron instability in a rotating plasma*, [2], section 7.4.

One more instability of this kind is:

(9) *lower-hybrid-drift instability in a plasma with hot ions*.

We shall discuss this below.

A main goal of this chapter is to study *electromagnetic current-driven instabilities* and the effect of a *magnetic-field inhomogeneity* on the *electrostatic current-driven instabilities*. In addition, to complete the picture, we shall also present results about *electrostatic instabilities* revealed in the *homogeneous magnetic field* approximation but which are not considered in [1, 2].

There are *two possible reasons* for small-scale instabilities in a plasma with a transverse current: first, the *usual drift effects* due to density and temperature inhomogeneity, and second, the *relative motion of electrons and ions* associated with the *current*. The drift effects result in instabilities only when the plasma temperature is nonvanishing. Therefore, we can rule out all the instabilities due to the usual drift effects assuming the plasma to be cold. We shall start the analysis of specific types of current-driven instabilities with just this simple situation.

The *cold-plasma approximation* is justified if the plasma pressure is sufficiently small, $\beta \ll 1$, the transverse phase velocity of perturbations is large compared with the ion thermal velocity, $\omega/k_\perp \gg v_{Ti}$, and the electron Larmor radius is small compared with the transverse wavelength, $k_\perp \rho_e \ll 1$. Then, in contrast to the case of unmagnetized plasma, it is not necessary to assume the current velocity denoted by V_j to be larger than the electron thermal velocity v_{Te}. Therefore, we shall assume below $V_j \ll v_{Te}$. At the same time, we *rule out*, from our discussion, *Buneman's instability* (i.e., instability (4), see above) irrespective of the magnetic field effects. Then, our goal in the *cold-plasma approximation* will be to investigate the *role of electromagnetic effects* in the *modified two-stream instability* (instability (1)) and in the *lower-hybrid-drift instability in a cold plasma* (instability (7)), and also to study the possibility of exciting other kinds of *electromagnetic instabilities*. We realize this programme in the *small-scale approximation* in section 8.1. The *large-scale perturbations* are studied in section 8.2.

In section 8.3 we discuss the role of a transverse current in the *lower-hybrid-drift instability in a finite-β plasma*. Recall that, in the absence of a transverse current, this instability was studied in section 6.3. At the same time, the topic of section 8.3 is associated with the above-mentioned instability of type (9).

In section 8.4 we consider the *electron-acoustic oscillation branches* and their *electromagnetic modification*. Then we deal with the *electrostatic current-driven electron-acoustic instability* (instability (2)) and the corresponding type of *electromagnetic current-driven instabilities*.

In section 8.5 perturbations with $\omega \ll k_z v_{Te}$ are examined assuming them to be of long wavelength and of low frequency, with respect to electrons. Section 8.6 is devoted to *perturbations of short wavelength, with respect to electrons*. In particular, in sections 8.5 and 8.6 the *ion-acoustic instabilities* (instabilities (3) and (5)) and the *electron-cyclotron instability* (instability (6)) are discussed. In addition, in section 8.6 we pay attention to some additional varieties of *essentially non-electrostatic current-driven instabilities*. In sections 8.5 and 8.6 we also discuss *gradient instabilities* in the absence of additional current. These concern the topic of Chapter 6, but, as mentioned in the introduction of Chapter 6, we have placed them in the present chapter to avoid duplicating similar calculations.

Finally in section 8.7 we study the *ion-cyclotron perturbations* in a *finite-β rotating plasma*. We then pay attention to an *electromagnetic generalization* of the instability of type (8).

The *electromagnetic transverse-current-driven instabilities* began to be intensively studied in [7.14]. In particular, in [7.14] *large-scale instabilities* excited in the front of a *shock wave* were discussed. In [7.14] it has been shown that a front of width δ smaller than c/ω_{pi} is unstable against *whistler-type perturbations*. Later on in [7.15] this problem has been generalized to the case of narrower current profiles, where $\delta \simeq c/\omega_{pe}$. The results of [7.15] have been corrected in [8.1].

Almost at the same time as [7.14], in [8.2, 7.11] instabilities related to the *small-scale whistler-type perturbations* and to the *fast magnetoacoustic waves* were studied. In [8.2] a strongly collisional plasma (*semiconductor plasma*) and in [7.11] a *collisionless plasma* were considered.

A wide range of problems in the theory of transverse-current-driven instabilities has been analysed in [6.40–6.42, 7.4–7.13, 8.3]. We shall discuss the specific results of these and some other papers below.

8.1   Electromagnetic modified two-stream instability

Small-scale current-driven instabilities in the cold-plasma approximation are described by the *local dispersion relation* (7.60) which in the *electrostatic approximation* reduces to equation (7.62). The latter describes the *electrostatic modified two-stream instability* and the *electrostatic lower-hybrid-drift instability* in a *cold plasma* mentioned in the introduction of this chapter as the instabilities of types (1) and (7) respectively. Specific results following from (7.62) have been given in section 13.1 of [1] and section 2.5 of [2].

According to section 7.7.3, the *electromagnetic effects* are important for $k_\perp \lesssim \omega_{pe}/c$. On the other hand, according to section 13.1

of [1], the characteristic $k_\perp$ for *the electrostatic modified two-stream instability* is of order ω_{LH}/V_j. Therefore, it is clear that the electrostatic approximation is invalid for $V_j \gtrsim c\omega_{LH}/\omega_{pe}$, i.e., for

$$V_j \gtrsim c_A. \tag{8.1}$$

Let us now consider the case where $V_j \gg c_A$ starting with (7.54). We neglect the density gradient, $\kappa_n = 0$, in equation (7.54) and solve this equation assuming the terms with k_z^2 and c_A^2 to be small. As a result, we find

$$\mathrm{Re}\,\omega \approx \omega_j$$
$$\gamma \simeq (c_A/V_j)^{2/3}\omega_j. \tag{8.2}$$

The condition of smallness of the terms with k_z^2 is

$$k_z \lesssim \kappa_B (c_A/V_j)^{1/3}. \tag{8.3}$$

The frequency and the growth rate (8.2) increase with increasing k_y. However, k_y is limited by the condition $k_y < \omega_{pe}/c$. Therefore, in the validity limits we find the estimate for the maximum growth rate

$$\gamma_{\max} \simeq (V_j/c_A)^{1/3}\omega_{LH}. \tag{8.4}$$

It can be seen that the growth rate is larger for $V_j > c_A$ than for $V_j < c_A$. However, one should allow for the assumptions $V_j < v_{Te}$ and $c_A > (M_e/M_i)^{1/2}v_{Te}$. Therefore,

$$\gamma_{\max} < (M_i/M_e)^{1/6}\omega_{LH}. \tag{8.5}$$

In allowing for the density gradient, $\kappa_n \neq 0$, in (7.54) it proves that an *additional destabilization* is associated with this gradient if

$$\partial \ln n_0 / \partial \ln B_0 > 0 \tag{8.6}$$

and some stabilization takes place in the opposite inequality. Condition (8.6) can also be presented in the form

$$\kappa_n V_j / \Omega_e < 0 \tag{8.7}$$

which coincides with equation (2.65) of [2].

Note that inequality (8.7) is a necessary condition for an electrostatic instability of perturbations with $k_z = 0$ (see section 2.5 of [2]). However, it is clear from the above that, due to *electromagnetic effects*, perturbations with $k_z = 0$ can be unstable even under the condition opposite to (8.7).

The results presented here have been obtained mainly in [7.4], see also [7.11, 7.12].

The instability of a plasma with $V_j > c_A$ discussed here can be called the *electromagnetic modified two-stream instability* (see, e.g., [8.4]).

The authors of [7.13] have arrived at an erroneous conclusion which states that the modified two-stream instability is impossible when $V_j > c_A$. This erroneous conclusion has been reproduced in [8.5-8.8] and some other papers. In section 7.7.3 we have elucidated the reason for this error: it is related to the fact that the authors of [7.13] have used the approximation of a homogeneous plasma and homogeneous magnetic field, even though one should take into account equation (7.22).

A possibility for the instability for $V_j > c_A$ has been pointed out in [7.4, 7.12, 8.9, 8.10].

Certain applied topics, where the notions of the modified two-stream instability are used, were mentioned in the introductions of Chapters 7 and 8. In addition, we note the problem of interaction between a *penetrating neutral gas and a magnetized plasma* [8.11–8.15].

8.2   Electromagnetic current-profile instability

Consider some consequences of (7.72)–(7.77). We assume for simplicity that $V_E \neq 0$ only in a certain region with the characteristic size δ smaller than the wavelength of perturbations considered, $k\delta \ll 1$. Then, using the *surface-wave method* (see section 1.7 of [1]), we find the following instability condition

$$2k < \int U(x,0)\mathrm{d}x \qquad (8.8)$$

where $U(x,0)$ is defined by (7.76) or (7.77) for $\omega = 0$. In particular, in this case for (7.76)

$$U(x,0) = \frac{\omega_{pe}^2}{c^2}\frac{V_E'}{\Omega_e} + \frac{\omega_{pe}^4 k_y^2 V_E^2}{\Omega_e^2 c^4 k_z^2}\frac{\kappa_n}{\kappa_B}. \qquad (8.9)$$

For sufficiently small k_z, a simpler instability condition follows from (7.74), (8.8) and (8.9) (cf. (8.6))

$$\frac{|k_y|}{2k_z^2}\int \mathrm{d}x\kappa_B\kappa_n > 1. \qquad (8.10)$$

For a significant variation of the density and the magnetic field, i.e., for $\kappa_n \simeq \kappa_B \simeq 1/\delta$, the growth rate is of order

$$\gamma \simeq (c/\omega_{pe}\delta)^2 \Omega_e. \tag{8.11}$$

This growth rate is attained at the limit of validity of the surface-wave approximation, i.e., for $k_y \simeq 1/\delta$. When $c/\omega_{pe} < \delta < c/\omega_{pi}$, the growth rate is within the limits of $\Omega_i < \gamma < \Omega_e$.

Note that the instability considered is purely electronic. It is essentially non-local, and in this meaning it is *similar to the large-scale Kelvin–Helmholtz instability* discussed in section 13.1. It can be called the *electromagnetic current-profile instability*.

Instability condition (8.10) has been obtained in [7.14]. The author of [7.15] has considered a *symmetric soliton* and has found that the instability condition following in this case from (7.75), (7.77) and (8.8) is expressed in terms of the *Mach number* and the ratio k_z/k_y. In calculating the growth rate he has not taken into account the antisymmetric part of the potential U which is proportional to ω. This error has been corrected in [8.1]. According to [8.1], the *dispersion relation for the electromagnetic current-profile instability in a soliton* has the form

$$[(k\delta_0)^2 + (\gamma/k_z\delta_0)^2]^{1/2} = I_\alpha(M) + G_\alpha(M)(\gamma/k_z\delta_0)^2. \tag{8.12}$$

Here

$$\begin{aligned} I_\alpha(M) &= I(M) + \alpha^2 R(M) \\ G_\alpha(M) &= G(M) + \alpha^2 Q(M) \end{aligned} \tag{8.13}$$

where $\alpha = k_y/k_z$, $\delta_0 = c/\omega_{pe}(0)$, $x = 0$ is the coordinate of the soliton centre, and $I(M), R(M), G(M)$ and $Q(M)$ are certain functions of the Mach number M whose explicit form is given in [7.15, 8.1].

Note also that the term with α^2 in expression (8.13) for $G_\alpha(M)$ is a consequence of allowing for the above-mentioned antisymmetric part of the potential $U(x,\omega)$.

8.3 Electromagnetic current-driven lower-hybrid-drift instability

8.3.1 *Oscillation branch in a cold-electron plasma*
Assuming $k_z = 0$, we start with equation (7.49). When $T_e \to 0$, it follows from (7.49) that

$$\omega = \omega_{ni} + \frac{\beta/2}{1 + \beta/2}(\omega_{Ti} + \omega_j). \tag{8.14}$$

It can be seen from a comparison of (8.14) with (6.18) that the *role of a transverse current* in a modification of the oscillation branch considered is similar to the *role* of a *temperature gradient* (cf. also the discussion in section 7.5.2).

8.3.2   Instability in a cold-electron plasma

Retaining the imaginary ion terms (see (7.37)), by means of (7.51) and (7.53), we find $\omega = \mathrm{Re}\,\omega + i\gamma$ instead of (8.14), where $\mathrm{Re}\,\omega$ is given by the right-hand side of (8.14), while

$$
\gamma = -\frac{\pi^{1/2}}{kv_{Ti}(1+\beta/2)^3}\left[\omega_{ni}\left(1+\frac{\beta}{2}\right)+\frac{\beta}{2}(\omega_{Ti}+\omega_j)\right]
$$
$$
\times\left[\frac{\omega_{Ti}}{2}\left(1+\frac{3}{2}\beta\right)-\omega_j\right].
\tag{8.15}
$$

It can be seen that an *instability is possible even when* $\nabla n_0 = \nabla T = 0$. In this case

$$
\mathrm{Re}\,\omega = \frac{\beta/2}{1+\beta/2}\omega_j
$$
$$
\gamma = \frac{\pi^{1/2}\beta\omega_j^2}{2kv_{Ti}(1+\beta/2)^3}.
\tag{8.16}
$$

It should also be noted that $\gamma = 0$ when $\omega_j = 0$ and $\nabla T = 0$. In this case, to calculate the growth rate, it is necessary to take into account the effects of order $k^2\rho_{ei}^2$ (cf. section 6.3.1).

The results of (8.16) are only valid if $k\rho_{ei} \lesssim 1$. Then, assuming $\gamma \gtrsim \Omega_i$, we find the lower estimate for the current velocity for $\beta \simeq 1$ (cf. (2.16) of [2]):

$$
V_j/v_{Ti} \gtrsim (M_e/M_i)^{1/4}.
\tag{8.17}
$$

For smaller values of V_j the instability should be considered as an *ion-cyclotron one*; this is meaningful in the case of a *rotating plasma*. Such a situation will be presented in section 8.7.

8.3.3   Stabilization in a hot-electron plasma with $\beta \gg 1$

Assume $\beta \gg 1, T_e \gg T_i$, and $\omega_d \gg (\omega_n, \omega_T, \omega_j)$. Then, neglecting resonant electrons, by means of (7.49) we find (cf. (6.59))

$$
\omega = \omega_{pi}^* + \omega_j.
\tag{8.18}
$$

In allowing for *resonant electrons*, we assume $\nabla T = 0$. Then we obtain for $\omega_j \ll \omega_{pe}^*$

$$
\gamma = -2\pi\omega_j^2/\beta\,|\,\omega_{De}\,|.
\tag{8.19}
$$

It can be seen that, as in the case of the *drift loss-cone oscillation branch* (see section 6.4.2), the perturbations considered are *damped* due to electrons. The ion resonance is unimportant since its contribution in (8.19) is small, of order $\rho_i\kappa_B$ (cf. section 6.4.2).

The authors of [8.16] have studied the effect of the magnetic field gradient on the lower-hybrid-drift instability in a plasma with a transverse current using the electrostatic approximation. However, such an approximation is invalid in this problem, so that the results of [8.16] appear doubtful.

8.4 Electrostatic and electromagnetic electron-acoustic instabilities

Consider the perturbations described by equation (7.58) for $k_z \neq 0$. When $\epsilon_{00}^{(i)}$ is of the form of (7.37), this equation reduces to

$$
1 - \frac{\omega_{ni}}{\omega} + \frac{\beta}{2}\left(1 - \frac{\omega_{pi}^* + \omega_j}{\omega}\right) + k_\perp^2\rho_{ei}^2 - \frac{k_z^2 T_i}{M_i\omega^2(1+q)}
$$
$$
+ \frac{i\pi^{1/2}}{k_\perp v_{Ti}}\left[\omega - \omega_j - \omega_{ni}\left(1 - \frac{\eta}{2}\right)\right] = 0. \tag{8.20}
$$

For $\nabla T = 0$ this result was originally obtained in [7.5].

When $\beta \ll 1$ and $q \ll 1$, we find from (8.20) the dispersion relation of the *electrostatic approximation*

$$
1 - \frac{\omega_{ni}}{\omega} + k_\perp^2\rho_{ei}^2 - \frac{k_z^2 T_i}{M_i\omega^2} + \frac{i\pi^{1/2}}{k_\perp v_{Ti}}\left[\omega - \omega_j - \omega_{ni}\left(1 - \frac{\eta}{2}\right)\right] = 0. \tag{8.21}
$$

In the approximation of a homogeneous plasma, equation (8.21) describes the current excitation of the *electron-acoustic oscillation branches*. Such an excitation has been considered in section 13.2 of [1]. It was originally studied in [8.17].

The *gradient terms* in (8.21) modify the electron-acoustic oscillation branches. When in (8.21) one takes into account also the effects of *finite electron temperature*, the modified electron-acoustic branches become *hydrodynamically unstable* (see section 3.2 of [2]). According to (8.21), when $\omega_j = 0$, the *kinetic gradient-driven electron-acoustic instability* is also possible, which is an extension of the kinetic instability considered in section 2.1.3 of [2] into the region $k_z \neq 0$.

Neglecting the usual gradient terms in (8.21), i.e., assuming $\omega_{ni} \to 0$, we have

$$
1 + k_\perp^2\rho_{ei}^2 - \frac{k_z^2 T_i}{M_i\omega^2} + \frac{i\pi^{1/2}}{k_\perp v_{Ti}}(\omega - \omega_j) = 0. \tag{8.22}
$$

This equation describes the *electrostatic current-driven electron-acoustic instability*. Using (8.22), one can find that condition (8.17) is also necessary for this instability. Then $k_\perp \rho_{ei} \simeq 1$.

In the *essentially electromagnetic limit*, $q \gg 1$, equation (8.20) reduces to

$$1 - \frac{\omega_{ni}}{\omega} + \frac{\beta}{2}\left(1 - \frac{\omega_{pi}^* + \omega_j}{\omega}\right) - \frac{k_z^2 c_A^2 z_i}{\omega^2} + \frac{i\pi^{1/2}}{k_\perp v_{Ti}}\left[\omega - \omega_j - \omega_{ni}\left(1 - \frac{\eta}{2}\right)\right] = 0.$$
$$(8.23)$$

Qualitatively, it hence follows that the results of the electrostatic approximation hold up to $\beta \simeq 1$. A quantitative analysis of equation (8.23) as well as of the more general dispersion relation (8.20) has been performed in [7.5].

8.5   Oblique current-driven instabilities with $k_\perp \rho_e < 1$ and $\omega < \Omega_e$

We now proceed with the analysis of equation (7.67) and of some generalizations of this equation.

8.5.1   *Electrostatic perturbations*
It follows from (7.67) in the *electrostatic approximation* that

$$1 + (k\lambda_{De})^2 \epsilon_{00}^{(i)} = 0. \tag{8.24}$$

This equation has solutions only for $|\omega - \omega_j| > k v_{Ti}$. In this case $\epsilon_{00}^{(i)}$ is of the form of (7.36). Then equation (8.24) describes the *shifted high-frequency ion-acoustic oscillations* with the dispersion law

$$(\omega - \omega_j)^2 = k^2 T_e / M_i \equiv k^2 c_s^2 \tag{8.25}$$

where c_s is the *ion sound velocity*. As is well-known, this result is true only if $T_e \gg T_i$.

In order to consider the *instabilities associated with oscillation branches* (8.25), the *electron Landau damping* must be taken into account, i.e., terms of order $\omega / k_z v_{Te}$. Then we arrive at the dispersion relation

$$1 - \frac{k^2 T_e}{M_i(\omega - \omega_j)^2} + i\frac{\pi^{1/2}}{|k_z| v_{Te}}\left[\omega - \omega_{ne}\left(1 - \frac{\eta}{2}\right)\right] = 0. \tag{8.26}$$

For $V_j = 0$ this equation describes the *gradient-driven high-frequency ion-acoustic instability* discussed in section 2.2 of [2]. For instability it is necessary that

$$\rho_e / a > (M_e / M_i)^{1/2}. \tag{8.27}$$

In neglecting the gradient terms, equation (8.26) describes the *current-driven high-frequency ion-acoustic instability* if

$$V_j \gtrsim c_s. \tag{8.28}$$

Such a type of instability, with the exception for $z_e \gtrsim 1$, was discussed in section 13.3 of [1]. The approximation $z_e \ll 1$ taken here is satisfied when a plasma is not too dense

$$\omega_{pe}/\Omega_e \lesssim V_i/c_s. \tag{8.29}$$

We shall return to the case where $z_e \gtrsim 1$ in section 8.6.

8.5.2   *Electromagnetic perturbations with $\omega > k_\perp v_{Ti}$*
For simplicity we assume $\kappa_n = \kappa_T = 0$. Then it follows from (7.67) that

$$\left((\omega - \omega_j)^2 - \frac{k^2 T_e}{M_i} \right) \left(\omega(\omega - \omega_j) - \frac{k^2 k_z^2 c^4 \Omega_e^2}{\omega_{pe}^4} \right) - \omega^2 c_A^2 k^2 = 0. \tag{8.30}$$

Comparing (8.30) with (8.25), it can be seen that, when *electromagnetic effects* are taken into account, we have, in addition to the high-frequency ion-acoustic oscillation branches, still *two oscillation branches* which, in connection with (7.57), can be called the *modified whistlers*. Note also that, when $(\omega, \omega_j) \gg k_\perp c_s$, equation (8.30) reduces to (7.54) with $\kappa_n = 0$. In addition, it should be noted that, in the limit of small k_z and $V_j = 0$, equation (8.30) yields

$$\omega^2 = (c_A^2 + c_s^2)k^2. \tag{8.31}$$

This corresponds to the *fast magnetoacoustic waves* in a *plasma with $\beta_e \simeq 1$*.

It is significant that, in contrast to (8.24), equation (8.30) has solutions with $\omega \ll k_\perp c_s$ (it is assumed that $T_e \gg T_i$, so that the starting assumption $\omega \gg k_\perp v_{Ti}$ is valid). For such solutions, equation (8.30) reduces to (cf. (6.78))

$$\omega^2 + \frac{\beta}{2}\omega(\omega - \omega_j) - k_z^2 \frac{T_e}{M_i} \frac{c^2 k_\perp^2}{\omega_{pe}^2} = 0. \tag{8.32}$$

The oscillation branches with $\omega \ll k_\perp c_s$ are of interest due to the fact that, in contrast to the high-frequency ion-acoustic oscillation branches, to achieve excitation condition (8.28) is not required. In order to study the possible instabilities related to the oscillation branches (8.32) it is necessary to allow for *resonant terms* due to *electrons* (of order $\omega/ \mid k_z \mid v_{Te}$) and due to *ions* (formally small as $\exp[-(\omega - \omega_j)^2/k_\perp^2 v_{Ti}^2])$. One can show that the *resonant interaction* of waves with electrons results in *damping*, while that with ions results in *excitation*.

In the adopted approximation $\mid \omega - \omega_j \mid \gg k_\perp v_{Ti}$, the ion growth rate is exponentially small. Therefore, the analysis of the case $\mid \omega - \omega_j \mid \ll k_\perp v_{Ti}$ is of interest. This is the goal of section 8.5.3.

8.5.3   *Electromagnetic perturbations with* $\omega \ll k_\perp v_{Ti}$
For $\mid \omega - \omega_j \mid \ll k_\perp v_{Ti}$ it follows from (7.67) that (cf. (6.78))

$$\omega^2 \left[1 + \frac{\beta}{2} \left(1 - \frac{\omega_j}{\omega} \right) + \frac{i\pi^{1/2}(\omega - \omega_j)}{k_\perp v_{Ti}} \right] - k_z^2 \frac{T_e + T_i}{M_e} \frac{c^2 k_\perp^2}{\omega_{pe}^2} = 0. \quad (8.33)$$

Here, as in section 8.5.2, we neglect gradient effects.

By means of (8.33) we make sure that the *resonant interaction* of the ions with the waves leads to an *instability* with $\gamma/\mathrm{Re}\,\omega \simeq V_j/v_{Ti}$. This ratio has to be large compared with $\omega/k_z v_{Te}$ since we have neglected the imaginary terms of the mentioned order. As a result, we find, instead of (8.27), a necessary instability condition

$$V_j/v_{Ti} \gtrsim (M_e/M_i)^{1/2}. \quad (8.34)$$

In obtaining this estimate, we assumed $\beta \simeq 1$ and $T_e \simeq T_i$.
Note that (8.34) is in accordance with [6.40].

8.5.4   *Allowance for the gradient effects*
When $V^* > V_j$, *gradient terms* must be added into (8.33). This can be done using the equations of section 7.7.4. Then, in particular, for $V_j = 0$ we obtain a dispersion relation of the form of (6.78) but with the addition of an imaginary contribution due to the ion resonance.

As a result, by analogy with (8.34), we find a condition for a *gradient-driven instability*

$$\rho_i/a \gtrsim (M_e/M_i)^{1/6}. \quad (8.35)$$

As in (8.34), this estimate concerns the case where $\beta \simeq 1$ and $T_e \simeq T_i$.

8.6   Instabilities with $k_\perp \rho_e > 1$

We now proceed to study instabilities with $k_\perp \rho_e > 1$. As found in section 7.6, they can be considered to be *electrostatic* and can be described by dispersion relation (7.39).

In neglecting the transverse current and equilibrium gradients, the electrostatic perturbations with $k_\perp \rho_e > 1$ can be related to the following oscillating branches: *electron-Langmuir waves, electron-cyclotron harmonics (electron Bernstein modes)*, and *high-frequency ion-acoustic waves*. For excitation of Langmuir waves, it is necessary that $V_j > v_{Te}$. This corresponds to Buneman's instability. As mentioned in the introduction to this chapter, we assume $V_j < v_{Te}$ and thereby *exclude* Buneman's instability. Therefore, our problem reduces to the study of the excitation of *electron-cyclotron harmonics* and *ion-acoustic waves*. Then, according to (7.38)–(7.40) (see also (7.44)), our starting dispersion relation is of the form of

$$1 + k^2 \lambda_{De}^2 + \frac{T_i}{T_e}(1 + Z(x)) - \sum_n \left\langle \hat{\imath}_e \frac{\omega J_n^2(\xi_e)}{\omega - n\Omega_e - k_z v_z - \omega_{de}} \right\rangle = 0$$

(8.36)

where $x = (\omega - \omega_j)/k v_{Ti}$. We take $V_j > v_{Ti}$ and, consequently, *neglect* the *gradient effects* related to the ions.

8.6.1 Plasma with $\beta \gtrsim 1$

When $\beta \gtrsim 1$, according to (7.23) we have $\omega_{de} \gtrsim \omega_j$. This excludes excitation of the electron-cyclotron harmonics. In this case equation (8.36) can be replaced by a simpler one

$$1 + k^2 \lambda_{De}^2 - \frac{k^2 T_e}{M_i(\omega - \omega_j)^2} + \frac{i\pi^{1/2}}{k v_{Te}}\left[\omega - \omega_{ne}\left(1 - \frac{\eta}{2}\right)\right] = 0. \quad (8.37)$$

Here we have taken $T_e \gg T_i$ since in the opposite case the dispersion relation has no suitable solutions. The transition from (8.36) to (8.37) is performed by a *summation over the harmonics* (cf. (7.45)).

When $\omega_j = 0$, i.e., in the absence of an additional transverse current, equation (8.37) describes a *gradient-driven instability*. Such an instability is possible only if

$$\rho_e/a > (M_e/M_i)^{1/2} \quad (8.38)$$

which is the same as (8.27). The estimate for the growth rate is

$$\gamma \simeq \omega_{pi}\rho_e/a. \quad (8.39)$$

This instability has been studied in detail in [8.18]. The role of the gradient effects in the perturbations with $k_\perp \rho_e > 1$ was also considered in [8.19, 8.20].

By allowing for V_j and in neglecting gradient effects, equation (8.37) reduces to a dispersion relation considered in section 3.4 of [1].

8.6.2   *Oblique waves in a plasma with $\beta \to 0$*

When $k_z v_{Te} \gtrsim \Omega_e$, one can sum in (8.36) over the harmonics for arbitrary β, including $\beta \to 0$. Then we again arrive at (8.37) and the remaining equations of section 8.6.1. Consequently, as in the case of section 8.6.1, the magnetization of the electrons in perturbations with the mentioned values of k_z does not play a role (cf. the discussion in section 7.6.3). This fact has been noted in [8.21].

8.6.3   *Electron-cyclotron harmonics*

As is clear from sections 8.6.1 and 8.6.2, in studying the instabilities of *electron-cyclotron harmonics*, we should take β and k_z to be sufficiently small. For simplicity we shall consider $k_z = 0$. We shall initially discuss the case where $\beta = 0$ and then that of finite β. Then our problem becomes similar to the problem of the *drift-cyclotron instability* discussed in section 6.2 (see also section 2.1 of [2]). For simplicity we consider the *ions* to be *cold*. The gradient terms are neglected.

Thus, by analogy with section 6.2, we start with the case where $\beta = 0$. It then follows from (8.36) that

$$1 + k^2 \lambda_{De}^2 - \frac{k^2 T_e}{M_i(\omega - \omega_j)^2} - \frac{(2\pi z_e)^{-1/2}\omega}{\omega - n\Omega_e} = 0. \qquad (8.40)$$

By analogy with section 2.1 of [2], we conclude that the maximum growth rate for a fixed n, as a function of the wave number k, is attained for the intersection of the *Doppler-shifted ion-acoustic branch* with the *branch of the cyclotron harmonics*, i.e., for

$$\omega \approx \omega_j - \omega_I \approx n\Omega_e \qquad (8.41)$$

where ω_I is the frequency of the ion-acoustic waves defined by

$$\omega_I^2 = k^2 T_e / M_i (1 + k^2 \lambda_{De}^2). \qquad (8.42)$$

Then

$$\gamma \approx \left(\frac{M_e}{8\pi M_i}\right)^{1/4} \frac{n^{1/2} \, |\, \Omega_e \,|}{(1 + k^2 \lambda_{De}^2)^{3/4}}. \qquad (8.43)$$

This result has been obtained in [8.3, 8.22, 8.23], see also [8.24–8.27]. In the papers mentioned one can also find some additional details following from (8.40), as well as from more complicated dispersion relations allowing for a *finite ion temperature*.

In accordance with section 6.2, the result of (8.43) should be modified for $\omega_{de} \gtrsim \gamma$, i.e. for (cf. (6.29))

$$\beta \gtrsim \left(\frac{M_e}{M_i}\right)^{1/4} \frac{1}{n^{1/2}(1 + k^2 \lambda_{De}^2)^{3/4}}. \qquad (8.44)$$

According to (8.36), in this case equation (8.40) is replaced by the following (see also (7.44))

$$1 + k^2\lambda_{De}^2 - \frac{k^2 T_e}{M_i(\omega - \omega_j)^2} - \frac{\omega}{\pi}\left\langle\frac{1}{\xi_e(\omega - n\Omega_e - \omega_{de})}\right\rangle = 0. \quad (8.45)$$

Then, similarly to (6.34), we find that the growth rate under condition (8.44) is of order

$$\gamma \simeq \frac{|\Omega_e|}{1 + k^2\lambda_{De}^2}\left(\frac{nM_e}{M_i\beta}\right)^{1/3}. \quad (8.46)$$

This result has been obtained in [8.3].

Note that, when both sides of inequality (8.44) are of the same order, equation (8.46) means qualitatively the same as (8.43) (cf. the analogous situation in section 6.2).

8.7 Instabilities in a rotating plasma

When $\gamma > \Omega_i$ or $\omega_{di} > \Omega_i$, the theory of instabilities in a *rotating plasma* reduces to that given above. In studying instabilities with $(\gamma, \omega_{di}) < \Omega_i$ one should modify the expression for $\epsilon_{00}^{(i)}$. Then we find that $\epsilon_{00}^{(i)}$ has a form similar to (7.40) but with the substitution $\omega \to \omega - \omega_j$, i.e.,

$$\epsilon_{00}^{(i)} = \frac{1}{k^2\lambda_{Di}^2}\left(1 - \sum_n\left\langle J_n^2(\xi_i)\frac{\omega - \omega_j - \omega_{ni}\{1 + \eta[(M_iv_\perp^2/2T_i) - 1]\}}{\omega - \omega_j - n\Omega_i - \omega_{di}}\right\rangle\right). \quad (8.47)$$

Here we have omitted the term with $\hat{l} - 1$ which is unimportant for *ion-cyclotron instabilities*.

Substituting (8.47) into respective dispersion relations, one can consider a great number of varieties of the ion-cyclotron instabilities. However, we shall dwell only upon one of these, namely, upon the *cyclotron modification* of relations (8.16) when condition (8.17) is violated. In this case the starting dispersion relation has the form

$$1 + k^2\rho_{ei}^2 + \frac{\beta}{2}\left(1 - \frac{\omega_j}{\omega}\right) - \frac{1}{\pi}\left\langle\frac{\omega - \omega_j}{\xi_i(\omega - \omega_j - n\Omega_i - \omega_{di})}\right\rangle = 0. \quad (8.48)$$

Here we have neglected the density and temperature gradients and have assumed the electrons to be cold. In contrast to section 8.3.2

but by analogy with section 6.3.1, we have allowed for the terms of order $k^2 \rho_{ei}^2$.

Note that equation (8.48) is similar to (8.45). In addition, there is an analogy of (8.48) with (6.28). Therefore, we shall give results following from (8.48) without additional calculations.

It is necessary for instability that

$$V_j / v_{Ti} \gtrsim (M_e / M_i)^{1/2}. \tag{8.49}$$

The real part of the oscillation frequency is approximately given by the first expression of (8.16). The term ω_{di} in (8.48) can be neglected when

$$\beta < (M_e / M_i)^{1/2}. \tag{8.50}$$

In this case

$$\gamma \simeq \beta^{1/2} (M_e / M_i)^{1/4} \Omega_i. \tag{8.51}$$

When a condition opposite to (8.50) takes place, i.e., when

$$\beta > (M_e / M_i)^{1/2} \tag{8.52}$$

equation (8.51) is replaced by

$$\gamma \simeq (\beta M_e / M_i)^{1/3} \Omega_i. \tag{8.53}$$

This growth rate is smaller than (6.34) which is explained by the smallness of the frequency of (8.16).

9 Hydrodynamic Description of Collisional Finite-β Plasma Perturbations

Instabilities in an inhomogeneous finite-β plasma can be studied on the basis of two approaches: (i) kinetic, and (ii) hydrodynamic. The kinetic approach has been given in Chapter 1. Now we consider the *hydrodynamic approach*.

As is well-known, in the hydrodynamic approach, the kinetic equation for each particle species reduces to an *equation set for some first moments* of the velocity distribution function of these particles. Such a procedure is usually due to two reasons: first, to allow for *collisions between particles*, and, second, to obtain a *clearer* picture of plasma processes. These reasons stimulated a great number of approaches to the simplification of the kinetic equation. This resulted in an equation set of *two-fluid magnetohydrodynamics* containing density, velocity, pressure and some other moments of the distribution function. The authors of [9.1] have analysed a suitable sub-set for the problem of instabilities in an inhomogeneous finite-β plasma, and have improved some of the most familiar hydrodynamic approaches allowing for the terms essential for this problem (see in detail below). The same viewpoint forms the basis of the present chapter.

As will be seen below, for a complete description of the instabilities in a *collisional finite-β plasma* it is necessary to deal with a rather large number of moments of the distribution function. Therefore, in the general statement of the problem, one of the main advantages of the hydrodynamic description, i.e., clarity, is in fact lost. In order to keep some element of simplicity of the hydrodynamics we begin our presentation by obtaining the simplest set of hydrodynamic equations suitable for studying the perturbations with $k_z = 0$ by neglecting dissipative, inertial and magneto-viscous effects, section

9.1. Such equations are found by using the *drift kinetic equation*. They are the basis of the so-called *drift hydrodynamics* of a finite-β plasma.

In contrast to section 9.1, the goal of sections 9.2 and 9.3 is to find the starting equations for the *complete description of the instabilities in a finite-β plasma*. Note that the problem of choosing transport equations adequately describing these instabilities is far from trivial. This problem has been proceeded by a simpler problem of describing the instabilities of a collisional plasma with zero-β which has been considered in [9.2]. Before paper [9.2] it was generally accepted that, in order to describe completely the instabilities of a *collisional plasma*, one could use the well-known *Braginskii equations* [9.3, 9.4] or the similar Kaufman equations [9.5]. The author of [9.2] has shown that, in fact, some problems of the theory of collisional plasma instabilities can be analysed by means of equations of the type of [9.3–9.5]. At the same time, according to [9.2], in some cases the use of those equations leads to incorrect results. This is explained by the fact that the totality of small parameters adopted in deriving equations of type [9.3–9.5] does not correspond to that adopted in the theory of gradient (drift) instabilities. Such a nonconformity has been discussed in Chapter 4 of [2]; therefore, we shall not consider it in detail here. However, we should mention the fact that in the theories of [9.3–9.5] the *transverse velocities* of the plasma components are assumed to be of zeroth order, while the *transverse heat fluxes* are of first order. As for the ordering adopted in the drift-wave theory, the transverse velocities and the transverse heat fluxes are assumed to be of the same order. This fact results, in particular, in difficulties in the use of the expressions for the *viscosity tensor* given in [9.3, 9.4].

In the above-mentioned papers the viscosity tensor is expressed in terms of the velocity derivatives and does not contain the derivatives of the heat flux, which is obviously unacceptable for the drift-wave theory. These difficulties can be overcome if, in deriving the transport equations, the transverse velocities and the transverse heat fluxes are assumed to be of the same order. Such an ordering can be made by two approaches: first, by assuming the both quantities to be of first order, and, second, by considering them to be of zeroth order. The first approach has been initially developed in [9.2], where the transport equations suitable for describing the gradient instabilities in a zero-β plasma in a homogeneous magnetic field have been obtained, and then generalized in [9.6] to the case of *arbitrary β* and *curvilinear magnetic fields*. The second approach goes back to the idea used by Grad [9.7] of a *13-moment set of transport equations* which has been realized for the case of a plasma in a magnetic field in [9.8].

Recall that in *Grad's approach* the zeroth order values are density, temperature and velocity vector (i.e., 5 quantities), as well as the heat flux vector (3 quantities) and the viscosity tensor (5 quantities since the viscosity tensor is symmetric and of zeroth spur). At the same time, it is clear that, generally speaking, there is no reason, to adopt Grad's equations as being *a priori* acceptable for the problem of gradient instabilities, since there is *no guarantee* that in some cases other moments of the distribution function will not be essential. Moreover, Grad's equations fail to reproduce Braginskii's results. Therefore, generally speaking, the approach of [9.2, 9.6] is more consistent than that of [9.7, 9.8] for the drift-wave theory. However, Grad's approach is more obvious since, in this approach, all the moments of the distribution function have a simple physical meaning.

The authors of [9.1, 9.9] have extended Grad's approach by allowing for some additional moments of the distribution function. They have augmented Grad's 13 moments by 10 new moments. The 8 new moments are a tensor of the second rank similar to the viscosity tensor and a vector similar to the heat flux. Allowing for these tensor and vector moments is necessary in order to reproduce Braginskii's results. In addition, according to [9.1, 9.9], one should take into account two scalar moments which are not present in either [9.3–9.5] or [9.7] but arise in the approach of [9.2]. The above-mentioned *multimoment equation set* is given in section 9.2.

When the cyclotron frequency is large compared with the collision frequency, the multimoment equation set of section 9.2 is significantly simplified. Then this set reduces to a form similar to Braginskii's equations but with *modified expressions* for the *viscosity tensor* and the *heat flux*. Such a reduction is presented in section 9.3.

The equations of section 9.3 show that, in a number of problems, besides the above-mentioned *influence of the heat conductivity on viscosity*, such a crossed effect as the *influence of the viscosity on heat conductivity* is also significant (this effect is not described by Braginskii's and Kaufman's equations). In addition, the equations of section 9.3 allow for, more correctly than Braginskii's equations, the so-called *gyro-relaxation effect*, i.e., a collisional thermal redistribution among the different degrees of freedom. This effect was originally considered in [9.10, 9.11] in connection with the problem of plasma heating by an oscillating magnetic field. The gyro-relaxation effect is also important for the problem of collisional finite-β plasma instabilities which will be discussed below. The simplest set of hydrodynamic equations allowing for effects of order ω/ν_i, including the gyro-relaxation effect, is given in section 9.4 (ν_i is the *ion collision frequency*).

In accordance with the above, the equations of section 9.4 have been found by simplifying the general set of transport equations derived by means of the moment method from Boltzmann's kinetic equation. In section 9.5 we show that precisely the same equations follow from the *drift kinetic equation with a collisional term*. This, on the one hand, allows one to treat the corresponding hydrodynamic equations in terms of *"drift"* notions, and, on the other hand, is a guarantee of the correctness of calculations made under a simplification of the general transport equations.

The *simplified hydrodynamic equations* of sections 9.1, 9.4 and 9.5 do not allow for *inertia and magneto-viscosity effects*, nor for the *transverse heat conductivity* and the *transverse viscosity*, and a series of other effects important in some problems on instabilities. These effects are discussed in the remaining sections of this chapter.

In section 9.6 we explain the specificity of allowing for *inertial* and *magneto-viscosity effects*, while in section 9.7 we treat the *transverse heat conductivity* and the *transverse viscosity*. Some other collisional effects will be considered in Chapter 10 when deriving the dispersion relations for separate types of perturbations.

Note also that the hydrodynamic description is sometimes used even when neglecting collisions. The approach of [9.12] called the CGL *approach* is an example of such a description. The CGL equations have been augmented by the magnetic viscosity in [9.13] (see also [9.14]). Such modified CGL equations have been used in [9.13] in the problem of inertial perturbations. However, the procedure of [9.13] is, generally speaking, risky since, in the problem on drift waves in a collisionless finite-β plasma, there is no small parameter, allowing one to reduce the kinetic equation to a moment equation set.

9.1   Drift hydrodynamics of a finite-β plasma

As in the previous chapters, we consider a plasma to be in an equilibrium magnetic field $\boldsymbol{B}_0 \parallel \boldsymbol{z}$. In this section we shall find starting equations for describing perturbations with $k_z = 0$ and $E_z = 0$. In contrast to the previous chapters, we now assume that the collision frequency, ν, of each particle species is large compared with the perturbation frequency, $\nu \gg \omega$.

We start with *Boltzmann's kinetic equation* which, for the above-mentioned assumptions, has the form

$$\frac{\partial f}{\partial t} + \boldsymbol{v}_\perp \cdot \nabla f + \frac{e}{M} \boldsymbol{E} \cdot \frac{\partial f}{\partial \boldsymbol{v}_\perp} = \Omega \frac{\partial f}{\partial \alpha} + C. \qquad (9.1)$$

Here f is the total distribution function, i.e., the sum of the equilibrium and perturbed distribution functions, and C is the *collisional term*. The quantity Ω is the cyclotron frequency defined by the total magnetic field which, in contrast to Chapter 1, is now designated by $\boldsymbol{B}$, so that $\boldsymbol{B} = \boldsymbol{B}_0 + \tilde{\boldsymbol{B}}$, where $\tilde{\boldsymbol{B}}$ is the perturbed magnetic field (cf. sections 6.6.2 and 7.8), and α is the angle in particle transverse velocity space (cf. section 1.2.4).

For a sufficiently *strong magnetic field* when $\Omega \gg \nu$, equation (9.1) can be solved by *expansion in series in the small parameter* $1/\Omega$. A corresponding procedure for the case of a homogeneous magnetic field and electrostatic electric field was presented in Chapter 4 of [2]. Generalizing this procedure for the case involved, we arrive at the following equation

$$\frac{\partial \bar{f}}{\partial t} + \boldsymbol{V}_E \cdot \nabla \bar{f} - \mathcal{E}_\perp \frac{\partial \bar{f}}{\partial \mathcal{E}_\perp} \nabla \cdot \boldsymbol{V}_E + \mathcal{E}_\perp \boldsymbol{e}_z \cdot \nabla \bar{f} \times \nabla \frac{1}{\Omega} = \bar{C}. \qquad (9.2)$$

Here $\bar{f}$ is the part of f independent of the angle α, $\bar{C}$ is the average of C over α, $\boldsymbol{V}_E = c\boldsymbol{E} \times \boldsymbol{B}/B^2$ is the particle drift velocity in the crossed fields $\boldsymbol{E}$ and $\boldsymbol{B}$ (the *crossed-field drift velocity*), and $\mathcal{E}_\perp = v_\perp^2/2$. Equation (9.2) is simply the *drift kinetic equation with the collisional term*.

Assuming the function $\bar{f}$ in the left-hand side of (9.2) to be Maxwellian, we take the moments of (9.2). *Neglecting the dissipative effects* which will be considered below, we find the equations for the total (in the above-mentioned meaning) density n and temperature T of each species:

$$\frac{\partial n}{\partial t} + \boldsymbol{V}_E \cdot \nabla n + n\nabla \cdot \boldsymbol{V}_E + \boldsymbol{e}_z \cdot \nabla \frac{p}{M} \times \nabla \frac{1}{\Omega} = 0 \qquad (9.3)$$

$$\frac{\partial p}{\partial t} + \boldsymbol{V}_E \cdot \nabla p + \gamma_0 p \nabla \cdot \boldsymbol{V}_E + \gamma_0 \boldsymbol{e}_z \cdot \nabla \frac{pT}{M} \times \nabla \frac{1}{\Omega} = 0 \qquad (9.4)$$

where $\gamma_0 = 5/3$ is the adiabatic exponent and $p = nT$.

9.2   Multimoment transport equation set

Equations (9.3) and (9.4) allow one to study only the particular case of perturbations of a *collisional finite-β plasma* neglecting all the dissipative effects. We shall now derive equations suitable for a general analysis of the problem of instabilities in such a plasma.

We start with the fact that sufficiently frequent collisions between particles lead to a velocity distribution close to the Maxwellian case, i.e.,

$$f = f_0(1 + \Phi) \tag{9.5}$$

where

$$f_0 = n(M/2\pi T)^{3/2} \exp[-M(\boldsymbol{v} - \boldsymbol{V})^2/2T] \tag{9.6}$$

as long as the value of Φ is small, $\Phi \ll 1$. The quantities $n, \boldsymbol{V}$ and T are the density, velocity and temperature of the corresponding plasma component, respectively, if Φ satisfies the condition

$$\int (\boldsymbol{v} - \boldsymbol{V})^k f_0 \Phi d\boldsymbol{v} \qquad k = 0, 1, 2. \tag{9.7}$$

As a function of the "random" velocity vector $\boldsymbol{w} \equiv \boldsymbol{v} - \boldsymbol{V}$, Φ can be written as a sum of scalar, vector and tensor parts

$$\Phi = \Phi_0 + w_i \Phi_i + (w_i w_k - \delta_{ik} w^2/3)\Phi_{ik} + \ldots \tag{9.8}$$

The ellipsis in the right-hand side represents tensors of the third and higher ranks which are neglected.

A general expression for the scalar Φ_0 satisfying the first and third conditions of (9.7) can be presented as a series of the *Sonine–Laguerre polynomials* $L_l^{(1/2)}(x)$, where $x \equiv Mw^2/2T$, starting with $l = 2$:

$$\Phi_0 = \sum_{l=2}^{\infty} a^{(l)} L_l^{(1/2)}(x). \tag{9.9}$$

Similarly, the vector Φ_i satisfies condition (9.7) if

$$\Phi_i = \sum_{l=1}^{\infty} a_i^{(l)} L_l^{(3/2)}(x). \tag{9.10}$$

Since we impose no restrictions on the tensor Φ_{ik}, it can be written as a full series

$$\Phi_{ik} = \sum_{l=0}^{\infty} a_{ik}^{(l)} L_l^{(5/2)}(x). \tag{9.11}$$

Series of the same type can be found for the tensors of higher rank omitted on the right-hand side of (9.8).

The coefficients $a^{(l)}, a_i^{(l)}$ and $a_{ik}^{(l)}$ are higher moments of the distribution function. The quantities $a_i^{(1)}$ and $a_{ik}^{(0)}$ have the simplest meaning. They are related to the *heat flux* $\boldsymbol{q}$ and the *viscosity tensor* π_{ik} by

$$a_i^{(1)} = -\frac{2}{5}\frac{M}{pT}q_i \tag{9.12}$$

$$a_{ik}^{(0)} = \frac{M}{2pT}\pi_{ik}. \tag{9.13}$$

This can be proved by means of (9.5), (9.10) and (9.11) and the definitions of $\boldsymbol{q}$ and π_{ik}:

$$q = \frac{M}{2}\int w^3 f \, d\boldsymbol{v} \tag{9.14}$$

$$\pi_{ik} = M \int (w_i w_k - \delta_{ik} w^2/3)f \, d\boldsymbol{v}. \tag{9.15}$$

Allowing for (9.12) and (9.13), we shall also introduce the quantities $\boldsymbol{q}^*$ and π_{ik}^* determined by the relations

$$a_i^{(2)} = -\frac{2}{5}\frac{M}{pT}q_i^* \tag{9.16}$$

$$a_{ik}^{(1)} = \frac{M}{2pT}\pi_{ik}^*. \tag{9.17}$$

Using *Boltzmann's kinetic equation*

$$\frac{\partial f}{\partial t} + \boldsymbol{v} \cdot \nabla f + \frac{e}{M}\left(\boldsymbol{E} + \frac{\boldsymbol{v} \times \boldsymbol{B}}{c}\right) \cdot \frac{\partial f}{\partial \boldsymbol{v}} = C \tag{9.18}$$

a series of equations can be obtained relating the quantities $n, \boldsymbol{V}, T, a_i^{(l)}, a_i^{(l)}, a_{ik}^{(l)}$, etc. This series can be truncated in different ways neglecting quantities with a sufficiently high number of the tensor indices. The different ways of truncating the series of equations lead to a closed transport equation set with different degree of accuracy suitable for describing specific phenomena. The examples of such a transport equation set suitable for studying the *electrostatic instabilities in a low-β plasma* were given in Chapter 4 of [2].

To obtain the correct limiting transition to Braginskii's results, it is necessary to take into account not only the values $\boldsymbol{q}$ and π (i.e., $a_i^{(1)}$ and $a_{ik}^{(0)}$), but also $a_i^{(2)}$ and $a_{ik}^{(1)}$. This means that in expressions (9.10) and (9.11) for Φ_i and Φ_{ik} the first two terms of the series must be considered (*approximation of two polynomials*).

In addition, the contribution of the terms with the scalar Φ_0 (see (9.9)) and the tensor terms of third and higher ranks must be estimated. It appears necessary to take Φ_0 into account. It is also necessary in expression (9.9) to consider two polynomials to describe correctly the quantitative effects related to Φ_0.

As a consequence, our problem consists in the construction of the 23 equations for $n, \boldsymbol{V}, T, a_{ik}^{(0)}, a_{ik}^{(1)}, a_i^{(1)}, a_i^{(2)}, a^{(2)}, a^{(3)}$.

The first five equations of this set, obtained by integrating equation (9.18) with the weights $1, \boldsymbol{v}, (\boldsymbol{v} - \boldsymbol{V})^2$, have the well-known form

$$\partial n / \partial t + \nabla \cdot (n\boldsymbol{V}) = 0 \tag{9.19}$$

$$Mn\frac{\mathrm{d}\boldsymbol{V}}{\mathrm{d}t} = -\nabla p - \nabla \cdot \boldsymbol{\pi} + en(\boldsymbol{E} + \frac{1}{c}\boldsymbol{V} \times \boldsymbol{B}) + \boldsymbol{R} \tag{9.20}$$

$$\frac{3}{2}\frac{\mathrm{d}p}{\mathrm{d}t} + \frac{5}{2}p\nabla \cdot \boldsymbol{V} = -\nabla \cdot \boldsymbol{q} - (\boldsymbol{\pi} \cdot \nabla) \cdot \boldsymbol{V} + Q. \tag{9.21}$$

Here $\mathrm{d}/\mathrm{d}t = \partial/\partial t + \boldsymbol{V} \cdot \nabla$; $\boldsymbol{R}$ is the *momentum transfer* (the *frictional force*)

$$\boldsymbol{R} = M \int (\boldsymbol{v} - \boldsymbol{V})C\mathrm{d}\boldsymbol{v} \tag{9.22}$$

and Q is the *heat transfer*,

$$Q = \frac{M}{2} \int (\boldsymbol{v} - \boldsymbol{V})^2 C\mathrm{d}\boldsymbol{v}. \tag{9.23}$$

In the approximation $\nu_i \gg (\mathrm{d}/\mathrm{d}t, \nabla \cdot \boldsymbol{V})$ the equations for the tensors $\pi_{\lambda\mu}$ and $\pi_{\lambda\mu}^*$ are:

$$\begin{aligned} \nu_i \left(\pi_{\lambda\mu} + \frac{3}{4}\pi_{\lambda\mu}^* \right) - \frac{5}{3} \ll \boldsymbol{\pi} \times \boldsymbol{\Omega}_i \gg_{\lambda\mu} = -\frac{5}{6}p_i W_{\lambda\mu}^{(1)} \\ \nu_i \left(\frac{3}{4}\pi_{\lambda\mu} + \frac{205}{48}\pi_{\lambda\mu}^* \right) - \frac{35}{6} \ll \boldsymbol{\pi} \times \boldsymbol{\Omega}_i \gg_{\lambda\mu} = -\frac{7}{6}p_i W_{\lambda\mu}^{(2)}. \end{aligned} \tag{9.24}$$

Here

$$W_{\lambda\mu}^{(1)} = 2 \ll \nabla \cdot \boldsymbol{V} \gg_{\lambda\mu} + \frac{4}{5p_i} \ll \nabla \cdot \boldsymbol{q} \gg_{\lambda\mu}$$

$$\begin{aligned} W_{\lambda\mu}^{(2)} = &\frac{2}{p_i} \ll \frac{M_i}{T_i}\boldsymbol{F}_i \cdot \boldsymbol{q} - \nabla \cdot \boldsymbol{q} - 2\boldsymbol{q} \cdot \nabla \ln T_i \\ &+ \nabla \cdot \boldsymbol{q}^* + \boldsymbol{q}^* \cdot \nabla \ln T_i \gg_{\lambda\mu} \end{aligned}$$

$$\boldsymbol{F} = e\boldsymbol{E}/M + \boldsymbol{V} \times \boldsymbol{\Omega} - \mathrm{d}\boldsymbol{V}/\mathrm{d}t \tag{9.25}$$

$$\boldsymbol{\Omega} = e\boldsymbol{B}/Mc$$

$$\ll \boldsymbol{\pi} \times \boldsymbol{B} \gg_{\lambda\mu} = \pi_{\sigma\lambda}\epsilon_{\mu\sigma\tau}B_\tau$$

$$\ll \boldsymbol{A} \cdot \boldsymbol{B} \gg_{\lambda\mu} = \frac{1}{2}(A_\lambda B_\mu + A_\mu B_\lambda - \frac{2}{3}\delta_{\lambda\mu}\boldsymbol{A} \cdot \boldsymbol{B})$$

where $\epsilon_{\mu\sigma\tau}$ is the antisymmetric unit tensor of the third rank.

In the same approximation the equations for the vectors q and q^* are

$$q \times \Omega_i - \nu_i \left(\frac{4}{5}q + \frac{3}{5}q^* \right) = \frac{5}{2}\frac{p_i}{M_i}\nabla T_i + \frac{5}{2}\nabla \left(\frac{T_i}{M_i}\chi \right) - \frac{7}{2}\nabla \cdot \left(\frac{T_i}{M_i}\pi^* \right)$$

$$- \frac{7}{2}\frac{\pi}{M_i} \cdot \nabla T_i + \frac{T_i}{M_i}\nabla \cdot \pi - \pi \cdot F_i$$

$$\nu_i \left(\frac{3}{5}q + \frac{9}{4}q^* \right) - \frac{7}{4}q^* \times \Omega_i = 0. \tag{9.26}$$

Here $\chi \equiv a^{(2)}/p_i$. We have already presented the equations for the ion values q and q^*. The equations for the electron values q and q^* have similar form, but the terms of order ν_e/Ω_e in those equations should be omitted.

Finally, we have the equations for χ and $\chi^* \equiv a^{(3)}/p_i$:

$$\chi + \frac{3}{4}\chi^* = -\frac{2}{3\nu_i} \left(\nabla \cdot q - \frac{M_i}{T_i}F_i \cdot q \right)$$

$$\frac{3}{4}\chi + \frac{31}{16}\chi^* = 0. \tag{9.27}$$

Thus we have the set of 23 equations, (9.19)–(9.21), (9.24), (9.26), and (9.27), for the 23 quantities: $n, T, V, q, q^*, \pi_{\lambda\mu}, \pi^*_{\lambda\mu}, \chi$, and χ^*.

9.3   Reducing the multimoment transport equation set by expansion in $1/B$

Let us reduce our equation set assuming the cyclotron frequency of each particle species Ω to be large compared with all the remaining characteristic frequencies including the collision frequency ν. In other words, we consider the value $1/B$ to be a small parameter and expand in a series in this small parameter.

First of all, we find from (9.27)

$$\chi = -\frac{31}{33\nu_i} \left(\nabla \cdot q - \frac{M_i}{T_i}F_i \cdot q \right). \tag{9.28}$$

(It is clear that in the transition from equation (9.27) to (9.28) the smallness of $1/B$ is not required.)

It follows from (9.26) that

$$q = \frac{5}{2}\frac{p_i}{M_i\Omega_i}h \times \nabla T_i - \frac{3.9p_i}{\nu_i}\nabla_{\parallel}T_i - \frac{2p_i\nu_i}{M_i\Omega_i^2}\nabla_{\perp}T_i$$
$$+ \frac{1}{\Omega_i}h \times \left[\frac{5}{2}\nabla\left(\frac{T_i}{M_i}\chi\right) - \frac{7}{2}\nabla\cdot\left(\frac{T_i}{M_i}\pi^*\right)\right.$$
$$\left.+ \frac{7}{2}\frac{\pi}{M_i}\cdot\nabla T_i + \frac{T_i}{M_i}\nabla\cdot\pi - \pi\cdot F_i\right]$$

$$q^* = -\frac{6}{7}\frac{\nu_i}{\Omega_i^2}\frac{p_i}{M_i}\nabla T_i + \frac{25}{24}\frac{p_i}{M_i\nu_i}\nabla_{\parallel}T_i$$

(9.29)

where $h = B/B$, $\nabla_{\parallel} = h(h\cdot\nabla)$, $\nabla_{\perp} = -h\times h\times\nabla$.

Using (9.24) we find that in the reference frame in which two axes are perpendicular and the third one is parallel to B (so that $e_3 = h$) the components of the tensor π are of the form

$$\pi_{33} = -\frac{p_i}{\nu_i}\left(0.96W_{33}^{(1)} - 0.24W_{33}^{(2)}\right)$$

$$\pi_{11} = -\frac{1}{2}\frac{p_i}{\nu_i}\left[0.96\left(W_{11}^{(1)} + W_{22}^{(1)}\right) - 0.24\left(W_{11}^{(2)} + W_{22}^{(2)}\right)\right]$$
$$-\frac{1}{2}\frac{p_i}{\nu_i}W_{12}^{(1)} - \frac{3}{20}\frac{p_i\nu_i}{\Omega_i^2}\left[W_{11}^{(1)} - W_{22}^{(1)} + \frac{3}{10}\left(W_{11}^{(2)} - W_{22}^{(2)}\right)\right]$$

$$\pi_{22} = -\frac{1}{2}\frac{p_i}{\nu_i}\left[0.96\left(W_{11}^{(1)} + W_{22}^{(1)}\right) - 0.24\left(W_{11}^{(2)} + W_{22}^{(2)}\right)\right]$$
$$-\frac{1}{2}\frac{p_i}{\nu_i}W_{12}^{(1)} + \frac{3}{20}\frac{p_i\nu_i}{\Omega_i^2}\left[W_{11}^{(1)} - W_{22}^{(1)} + \frac{3}{10}\left(W_{11}^{(2)} - W_{22}^{(2)}\right)\right]$$

$$\pi_{12} = \pi_{21} = \frac{1}{4}\frac{p_i}{\Omega_i}\left(W_{11}^{(1)} - W_{22}^{(1)}\right) - \frac{3}{10}\frac{p_i\nu_i}{\Omega_i^2}\left(W_{12}^{(1)} + \frac{3}{10}W_{12}^{(2)}\right)$$

$$\pi_{13} = \pi_{31} = -\frac{p_i}{\Omega_i}W_{23}^{(1)} - \frac{6}{5}\frac{p_i\nu_i}{\Omega_i^2}\left(W_{13}^{(1)} - \frac{3}{10}W_{13}^{(2)}\right)$$

$$\pi_{23} = \pi_{32} = \frac{p_i}{\Omega_i}W_{13}^{(1)} - \frac{6}{5}\frac{p_i\nu_i}{\Omega_i^2}\left(W_{23}^{(1)} + \frac{3}{10}W_{23}^{(2)}\right).$$

(9.30)

As for the components of the tensor π^*, only the following ones will be necessary for us:

$$\pi_{33}^* = \frac{p_i}{\nu_i}\left(0.17W_{33}^{(1)} - 0.31W_{33}^{(2)}\right)$$

$$\pi_{11}^* = \pi_{22}^* = -\pi_{33}^*/2.$$

(9.31)

These expressions are required for calculating the heat flux (see (9.29)).

As a result, our *multimoment transport equation set* has been reduced to *Braginskii type equations* but with *different expressions for the viscosity and heat flux*.

9.4 The simplest hydrodynamic equation set allowing for dissipative effects of order ω/ν_i

As in section 9.1, we assume that the magnetic field is directed along the z-axis of the Cartesian coordinate system, $\boldsymbol{B} \parallel \boldsymbol{z}$, the velocities of the plasma components $\boldsymbol{V}$ and the electric field $\boldsymbol{E}$ are perpendicular to the magnetic field, $\boldsymbol{V} \perp \boldsymbol{z}, \boldsymbol{E} \perp \boldsymbol{z}$, and all the equilibrium and perturbed quantities are independent of z, $\partial/\partial z = 0$. We take into account that the values of type $(\rho_i \nabla)^2$, $\nu_i^{-1}\partial/\partial t$, and ν_i/Ω_i are small parameters. Using the equation of motion (9.20) and the expression for the heat flux (9.29), we find in the *zeroth approximation* with respect to the above-mentioned small parameters:

$$\boldsymbol{V}_\perp^{(0)} = \boldsymbol{e}_z \times (\nabla p/n - e\boldsymbol{E})/M\Omega \qquad (9.32)$$

$$\boldsymbol{q}_\perp^{(0)} = \frac{5}{2}\frac{p}{M\Omega}\boldsymbol{e}_z \times \nabla T. \qquad (9.33)$$

The viscosity tensor in this approximation vanishes. Substituting equations (9.32) and (9.33) into (9.19) and (9.21) and neglecting the heat transfer Q, we arrive at (9.3) and (9.4).

Note also that in the assumed approximation the transverse current density, according to (9.32), is of the form

$$\boldsymbol{j}_\perp^{(0)} = \sum en\boldsymbol{V}_\perp^{(0)} = \frac{c}{B}\boldsymbol{e}_z \times \nabla(p_e + p_i). \qquad (9.34)$$

Let us now augment equations (9.3), (9.4), and (9.34) by *dissipative terms* of order $\nu_i^{-1}\partial/\partial t$. For this purpose we represent $\boldsymbol{V}_\perp$ and $\boldsymbol{q}_\perp$ in the form

$$\begin{aligned} \boldsymbol{V}_\perp &= \boldsymbol{V}_\perp^{(0)} + \boldsymbol{V}_\perp^{(1)} \\ \boldsymbol{q}_\perp &= \boldsymbol{q}_\perp^{(0)} + \boldsymbol{q}_\perp^{(1)} \end{aligned} \qquad (9.35)$$

where $\boldsymbol{V}_\perp^{(1)}, \boldsymbol{q}_\perp^{(1)}$ are determined by the derivatives of the viscosity tensor $\boldsymbol{\pi}$ and the related tensor $\boldsymbol{\pi}^*$ as well as by the derivatives of the *scalar moment* χ. In calculating $\boldsymbol{\pi}, \boldsymbol{\pi}^*$ and χ we take $\boldsymbol{V}$ and $\boldsymbol{q}$ in the form of (9.32) and (9.33). By means of (9.28), (9.30) and (9.31) we then find

$$\chi = -2.35 p_i \boldsymbol{b} \cdot \nabla T_i / M_i \nu_i$$

$$\pi_{xx} = \pi_{yy} = -\frac{\pi_{zz}}{2} = -\frac{p_i}{3\nu_i}\left[0.96\left(\nabla \cdot \boldsymbol{V}_E + \frac{\boldsymbol{b} \cdot \nabla p_i}{M_i n_i}\right) + \frac{1.55}{M_i}\boldsymbol{b} \cdot \nabla T_i\right]$$

$$\pi_{\lambda\mu} = 0 \qquad \lambda \neq \mu$$

$$\pi^*_{xx} = \pi^*_{yy} = -\frac{\pi^*_{zz}}{2} = \frac{p_i}{3\nu_i}\left[0.17\left(\nabla \cdot \boldsymbol{V}_E + \frac{\boldsymbol{b} \cdot \nabla p_i}{M_i n_i}\right) + \frac{0.96}{M_i}\boldsymbol{b} \cdot \nabla Ti\right]$$

$$\pi^*_{\lambda\mu} = 0 \qquad \lambda \neq \mu.$$

$$(9.36)$$

Here $\boldsymbol{b} \equiv \nabla\Omega_i^{-1} \times \boldsymbol{e}_z$. In addition, we have used the fact that in the assumed approximation $\boldsymbol{F}_i = \boldsymbol{F}_i^{(0)} = \nabla p_i / M_i n_i$.

It follows from (9.20), (9.29) and (9.36) that

$$\boldsymbol{V}_{\perp i}^{(1)} = -\frac{1}{3nM\Omega}\boldsymbol{e}_z \times \nabla\frac{p}{\nu}\left[0.96\left(\nabla \cdot \boldsymbol{V}_E + \frac{\boldsymbol{b} \cdot \nabla p}{Mn}\right) + 1.55\frac{\boldsymbol{b} \cdot \nabla T}{M}\right]$$

$$\boldsymbol{q}_{\perp i}^{(1)} = -\frac{1}{3M\Omega}\boldsymbol{e}_z \times \left\{\nabla\frac{pT}{\nu}\left[1.55\left(\nabla \cdot \boldsymbol{V}_E + \frac{\boldsymbol{b} \cdot \nabla p}{Mn}\right) + 22.51\frac{\boldsymbol{b} \cdot \nabla T}{M}\right]\right.$$

$$\left. + \frac{p}{\nu}\left[0.96\left(\nabla \cdot \boldsymbol{V}_E + \frac{\boldsymbol{b} \cdot \nabla p}{Mn}\right) + 1.55\frac{\boldsymbol{b} \cdot \nabla T}{M}\right]\left(\frac{5}{2}\nabla T - \frac{\nabla p}{n}\right)\right\}.$$

$$(9.37)$$

For simplicity we have omitted the ion subscripts in the right-hand sides of these equations.

Using (9.37) we find, instead of (9.3) and (9.4), the following equations

$$\partial n/\partial t + \ldots + \Delta_n = 0 \qquad (9.38)$$

$$\partial p_i/\partial t + \ldots + \Delta_p = 0. \qquad (9.39)$$

Here the ellipses represent all the terms in the left-hand sides of equations (9.3) and (9.4), except for $\partial n/\partial t$ and $\partial p_i/\partial t$. The quantities Δ_n and Δ_p are defined by

$$\Delta_n = -\frac{1}{3M}\boldsymbol{b} \cdot \nabla\frac{p}{\nu}\left[0.96\left(\nabla \cdot \boldsymbol{V}_E + \frac{\boldsymbol{b} \cdot \nabla p}{Mn}\right) + 1.55\frac{\boldsymbol{b} \cdot \nabla T}{M}\right]$$

$$\Delta_p = -\frac{7}{9}\boldsymbol{b} \cdot \nabla\frac{pT}{M\nu}\left[1.13\left(\nabla \cdot \boldsymbol{V}_E + \frac{\boldsymbol{b} \cdot \nabla p}{Mn}\right) + 7.54\frac{\boldsymbol{b} \cdot \nabla T}{M}\right]$$

$$ - \frac{2}{9}\frac{p}{\nu}\nabla \cdot \boldsymbol{V}_E\left[0.96\left(\nabla \cdot \boldsymbol{V}_E + \frac{\boldsymbol{b} \cdot \nabla p}{Mn}\right) + 1.55\frac{\boldsymbol{b} \cdot \nabla T}{M}\right].$$

$$(9.40)$$

Similar additions arise in the expression for $j_\perp$, so that (9.34) is now replaced by

$$j_\perp = \frac{c}{B} e_z \times \nabla(p_i + p_e) - \frac{1}{3}\frac{c}{B} e_z$$
$$\times \nabla \frac{p_i}{\nu_i}\left[0.96\left(\nabla \cdot V_E + \frac{b \cdot \nabla p_i}{M_i n}\right) + 1.55\frac{b \cdot \nabla T_i}{M_i}\right].$$
$$(9.41)$$

Physically, the terms with $1/\nu_i$ correspond to the *gyro-relaxation effect* mentioned in the introduction to this chapter.

9.5   Agreement between the moment method and the drift kinetic equation method

The equation set (9.38)–(9.41) can also be derived by means of the *drift kinetic equation* (9.2). According to [9.9], this derivation is performed in the following manner.

We assume $\bar f = F + f_1$, where $F \sim \exp(-Mv^2/2T)$ is the Maxwellian distribution function satisfying the condition $C(F) = 0$ when the electrons and the ions have the same temperatures. Substituting this value of $\bar f$ into (9.2), we arrive at the following equation for f_1

$$\left[-\frac{1}{2}A_0\left(x_\parallel - \frac{x}{3}\right) + \frac{1}{2}A_1\left(x_\parallel - \frac{x}{3}\right)L_1^{(5/2)}(x)\right.$$
$$\left. + \frac{4}{3}A_1 L_2^{(1/2)}(x)\right]F = \bar C(f_1).$$
$$(9.42)$$

Here $x = Mv^2/2T, x_\parallel = Mv_z^2/2T$, and

$$A_0 = \nabla \cdot V_E + \frac{b}{M}\cdot\left(\frac{\nabla p}{n} + \nabla T\right)$$
$$A_1 = b \cdot \nabla T/M.$$
$$(9.43)$$

The solution of (9.42) can be represented in the form $f_1 = \Phi F$, where

$$\Phi = \frac{1}{p}\left[\sum_{m=2}^{\infty} a_m^{(1)} L_m^{(1/2)}(x) + \frac{1}{2}\left(x_\parallel - \frac{x_\perp}{2}\right)\sum_{m=0}^{\infty} a_m^{(2)} L_m^{(5/2)}(x)\right] \quad (9.44)$$

where $x_\perp \equiv Mv_\perp^2/2T$. We restrict ourselves to only the first two terms of the series in each of the sums of (9.44). Multiplying (9.42) successively by $L_2^{(1/2)}$, $L_3^{(1/2)}$, $x_\parallel - x/3$, and $(x_\parallel - x/3)L_1^{(5/2)}$, and integrating over velocities, we find the following equation set for the coefficients $a_m^{(l)}$

$$a_2^{(1)} + \frac{31}{12}a_3^{(1)} = 0$$

$$a_2^{(1)} + \frac{3}{4}a_3^{(1)} = -\frac{5}{3}\frac{p_i}{\nu_i}A_1$$

$$a_0^{(2)} + \frac{3}{4}a_1^{(2)} = \frac{5}{9}\frac{p_i}{\nu_i}A_0$$

$$a_0^{(2)} + \frac{205}{36}a_1^{(2)} = -\frac{70}{27}\frac{p_i}{\nu_i}A_1.$$

(9.45)

Hence, it follows that

$$a_2^{(1)} = -\frac{155}{66}\frac{p_i}{\nu_i}A_1$$

$$a_3^{(1)} = \frac{10}{11}\frac{p_i}{\nu_i}A_1$$

$$a_0^{(2)} = \frac{2}{3}\frac{p_i}{\nu_i}\left(\frac{1025}{1068}A_0 + \frac{105}{178}A_1\right)$$

$$a_1^{(2)} = -\frac{2}{3}\frac{p_i}{\nu_i}\left(\frac{15}{89}A_0 + \frac{70}{89}A_1\right).$$

(9.46)

Taking equation (9.44) and the above relation $\bar{f} = F + f_1$ into account, and integrating (9.2) over velocities with weights 1 and $Mv^2/2$, respectively, we obtain equations for the density and pressure of the form of (9.38) and (9.39) with

$$\Delta_n = -b \cdot \nabla a_0^{(2)}/2M \qquad (9.47)$$

$$\Delta_p = b \cdot \nabla \frac{T}{M}\left(\frac{5}{3}a_2^{(1)} - \frac{7}{6}a_0^{(2)} + \frac{7}{6}a_1^{(2)}\right) - \frac{1}{2}a_0^{(2)}\nabla \cdot V_E. \quad (9.48)$$

Using (9.46) one can see that expressions (9.47) and (9.48) are identical to (9.40).

9.6   Hydrodynamic description of inertial perturbations, neglecting dissipative effects

We add together the ion and electron equations of motion (9.20), use Maxwell's equations to express the current in terms of the derivatives of the magnetic field, and then take the curl of the resulting equation.

The z-component of this equation, called the *inertial equation* [2.1, 2.2], means

$$\text{curl}_z(M_i n \, d\mathbf{V}_i/dt + \nabla \cdot \boldsymbol{\pi}) = 0. \tag{9.49}$$

This is the *main starting equation* of the hydrodynamic theory of the *drift-Alfven perturbations* with $\partial/\partial z = 0$. It can be seen that the drift-Alfven perturbations depend essentially on the *transverse inertia* of the ions. Thereby we explain the meaning of the term *"inertial perturbations"* equivalent to the term *"Alfven perturbations"*.

All the terms of equation (9.49) are formally small compared with those contained in equations (9.3) and (9.4) for the inertialess perturbations. This can be seen at least from comparison of (9.49) with (9.20). On the other hand, according to section 9.1, equations (9.3) and (9.4) correspond to the *standard drift kinetic equation*. Therefore, it is clear that the Alfven perturbations can not be studied in the scope of this equation. Just this fact has caused the various modifications of the drift kinetic equation mentioned in the introduction to the present chapter.

Note that we could follow, instead of the above procedure for obtaining equation (9.49), another one based on the *current closure condition*

$$\text{div} \, \mathbf{j}_\perp = 0. \tag{9.50}$$

Neglecting the *inertial* and *magneto-viscosity effects*, i.e., taking $\mathbf{j}_\perp$ of the form of (9.34) or (9.41), we could obtain that (9.50) is equivalent to

$$\mathbf{e}_z \cdot \nabla p_\perp \times \nabla B = 0. \tag{9.51}$$

Here

$$p_\perp = p_e + p_i$$

when neglecting the terms of order ω/ν and

$$p_\perp = p_e + p_i + \pi_\perp \tag{9.52}$$

when allowing for these terms, where $\pi_\perp = \pi_{xx} = \pi_{yy}$, and π_{xx} and π_{yy} are defined by the first equation of (9.36). On the other hand, substituting $\mathbf{j}_\perp$ in the form of (9.34) or (9.41) into Maxwell's equation ($\text{curl}_\perp \mathbf{B} = 4\pi \mathbf{j}_\perp/c$ for $\mathbf{B} = B\mathbf{e}_z$) we find

$$\nabla(p_\perp + B^2/8\pi) = 0. \tag{9.53}$$

However, under condition (9.53), equation (9.51) is satisfied identically. Consequently, in the case being considered, equation (9.50) yields no information. To make equation (9.50) informative we must take into account the above-mentioned *inertial and magneto-viscosity effects* in calculating $\mathbf{j}_\perp$. For this purpose, in the case where

$\omega/\nu_i \to 0$, the ion velocity must be presented (in contrast to (9.32)) in the form

$$V_\perp = V_\perp^{(0)} + \delta V_\perp \qquad (9.54)$$

where, according to (9.20)

$$\delta V_\perp = e_z \times (Mnd V^{(0)}/dt + \nabla \cdot \pi)/Mn\Omega. \qquad (9.55)$$

The ion subscripts are omitted for simplicity. In the case of finite ω/ν_i the right-hand side of (9.55) must be augmented by $V_\perp^{(1)}$ (see (9.35) and (9.37)).

When $V_\perp$ is of the form of (9.54), equation (9.34) is replaced by

$$j_\perp = \frac{c}{B} e_z \times \nabla(p_e + p_i) + \frac{e}{M\Omega} e_z \times \left(Mn\frac{d V^{(0)}}{dt} + \nabla \cdot \pi \right). \qquad (9.56)$$

Substituting (9.56) into (9.50) we arrive at equation (9.49).

One should also keep in mind that equation (9.50) can be considered as the result of adding together the ion and electron continuity equations. Therefore, in studying the inertial perturbations by means of both continuity equations, the ion continuity equation (9.3) must be augmented by the term $\nabla \cdot (n\delta V_\perp)$.

Since equation (9.49) contains terms as small as ω/Ω_i and $\rho_i^2 \nabla^2$, the velocity $V_\perp$ and the heat flux $q_\perp$ can be substituted in it under the zeroth approximation for the given parameters, i.e., in the form of (9.32) and (9.33). Then neglecting the small dissipative terms which will be taken into account below, the components of the viscosity tensor contained in (9.49), according to (9.30), are given by

$$\pi_{xx}^{(0)} = -\pi_{yy}^{(0)} = -\frac{p_i}{2\Omega_i} \left[\frac{\partial V_x^{(0)}}{\partial y} + \frac{\partial V_y^{(0)}}{\partial x} + \frac{2}{5p_i} \left(\frac{\partial q_x^{(0)}}{\partial y} + \frac{\partial q_y^{(0)}}{\partial x} \right) \right]$$

$$\pi_{xy}^{(0)} = \pi_{yx}^{(0)} = \frac{p_i}{2\Omega_i} \left[\frac{\partial V_x^{(0)}}{\partial x} - \frac{\partial V_y^{(0)}}{\partial y} + \frac{2}{5p_i} \left(\frac{\partial q_x^{(0)}}{\partial x} - \frac{\partial q_y^{(0)}}{\partial y} \right) \right].$$
$$(9.57)$$

As mentioned above, equation (9.49) is valid for $\partial/\partial z = 0$. When $\partial/\partial z \neq 0$, equation (9.49) is replaced by

$$\text{curl}_z \left(M_i n \frac{d V^{(i)}}{dt} + \nabla \cdot \pi + \frac{1}{4\pi} B \times \nabla \times B \right) = 0. \qquad (9.58)$$

Similarly to (9.49), this equation can be found by adding together the ion and electron equations of motion (9.20) and by subsequently taking curl$_z$ of the resulting sum.

9.7 Allowance for the transverse heat conductivity and transverse viscosity in perturbations with $\partial/\partial z = 0$

Consider a scheme which allows for the *dissipative effects* of order $(k_\perp \rho_i)^2 \nu_i/\omega$ due to the *transverse heat conductivity* and the *transverse viscosity*.

In the approximation required for us, the heat flux vector q does not contain the additional ("unusual") terms given in section 9.3 (see (9.29)) and is defined by the well-known expression obtained in [9.3, 9.4] (cf. (9.33)):

$$q_i = \frac{5}{2}\frac{p_i}{M_i\Omega_i}e_z \times \nabla T_i - 2\frac{p_i\nu_i}{M_i\Omega_i^2}\nabla_\perp T_i. \qquad (9.59)$$

As for the expression for the viscosity tensor π, it differs from that generally used

$$\pi_{xx} = -\pi_{yy} = -\frac{p_i}{2\Omega_i}W_{xy}^{(1)} - \frac{3}{20}\frac{p_i\nu_i}{\Omega_i^2}$$

$$\times \left[W_{xx}^{(1)} - W_{yy}^{(1)} + \frac{3}{10}\left(W_{xx}^{(2)} - W_{yy}^{(2)}\right)\right] \qquad (9.60)$$

$$\pi_{xy} = \pi_{yx} = \frac{p_i}{4\Omega_i}\left(W_{xx}^{(1)} - W_{yy}^{(1)}\right) - \frac{3}{10}\frac{p_i\nu_i}{\Omega_i^2}\left(W_{xy}^{(1)} + \frac{3}{10}W_{xy}^{(2)}\right).$$

Here, as follows from (9.25),

$$W_{\lambda\mu}^{(1)} = \frac{\partial V_\lambda}{\partial x_\mu} + \frac{\partial V_\mu}{\partial x_\lambda} - \frac{2}{3}\delta_{\lambda\mu}\nabla\cdot V - \frac{2}{5}W_{\lambda\mu}^{(2)}$$

$$W_{\lambda\mu}^{(2)} = -\frac{1}{p_i}\left(\frac{\partial q_\lambda}{\partial x_\mu} + \frac{\partial q_\mu}{\partial x_\lambda} - \frac{2}{3}\delta_{\lambda\mu}\nabla\cdot q\right). \qquad (9.61)$$

As in deriving (9.36), we have taken into account that $F_i = F_i^{(0)} = \nabla p_i/M_i n$.

Expressions (9.60) reduce to those derived by Braginskii [9.3, 9.4] when the terms with $W_{\lambda\mu}^{(2)}$ are omitted. However, such an approximation is not justified in the problems considered since $q_\perp$ and $V_\perp$ are values of the same order: they are proportional to the *spatial derivatives of the temperature and the pressure*.

Respectively, expressions (9.59) and (9.60) differ from (9.33) and (9.57) by terms of order ν_i/Ω_i. Due to the *law of conservation of momentum* under collisions between the ions, the terms of order ν_i/Ω_i do not appear in the expression for the velocity, so that one can take $V_\perp = V_\perp^{(0)}$ in calculating the viscosity tensor, where $V_\perp^{(0)}$ is given by (9.32). The same accuracy is sufficient for substituting $V_\perp$ into the inertial term of (9.49) since it even contains the small factor of order ω.

10 Starting Equations for Basic Types of Hydrodynamic Perturbations

Using the results of Chapter 9, in the present chapter we shall find the starting equations for a number of specific types of *hydrodynamic perturbations*. Thereby we shall complete the preparation of everything required for the analysis of the stability of such perturbations which will then be made in the successive chapters.

In the present chapter we consider mainly *small-scale perturbations*. As in Chapter 1, we assume that they can be described within the scope of the local approach of the WKB approximation. Accordingly, our goal will be the derivation of the local dispersion relations for such perturbations. In addition, a part of the present chapter is devoted to *large-scale perturbations*, which we shall discuss in detail below.

According to Chapter 2, an important property of low-frequency long-wavelength perturbations of a collisionless plasma is the *splitting* of their dispersion relation into two dispersion relations, one of which corresponds to the Alfven perturbations (inertial), while the second corresponds to magnetoacoustic and other perturbations (also called inertialess). This fact is important for both the physical viewpoint (to understand the nature of instabilities) and the formal viewpoint, since it substantially facilitates the analysis of the perturbations. Therefore it appears to be important to find out whether the splitting takes place in a *collisional plasma*. The study of this problem is one of the objectives of the present chapter. As a result, we show that the *splitting does take place*. In addition, our analysis allows one to give a *hydrodynamic interpretation* of the splitting and justifies the use of the terms inertial and inertialess perturbations.

Our presentation starts with the derivation of the dispersion rela-

156

tions for the simplest case of perturbations with $k_z = 0$, neglecting dissipative effects. This is the goal of section 10.1. In such an approximation the inertialess and inertial perturbations come to be mutually independent. Then, in section 10.2 we explain how the dispersion relations for the above-mentioned perturbations are modified when allowing for the dissipative effects of orders ω/ν_i, $\nu_i k_\perp^2 \rho_i^2/\omega$ and ν_{ie}/ω, and find the dispersion relation for the transverse drift waves ($k_z = 0$).

Section 10.3 is devoted to deriving the dispersion relations for perturbations with $k_z \neq 0$. There we consider the perturbations with $k_z \simeq (\nu_e \omega)^{1/2}/v_{Te}$, those with $k_z \simeq \omega/v_{Ti}$ in a *weakly collisional high-β plasma* and those in a *strongly collisional plasma*.

Finally, in sections 10.4 and 10.5 we deal with the perturbations in a *plasma flow with an inhomogeneous velocity profile* (the so-called *shear-velocity flow*). In section 10.4 we derive the dispersion relations for the inertialess perturbations in such a plasma flow. Section 10.5 concerns the inertial (Alfven) perturbations. As in sections 10.1–10.3, we discuss small-scale perturbations in section 10.4. In section 10.5 both small-scale and large-scale perturbations are studied.

A more detailed analysis of the problems considered in the present chapter can be found in [2.1, 2.2, 9.1, 9.9, 10.1–10.6]. The results of these papers are used in our presentation.

The fact mentioned above, namely, the *splitting of the oscillation branches in a collisional plasma* into inertial and inertialess ones, is presented in section 10.1 and in some subsequent sections. Our analysis reveals a hydrodynamic feature of this effect: the set of hydrodynamic equations, obtained by neglecting the inertial and magneto-viscous effects, allows one to obtain a dispersion relation in spite of the fact that the number of hydrodynamic variables included in these equations (density, temperature, etc.) exceeds the number of equations. In this connection, the procedure of deriving the hydrodynamic dispersion relation is based upon the transition to the so-called *"canonical variables"*, the number of which coincides with the number of equations. Such a procedure for deriving the dispersion relation was initially used in [2.1, 2.2].

The dispersion relation for inertialess perturbations given in section 10.1, was initially obtained in [2.1] in the approximation of straight magnetic field lines. As will be explained in Chapter 15, the same dispersion relation is obtained in the case when transverse inertia and magnetic viscosity are small compared with the effects due to the *magnetic-field curvature*. For this case the dispersion relation for the inertialess perturbations has been found in [3.1], where such perturbations have been termed *entropy waves*. Following [3.1], we use the term *"magneto-drift entropy perturbations"* for the inertialess perturbations with $k_z = 0$.

The authors of [10.7] studied the *non-linear regime of the entropy waves*. They have shown that the set of non-linear equations for the entropy waves, as well as that of linear equations, is autonomic. Autonomy means that, in analysing these equations, the additional (i.e., inertial) equation is not required since, similarly to section 10.1, one may express the hydrodynamic variables in terms of a smaller number of canonical variables.

One more remark should be made concerning the perturbations in a *collisional plasma flow with an inhomogeneous velocity profile* (sections 10.4 and 10.5). Note that the dispersion relations for the collisional plasma flow were initially obtained in [10.8]. However, the results of [10.8] are inaccurate since, firstly, in the case of the inertialess perturbations, the viscosity tensor has been incorrectly linearized, and, second, in the case of Alfven perturbations, this tensor has been taken in an inadequate form.

10.1   Dispersion relations for perturbations with $k_z = 0$, neglecting dissipative effects

10.1.1 *Inertialess perturbations*
Linearizing (9.3) and (9.4), we find

$$-\mathrm{i}\omega \frac{\tilde{n}}{n_0} + \mathrm{i}\omega_\mathrm{D} \frac{\tilde{p}}{p_0} + \frac{cE_y}{B_0}(\kappa_n - \kappa_B) + \mathrm{i}\frac{\tilde{B}_z}{B_0}(\omega - \omega_p^*) = 0 \quad (10.1)$$

$$-\mathrm{i}\frac{\tilde{p}}{p_0}(\omega - 2\gamma_0\omega_\mathrm{D}) - \mathrm{i}\gamma_0 \frac{\tilde{n}}{n_0}\omega_\mathrm{D} + \frac{cE_y}{B_0}(\kappa_p - \gamma_0\kappa_B)$$

$$+ \mathrm{i}\gamma_0 \frac{\tilde{B}_z}{B_0}(\omega - 2\omega_p^* + \omega_n) = 0. \quad (10.2)$$

Here the subscript zero denotes the equilibrium quantities, while a tilde denotes perturbations, $\kappa_p = \partial \ln p/\partial x$. The remaining notations are clear from the previous chapters. Both equations (10.1) and (10.2) are associated with each particle species. For simplicity we omit the subscripts of the particle species.

From (10.1) and allowing for the *quasineutrality condition* $n_e = n_i$ we obtain

$$4\pi(\tilde{p}_i + \tilde{p}_e) + \tilde{B}_z B_0 = 0. \quad (10.3)$$

This is the so-called *perturbed pressure-balance equation*.

By means of (10.3) we exclude the field $\tilde{B}_z$ from (10.1) and (10.2). Following [3.1, 2.1] we introduce the variables X, Y, Z (*canonical variables*) defined by the relations

$$X = \left(1 + \frac{T_{0i}}{T_{0e}}\right)^{-1}\left(\frac{\tilde{p}_i}{p_{0i}} - \frac{\tilde{p}_e}{p_{0e}}\right)$$

$$Y = \frac{\omega}{\kappa_p}\frac{\tilde{p}_i + \tilde{p}_e}{p_{0i} + p_{0e}} + \frac{icE_y}{B_0} \qquad (10.4)$$

$$Z = \frac{\tilde{n}}{n_0} - \frac{\kappa_n}{\kappa_p}\frac{\tilde{p}_i + \tilde{p}_e}{p_{0i} + p_{0e}}.$$

Then, from (10.2) and (10.1) we find

$$\omega_{Di}X + (\kappa_B - \kappa_n)Y - \omega Z = 0$$
$$\omega(T_i/T_e)X + \gamma_0\omega_{Di}Z = 0 \qquad (10.5)$$
$$[\omega(1 - T_e/T_i) - 2\gamma_0\omega_{Di}]X + (\kappa_p - \gamma_0\kappa_B)Y = 0.$$

Here for simplicity we have omitted the subscript zero at T_e and T_i.

Equating the determinant of (10.5) to zero, we arrive at the dispersion relation [3.1, 2.1]

$$\omega^2 + \gamma_0\omega\left(1 - \frac{T_e}{T_i}\right)\omega_{Di}\frac{\kappa_B - \kappa_n}{\kappa_p - \gamma_0\kappa_B} + \gamma_0\omega_{Di}^2\frac{T_e}{T_i}\left(1 + 2\gamma_0\frac{\kappa_B - \kappa_n}{\kappa_p - \gamma_0\kappa_B}\right) = 0.$$
$$(10.6)$$

This dispersion relation is the *hydrodynamic analogue* of the kinetic dispersion relation (2.14) examined in section 3.1. The analysis of (10.6) will be performed in section 11.1.

Note that the variables X, Y, and Z defined by relations (10.4) are constructed by means of the four perturbed quantities : $\tilde{p}_i, \tilde{p}_e, \tilde{n}$ and E_y. Therefore, it is clear that the equation set (10.1)–(10.3) is *incomplete*; it should be supplemented by one more equation connecting the above-mentioned quantities. One can take equation (9.49) as the supplementary equation.

Nevertheless, this incomplete equation set allows one to obtain dispersion relation (10.6). This reflects the splitting of low-frequency long-wavelength oscillation branches into two mutually independent kinds (this fact was noted in Chapter 2 and in the introduction of Chapter 9). In fact, it follows from the previous equations that the perturbations described by dispersion relation (10.6) can be studied by means of the hydrodynamic equations (9.3) and (9.4) derived by neglecting the *plasma transverse inertia*. This provides the basis for calling perturbations of type (10.6) inertialess, which we have done since Chapter 2.

Conversely, the supplementary equation (9.49), as mentioned above, depends very substantially on the *transverse inertia*. This equation can be satisfied in the following two cases: (i) when $(X, Y, Z) \neq 0$, i.e., under condition (10.6); in this case it will give the missing connection between the quantities contained in (10.1)–(10.3); and (ii) when $X = Y = Z = 0$; resulting in some other dispersion relation which will only correspond to *inertial perturbations.*

Hence, the given considerations lead to the conclusion of the existence of *two mutually independent types of perturbations in a collisional plasma*, analogous to the case of a collisionless plasma.

10.1.2 Inertial perturbations

Using equations (9.32), (9.33) and (9.57), we linearize (9.49). As a result, we arrive at the equation

$$
\begin{aligned}
&\frac{3}{2}\omega_{\text{Di}}\frac{\tilde{n}}{n_0} - \frac{1}{2}(\omega - 2\omega_{\text{pi}}^* + \omega_{\text{ni}})\frac{\tilde{B}_z}{B_0} \\
&+ (\omega - 3\omega_{\text{Di}})\frac{\tilde{p}_i}{p_0} + \mathrm{i}\frac{\kappa_p}{\omega_{\text{pi}}^*}(\omega - \frac{3}{2}\omega_{\text{Di}})\frac{cE_y}{B_0} = 0.
\end{aligned}
\tag{10.7}
$$

In terms of the *canonical variables* X, Y, Z defined by relations (10.4) (for $T_{0e} = T_{0i} \equiv T$) this equation implies

$$
\begin{aligned}
&[\omega^2 - \omega(\omega_{\text{pi}}^* + \omega_{\text{Di}}) + \omega_{\text{pi}}^*\omega_{\text{Di}}\sigma]\frac{\tilde{B}_z}{B_0} - \omega_{\text{Di}}(\omega - 3\omega_{\text{Di}})X \\
&- \kappa_B(\omega - \frac{3}{2}\omega_{\text{Di}})Y - \frac{3}{2}\omega_{\text{Di}}^2 Z = 0
\end{aligned}
\tag{10.8}
$$

where σ is defined by (2.10). At the same time, we have equations (10.5) so that we now have the complete equation set.

In section 10.1.1 we found the dispersion relation (10.6) assuming X, Y, Z to be finite. At the same time, according to the discussion above, equations (10.5) can also be satisfied for $X = Y = Z = 0$. In this case, under the condition

$$
\tilde{B}_z \neq 0
\tag{10.9}
$$

the dispersion relation (4.1) is obtained from (10.8) corresponding to the *Alfven perturbations.* Thus, it is clear that the dispersion relation for the inertial perturbations in a collisional plasma is the same as that in the case of a collisionless one [2.1, 2.2].

Taking into account (10.3) and (10.4), from the condition $X = Y = Z = 0$ we find that, in the inertial perturbations,

$$\frac{\tilde{n}}{n_0} = \frac{\kappa_n}{\kappa_B}\frac{\tilde{B}_z}{B_0}$$

$$\frac{\tilde{T}_i}{T_{0i}} = \frac{\tilde{T}_e}{T_{0e}} = \frac{\kappa_T}{\kappa_B}\frac{\tilde{B}_z}{B_0} \qquad (10.10)$$

$$-i\omega \tilde{B}_z/B_0 = \kappa_B c E_y/B_0$$

where $\kappa_T = \partial \ln T/\partial x$. The last equality in (10.10) can be represented in the form

$$\nabla \cdot \boldsymbol{V}_{\mathrm{E}} = 0 \qquad (10.11)$$

where, as in section 9.1, $\boldsymbol{V}_{\mathrm{E}} = c\boldsymbol{E} \times \boldsymbol{B}/B^2$ is the *crossed-field drift velocity*.

Therefore, it is clear, that the Alfven perturbations in an inhomogeneous plasma are characterized by the crossed-field drift velocity corresponding to the *incompressible plasma motion* (see (10.11)). Due to the equilibrium magnetic-field inhomogeneity such a motion leads to perturbations of the z-component of the total magnetic field, i.e., to *compression and expansion of this field* (see (10.9)). This is in contrast to the case of a homogeneous magnetic field where the Alfven perturbations do not result in the compression and expansion of the field lines. According to (10.10), the density, temperature and plasma pressure are perturbed together with the magnetic field.

10.2 Dispersion relations for perturbations with $k_z = 0$ allowing for dissipative effects

10.2.1 Effects of order ω/ν_i

Let us linearize equations (9.38)–(9.40) using the quasineutrality condition (cf. section 10.1). Then (10.1) and (10.2) are replaced by

$$-i\omega \frac{\tilde{n}}{n_0} + i\omega_{\mathrm{Di}}\frac{\tilde{p}_i}{p_{0i}} + (\kappa_n - \kappa_B)\frac{cE_y}{B_0} + i(\omega - \omega_{pi}^*)\frac{\tilde{B}_z}{B_0}$$

$$+ i\frac{0.96}{5}\frac{\omega_{\mathrm{Di}}}{\nu_i}\left[-i\omega \frac{\tilde{p}_i}{p_{0i}} + \kappa_p \frac{cE_y}{B_0} + i\frac{42}{41}\left(\omega_{\mathrm{Ti}}\frac{\tilde{B}_z}{B_0} - \omega_{\mathrm{Di}}\frac{\tilde{T}_i}{T_{0i}}\right)\right] = 0$$

$$(10.12)$$

$$-\mathrm{i}(\omega - 2\gamma_0\omega_{\mathrm{Di}})\frac{\tilde{p}_{\mathrm{i}}}{p_{0\mathrm{i}}} + (\kappa_p - \gamma_0\kappa_B)\frac{cE_y}{B_0} - \mathrm{i}\gamma_0\omega_{\mathrm{Di}}\frac{\tilde{n}}{n_0}$$

$$+ \mathrm{i}\gamma_0(\omega - 2\omega_{p\mathrm{i}}^* + \omega_{n\mathrm{i}})\frac{\tilde{B}_z}{B_0} + \mathrm{i}1.13\frac{7}{15}\frac{\omega_{\mathrm{Di}}}{\nu_{\mathrm{i}}}\left(-\mathrm{i}\omega\frac{\tilde{p}_{\mathrm{i}}}{p_{0\mathrm{i}}} + \kappa_p\frac{cE_y}{B_0}\right)$$

$$+ 6.41\frac{7}{9}\frac{\omega_{\mathrm{Di}}}{\nu_{\mathrm{i}}}\left(\omega_{\mathrm{Di}}\frac{\tilde{T}_{\mathrm{i}}}{T_{0\mathrm{i}}} - \omega_{T\mathrm{i}}\frac{\tilde{B}_z}{B_0}\right) = 0$$

$$(10.13)$$

$$\frac{\tilde{B}_z}{B_0} = -\frac{\beta}{2}\frac{\tilde{p}_{\mathrm{i}\perp} + \tilde{p}_{\mathrm{e}}}{p_{0\mathrm{i}} + p_{0\mathrm{e}}}. \qquad (10.14)$$

The quantity $\tilde{p}_{\mathrm{i}\perp}$ in the last equation means

$$\tilde{p}_{\mathrm{i}\perp} = \tilde{p}_{\mathrm{i}} - \frac{\pi_{zz}}{2} = \tilde{p}_{\mathrm{i}} - \frac{0.96}{5}\frac{p_{0\mathrm{i}}}{\nu_{\mathrm{i}}}\left[-\mathrm{i}\omega\frac{\tilde{p}_{\mathrm{i}}}{p_{0\mathrm{i}}} + \kappa_p\frac{cE_y}{B_0}\right.$$

$$\left. + \mathrm{i}\frac{42}{41}\left(\omega_{T\mathrm{i}}\frac{\tilde{B}_z}{B_0} - \omega_{\mathrm{Di}}\frac{\tilde{T}_{\mathrm{i}}}{T_{0\mathrm{i}}}\right)\right]. \qquad (10.15)$$

As before, the electron density and the electron pressure are determined by equations (10.1) and (10.2).

Assuming $T_{0\mathrm{e}} = T_{0\mathrm{i}}$ for simplicity, we introduce the following variables similar to (10.4):

$$X_\perp = \frac{\tilde{p}_{\mathrm{i}\perp} - \tilde{p}_{\mathrm{e}}}{p_0}$$

$$Y_\perp = \frac{\omega}{\kappa_p}\frac{\tilde{p}_{\mathrm{i}\perp} + \tilde{p}_{\mathrm{e}}}{p_0} + \mathrm{i}c\frac{E_y}{B_0} \qquad (10.16)$$

$$Z_\perp = \frac{\tilde{n}}{n_0} - \frac{\kappa_n}{\kappa_p}\frac{\tilde{p}_{\mathrm{i}\perp} + \tilde{p}_{\mathrm{e}}}{p_0}.$$

Here $p_0 = p_{0\mathrm{e}} + p_{0\mathrm{i}}$. In terms of these variables the set (10.12)–(10.14) takes the form (cf. (10.5))

$$\omega_{\mathrm{Di}}X_\perp + (\kappa_B - \kappa_n)Y_\perp - \omega Z_\perp = 0$$

$$[\omega + \epsilon(A + D)]X_\perp + (\kappa_p/\omega)\epsilon AY_\perp + (\gamma_0\omega_{\mathrm{Di}} - \epsilon D)Z_\perp = 0$$

$$[-2\gamma_0\omega_{\mathrm{Di}} + \epsilon(A + D)]X_\perp + [\kappa_p - \gamma_0\kappa_B + (\kappa_p/\omega)\epsilon A]Y_\perp - \epsilon AZ_\perp = 0.$$

$$(10.17)$$

Here

$$\epsilon = \mathrm{i}\frac{0.96}{10}\frac{\omega}{\nu_{\mathrm{i}}}$$

$$A = \omega - 0.59\omega_{\mathrm{Di}} \qquad (10.18)$$

$$D = \frac{42}{41}\frac{\omega_{\mathrm{Di}}}{\omega}(\omega + 22.02\omega_{\mathrm{Di}}).$$

The determinant of equation set (10.17) yields the dispersion relation. When $\beta \ll 1$, this dispersion relation is of the form of

$$
\omega^2 - \gamma_0 \omega_{\mathrm{Di}}^2 \left(2\gamma_0 \frac{\kappa_n}{\kappa_p} - 1 \right) + \epsilon \left[(A + D) \left(\omega + \gamma_0 \omega_{\mathrm{Di}} \frac{\kappa_n}{\kappa_p} \right) \right.
$$
$$
\left. + A \left(2\gamma_0 \omega_{\mathrm{Di}} + \omega + \frac{\gamma_0 \omega_{\mathrm{Di}}}{\omega} \right) + D \left((2\gamma_0 \omega_{\mathrm{Di}} + \omega) \frac{\kappa_n}{\kappa_p} - \omega_{\mathrm{Di}} \right) \right] = 0.
$$

$$(10.19)$$

Equations (10.17)–(10.19) have been obtained in [9.1, 9.9]. The analysis of equation (10.19) will be performed in section 11.1.2. In addition, we shall use equation set (10.17) in section 12.3.1.

The dispersion relation (10.19) takes into account the terms of order ω/ν_i corresponding to the *longitudinal viscosity and related effects*. When one takes into account similar terms in deriving the dispersion relation for the *inertial perturbations*, it can be seen that all these terms are expressed only in terms of the variables $X_\perp, Y_\perp, Z_\perp$ (see (10.16)). This means that the dissipative effects of order ω/ν_i do not appear in the dispersion relation for the inertial perturbations. In other words, even when the terms of order ω/ν_i are allowed for, equation (4.1) still holds.

10.2.2 *Effects of order $\nu_i k_\perp^2 \rho_i^2 / \omega$*

Using (9.59)–(9.61), we find that allowing for terms of order $\nu_i k_\perp^2 \rho_i^2 / \omega$ modifies the ion *heat-balance equation* (10.2) and the *inertial equation* (10.7) which now take the form, respectively:

$$
-i \frac{\tilde{p}_i}{p_{0i}} (\omega - 2\gamma_0 \omega_{\mathrm{Di}} + i\alpha) - i \frac{\tilde{n}}{n_0} (\gamma_0 \omega_{\mathrm{Di}} - i\alpha)
$$
$$
+ \frac{cE_y}{B_0} (\kappa_p - \gamma_0 \kappa_B) + i\gamma_0 \frac{\tilde{B}_z}{B_0} (\omega - 2\omega_{pi}^* + \omega_{ni}) = 0
$$

$$(10.20)$$

$$
\frac{\tilde{n}}{n_0} \left(3\omega_{\mathrm{Di}} - \frac{57}{80} i\alpha \right) - \frac{\tilde{B}_z}{B_0} (\omega - 2\omega_{pi}^* + \omega_{ni})
$$
$$
+ \frac{\tilde{p}_i}{p_{0i}} \left(2(\omega - 3\omega_{\mathrm{Di}}) + \frac{93}{80} i\alpha \right)
$$
$$
+ \frac{icE_y}{B_0} \frac{\kappa_p}{\omega_{pi}^*} \left(2\omega - 3\omega_{\mathrm{Di}} + \frac{9}{20} i\alpha \right) = 0
$$

$$(10.21)$$

where $\alpha = (4/3)\nu_i k_\perp^2 \rho_i^2$. These equations are supplemented by the *continuity equation* (10.1), by the *electron heat-balance equation* (10.2) and by the perturbed *pressure-balance equation* (10.3) which are unchanged. From (10.20), (10.21), and (10.1)–(10.3) after transforming to the variables X, Y, Z we find the dispersion relation [10.5]

$$(Q_1^{(0)} + \delta Q_1)(Q_2^{(0)} + \delta Q_2) = -\frac{i\alpha\kappa_T}{2\kappa_p(2 + \gamma_0\beta)}$$
$$\times \{2\omega^3 + \omega^2[\omega_{\mathrm{Di}}(2\gamma_0 - 3) + 2\omega_{pi}^*]$$
$$+ \omega\omega_{\mathrm{Di}}[3\omega_{\mathrm{Di}} + \omega_{ni}(2\gamma_0 - 3) - 6\omega_{pi}^*]$$
$$+ 3\omega_{\mathrm{Di}}^2(2\gamma_0\omega_{\mathrm{Di}} + 4\omega_{pi}^* - 4\gamma_0\omega_{ni})\}. \tag{10.22}$$

Here $Q_1^{(0)}$ and $Q_2^{(0)}$ represent the left-hand sides of equations (4.1) and (10.6) (for $T_i = T_e$), respectively, while δQ_1 and δQ_2 are of the form

$$\delta Q_1 = \frac{9}{40}i\alpha\left(\omega - \frac{31}{12}\omega_{pi}^* + \frac{19}{12}\omega_{ni}\right) \tag{10.23}$$

$$\delta Q_2 = \frac{i\alpha}{2 + \gamma_0\beta}\left[\omega\left(1 + \frac{\beta}{2}(\gamma_0 + 1) + \frac{\kappa_n}{\kappa_p}\right)\right.$$
$$\left. + \omega_{\mathrm{Di}}(\gamma_0\beta - 1) - \frac{3}{2}\gamma_0\beta\omega_{ni}\right]. \tag{10.24}$$

The dispersion relation (10.22) will be studied in section 11.1.3.

10.2.3 Allowance for the heat transfer
Above we neglected the heat transfer Q in the heat-balance equations (9.21). According to [9.3, 9.4]

$$Q_e = -Q_i = -(3/2)\nu_{ie}n(T_e - T_i) \tag{10.25}$$

where $\nu_{ie} = 2M_e/M_i\tau_e$, and τ_e is the characteristic relaxation time of the electron gas. (The explicit definition of τ_e is given in [9.3, 9.4].)
When $T_{0e} = T_{0i}$, allowance for Q leads to the following modification of dispersion relation (10.6) for the *inertialess perturbations* [3.1]:

$$\omega(\omega + i\nu_{ie}) + \gamma_0\omega_{\mathrm{Di}}^2\left(1 + 2\gamma_0\frac{\kappa_B - \kappa_n}{\kappa_p - \gamma_0\kappa_B}\right) = 0. \tag{10.26}$$

Since Q is expressed only in terms of X, the heat transfer contribution into the dispersion relation for the inertial perturbations vanishes.
We shall examine (10.26) in section 11.1.4.

10.2.4 Dispersion relation for transverse drift waves

Consider the perturbations with $k_z = 0$ and $E \parallel B_0$. In this case $B = B_0 e_z + \tilde{B}_\perp$, where $\tilde{B}_\perp = (\tilde{B}_x, \tilde{B}_y, 0)$. We start with the z-component of the electron motion equation (9.20) written in the form

$$e_e n_0 (E_z - V_{pe}^* \tilde{B}_x / c) + \tilde{R}_{ze} = 0. \tag{10.27}$$

Here $V_{pe}^* = c T_e \kappa_p / e_e B_0$ is the *electron pressure-gradient drift velocity* and $\tilde{R}_{ze}$ is the z-component of the *electron momentum transfer* $\tilde{R}_e$. According to [9.3, 9.4]

$$R_e = -s n \nabla_\parallel T_e - \nu_e n M_e (V_{\parallel e} - V_{\parallel i}) e_0 \tag{10.28}$$

where $s = 0.71, V_\parallel = V \cdot e_0, e_0 = B/B \equiv h, \nu_e$ is the *electron collision frequency* $(\nu_e \simeq 1/\tau_e)$, and $\nabla_\parallel = e_0(e_0 \cdot \nabla)$. Linearizing (10.28), we take into account

$$e_0 \cdot \nabla = \frac{\partial}{\partial z} + \frac{1}{B_0} \tilde{B}_\perp \cdot \nabla. \tag{10.29}$$

Therefore,

$$\tilde{R}_{ze} = -s n_0 \tilde{B}_x (\partial T_{0e}/\partial x)/B_0 - \nu_e n_0 M_e V_{ze}. \tag{10.30}$$

We neglect the ion longitudinal velocity. Besides, it is clear that we have neglected the electron longitudinal inertia in (10.27).

We supplement equations (10.27) and (10.30) by Maxwell's equations

$$\text{curl}_z \tilde{B}_\perp = 4\pi e_e n_0 \tilde{V}_{ze}/c$$
$$E_z = \omega \tilde{B}_x / c k_y. \tag{10.31}$$

Taking into account $\nabla \cdot \tilde{B}_\perp = 0$, from the first equation of (10.31) we find

$$\tilde{V}_{ze} = -i k_\perp^2 c \tilde{B}_x / 4\pi e_e n_0 k_y. \tag{10.32}$$

As a result, the dispersion relation is obtained (cf. (3.16))

$$\omega = \omega_{pe}^* + s \omega_{Te} - i c^2 k_\perp^2 \nu_e / \omega_{pe}^2. \tag{10.33}$$

The oscillation branch (10.33) will be discussed in section 11.2.

10.3   Starting equations for oblique perturbations

10.3.1 Perturbations with $k_z \simeq (\nu_e\omega)^{1/2}/v_{Te}$

In contrast to sections 10.1 and 10.2, we now consider $k_z \neq 0$ (*oblique perturbations*) and allow for *finite electron heat conductivity* along the lines of the total magnetic field, i.e., take $\boldsymbol{q}_e$ of the form [9.3, 9.4]

$$\boldsymbol{q}_e = -3.16\frac{p_e}{M_e\nu_e}\nabla_{\|}T_e + \frac{5}{2}\frac{p_e}{M_e\Omega_e}\boldsymbol{e}_0 \times \nabla T_e. \qquad (10.34)$$

We neglect the ion longitudinal motion, $V_{zi} = 0$. Taking into account (10.32), we neglect the contribution of V_{ze} in the electron equations of motion and of heat balance, (9.20) and (9.21), as being as small as $k_\perp^2$. Therefore, the linearized electron continuity equation is the same as that for $k_z = 0$ and is of the form of (10.1). Using the z-component of the electron motion equation (9.20) (cf. (10.27)), we express the magnetic field perturbation $\tilde{B}_x$ in terms of other perturbed quantities:

$$\tilde{B}_x = \frac{ik_z cT_{0e}k_y}{e_e\tilde{\omega}}\left(\frac{icE_y}{B_0}\frac{\kappa_n}{\omega_{ni}} - \frac{\tilde{p}_e}{p_{0e}}(1+s) + s\frac{\tilde{n}}{n_0}\right). \qquad (10.35)$$

Here $\tilde{\omega} = \omega - \omega_{pe}^* - s\omega_{Te}$ (cf. (10.33)).

Using (10.34) and (10.35), we reduce the linearized electron heat-balance equation (9.21) to the form (cf. (10.2))

$$-i\frac{\tilde{p}_e}{p_{0e}}\left(\omega - 2\gamma_0\omega_{De} + i\Delta\frac{\omega - \omega_{ne}}{\tilde{\omega}}\right) - i\frac{\tilde{n}}{n_0}\left(\gamma_0\omega_{De} - i\Delta\frac{\omega - \omega_{pe}^*}{\tilde{\omega}}\right)$$

$$+ \frac{cE_y}{B_0}\left(\kappa_p - \gamma_0\kappa_B + i\Delta\frac{\kappa_T}{\tilde{\omega}}\right) + i\gamma_0\frac{\tilde{B}_z}{B_0}(\omega - 2\omega_{pe}^* + \omega_{ne}) = 0. \qquad (10.36)$$

Here $\Delta = (2/3)3.16k_z^2T_0/M_e\nu_e$. It is assumed that $T_{0e} = T_{0i} \equiv T_0$.

Introducing the variables X, Y, Z given by (10.4), we arrive at the set of three equations similar to (10.5):

$$\omega_{Di}X + (\kappa_B - \kappa_n)Y - \omega Z = 0$$

$$X(\omega - 2\gamma_0\omega_{Di}) + Y(\kappa_p - \gamma_0\kappa_B) + \gamma_0\omega_{Di}Z = 0$$

$$X\left(\omega + \frac{i\Delta}{2\tilde{\omega}}(\omega - \omega_{ne})\right) - iY\frac{\Delta}{2\tilde{\omega}}\kappa_T + Z\left(-\gamma_0\omega_{De} + \frac{i\Delta}{2\tilde{\omega}}(\omega - \omega_{pe}^*)\right) = 0. \qquad (10.37)$$

Hence the dispersion relation follows [10.1]

$$\left(1 + \frac{\gamma_0 \beta}{2}\right)\left[x^2 + \gamma_0 \frac{\beta^2 y^2}{4}\left(1 - \frac{2\gamma_0(1 + y\beta/2)}{y + \gamma_0\beta/2}\right)\right][x + y(1 + s) - s]$$

$$+ i\lambda\left\{x^2\left(1 + \frac{2}{5}\gamma_0\beta\right) + x\left[1 + \frac{\gamma_0\beta}{2} + \frac{y\gamma_0\beta}{2}\left(1 + \frac{\beta}{2}\right)\right]\right.$$

$$\left. + \frac{\gamma_0\beta}{4}\left(1 + 2y + \beta y^2 - \frac{y^2}{\gamma_0}\right)\right\} = 0.$$

$$(10.38)$$

Here $x = \omega/\omega_{ni}$, $y = \kappa_p/\kappa_n$, and $\lambda = \Delta/\omega_{ni}$.

We shall examine equation (10.38) in section 11.2.

10.3.2 *Perturbations with $k_z \simeq \omega/v_{Ti}$ in a weakly collisional plasma*

Following [2.2], we shall now consider the perturbations with $k_z \simeq \omega/v_{Ti}$. We neglect the heat transfer assuming $\omega \gg \nu_{ie}$. This condition corresponds to the so-called *weakly collisional plasma regime*. Another limiting case, $\omega \ll \nu_{ie}$, corresponding to the strongly collisional plasma regime will be considered in section 10.3.4.

When $\omega \gg \nu_{ie}$, the leading term in the electron heat-balance equation (9.21) is due to the heat flux related to the longitudinal electron heat conductivity. This heat flux is described by the first term of the right-hand side of (10.34). Then (9.21) reduces to

$$e_0 \cdot \nabla T_e = 0. \qquad (10.39)$$

Linearizing (10.39) allowing for (10.29), we find

$$\tilde{T}_e/T_0 = i\kappa_T \tilde{B}_x/k_z B_0. \qquad (10.40)$$

In contrast to section 10.3.1, the ion longitudinal motion is now important, $\tilde{V}_{zi} \neq 0$. Then Maxwell's equation for the field $\tilde{B}_x$(10.31) contains the difference $\tilde{V}_{ze} - \tilde{V}_{zi}$ instead of $\tilde{V}_{ze}$. Equation (10.32) is also modified in a similar manner. From that equation it then follows that the difference $\tilde{V}_{ze} - \tilde{V}_{zi}$ is small, of the order of $k_\perp^2$. Therefore, one can assume

$$\tilde{V}_{ze} = \tilde{V}_{zi} \equiv V_z \qquad (10.41)$$

in the electron and ion continuity equations and in the ion heat-balance equation.

In allowance for $V_z \neq 0$ the term with $ik_z V_z$ is added to the left-hand side of equation (10.1) (i.e., of the linearized continuity equation), while the addition $i\gamma_0 k_z V_z$ appears on the left-hand side of equation (10.2) for the ions (i.e, the linearized ion heat-balance equation). In addition, we have the inertial equation (9.58). Linearizing the last term in the brackets of (9.58), we find in the small-scale perturbation approximation:

$$\operatorname{curl}_z(M_i n_0 dV_i/dt + \nabla \cdot \pi) = B_0 \tilde{B}_x k_\perp^2 k_z / 4\pi k_y. \qquad (10.42)$$

The left-hand side of this equality is calculated precisely in the same manner as in section 10.1. As a result, the inertial equation (10.42) reduces to the equation of form (10.7) with the additional term

$$-i k_z c_A^2 (\kappa_p/\omega_{pi}^*) \tilde{B}_x / B_0$$

in the right-hand side.

We supplement the above-mentioned equations by the *pressure-balance equation* (10.3) which is unchanged. It is obtained, e.g., from Maxwell's equation for the field $\tilde{B}_z$,

$$\partial \tilde{B}_z / \partial y \approx 4\pi j_y / c \qquad (10.43)$$

and by using the expression for the current density (9.34).

In addition, we also have the electron and ion longitudinal motion equations (9.20) and the second equation of (10.31) relating E_z and $\tilde{B}_x$. One can consider $R_e = 0$ in the electron longitudinal motion equation; this is a consequence of the approximate equalities (10.39) and (10.41). Therefore, using the second equation of (10.31), from the electron longitudinal motion equation we find (cf. (10.35))

$$\frac{\tilde{B}_x}{B_0} = \frac{k_z}{\omega - \omega_{ne}}\left(-c\frac{E_y}{B_0} + i\frac{\tilde{n}}{n_0}\frac{\kappa_p}{\omega_{pe}^*}\right). \qquad (10.44)$$

By means of this relation one can exclude $\tilde{B}_x$ from all the remaining equations.

In the ion longitudinal motion equation we allow for the *oblique (magnetic) viscosity*. Then, using the electron longitudinal motion equation we have the following starting equation

$$M_i n dV_{zi}/dt = -\partial(p_e + p_i)/\partial z - \partial \pi_{z\alpha}/\partial x_\alpha - e_i n_0(V_{pi}^* - V_{pe}^*)\tilde{B}_x/c \qquad (10.45)$$

where $V_{pi}^* = cT_{0i}\kappa_p/e_i B_0$ is the *ion pressure-gradient drift velocity*. Here it is assumed that $\pi_{zz} \to 0$. According to (9.30), to the required accuracy the components π_{xz} and π_{yz} have the form

$$\pi_{xz} = -\frac{p_i}{\Omega_i}\left[\frac{\partial V_y}{\partial z} + \frac{\partial V_z}{\partial y} + \frac{2}{5p_i}\left(\frac{\partial q_y}{\partial z} + \frac{\partial q_z}{\partial y}\right)\right]$$

$$\pi_{yz} = \frac{p_i}{\Omega_i}\left[\frac{\partial V_x}{\partial z} + \frac{\partial V_z}{\partial x} + \frac{2}{5p_i}\left(\frac{\partial q_x}{\partial z} + \frac{\partial q_z}{\partial x}\right)\right]. \qquad (10.46)$$

In the right-hand side of these equations we have omitted the ion subscripts for V and q. It is assumed that $V_\perp$ and $q_\perp$ are of the form of (9.32) and (9.33).

Using equations (10.44) and (10.46), we reduce the linearized equation (10.45) to the form

$$(\omega - \omega_{\mathrm{Di}})V_z = -k_z c_A^2 \left[\frac{\tilde{B}_z}{B_0} + \frac{1}{\omega - \omega_{ne}} \left(i\kappa_B \frac{cE_y}{B_0} - \omega_{\mathrm{De}} \frac{\tilde{n}}{n} \right) \right]. \quad (10.47)$$

Now we make the transition to the variables X, Y, Z and $\tilde{B}_z$. Then, as before, we find that $\tilde{B}_z$ is contained only in the inertial equation, whilst in all the remaining equations only X, Y, Z appear. Therefore, considering $(X, Y, Z) = 0$, from the inertial equation we obtain the dispersion relation for the Alfven perturbations which coincides with (4.10). On the other hand, for the perturbations with $(X, Y, Z) \neq 0$ the following dispersion relation is obtained [2.2]

$$x^2 \left(x + \frac{\beta}{2} \right) \left(1 + \frac{2\beta}{3} \right) + x \left(x + \frac{\beta}{2} \right) \left[\left(1 + \frac{5\beta}{6} \right) y + \frac{5}{6}\beta + \frac{5}{12}\beta^2 \right]$$
$$+ \frac{5}{12}\beta \left(x + \frac{\beta}{2} \right) \left(y^2 + 2y + \beta - \frac{3}{5} \right)$$
$$- \alpha \left(\frac{8}{3}x + \frac{5}{3}y + \frac{5}{3}\beta - 1 \right) = 0.$$

$$(10.48)$$

Here $x = \omega/\omega_{pi}^*$, $y = (1 + \eta)^{-1}$, and $\alpha = k_z^2 T_0/M_i \omega_{pi}^{*2}$.
We shall examine equation (10.48) in section 11.3.

10.3.3 Perturbations in a high-β plasma with $k_z > \omega/v_{\mathrm{Ti}}$

Let us augment the equations of section 10.3.2 by the longitudinal ion viscosity, $\pi_{zz} \neq 0$, and by the longitudinal ion heat flux described by the second term of the right-hand side of the first equation of (9.29). When allowing for π_{zz} it is more convenient to use, instead of the perturbed ion pressure $\tilde{p}_i$, the quantity $\tilde{p}_{i\perp}$ given by the first equality of (10.15). In addition, when $\pi_{zz} \neq 0$, it is necessary to take into account the modification of the perturbed pressure-balance equation described by (10.14). In calculating $\pi_{z\alpha}, \alpha = (x, y, z)$, we use equations (9.30) taking into account the contribution of both the longitudinal heat flux and the part of the vector q^* proportional to $1/\nu_i$ (see the second equation of (9.29)) in these viscosity tensor components, approximating $25/24$ to unity for simplicity. The remaining assumptions are the same as in section 10.3.2.

We write the linearized equations in terms of the variables $X_\perp, Y_\perp, Z_\perp$ defined by (10.16). As a result, we arrive at the following equation set [10.2]:

the electron continuity equation

$$\omega_{\mathrm{Di}} X_\perp + (\kappa_B - \kappa_n) Y_\perp - \omega Z_\perp + k_z V_z = 0 \qquad (10.49)$$

the electron longitudinal motion equation

$$(\omega - \omega_{ne}) X_\perp + \kappa_T Y_\perp + (\omega - \omega_{pe}^*) Z_\perp = 0 \qquad (10.50)$$

the ion heat-balance equation

$$(\omega - 2\gamma_0 \omega_{\mathrm{Di}} - 2\mathrm{i}\Delta_1) X_\perp + (\kappa_p - \gamma_0 \kappa_B) Y_\perp + \gamma_0 \omega_{\mathrm{Di}} Z_\perp - \gamma_0 k_z V_z = 0 \qquad (10.51)$$

the ion longitudinal motion equation

$$(\omega - \omega_{\mathrm{Di}}) k_z V_z = \frac{3}{2} \frac{k_z^2}{M_{\mathrm{i}} n_0} \pi_{zz} + 2 \frac{\kappa_p}{\kappa_T} \frac{k_z^2 T_0}{M_{\mathrm{i}}} (X_\perp + Z_\perp) - \mathrm{i} \frac{6}{5} \Delta_1 \omega_{\mathrm{Di}} X_\perp \qquad (10.52)$$

the linearized form of the first equation of (9.30) for the components of the viscosity tensor

$$\pi_{zz} = -\mathrm{i} 0.96 \frac{p_{0\mathrm{i}}}{\nu_{\mathrm{i}}} \left[\frac{4}{3} k_z V_z - \left(\frac{26}{15} \omega_{\mathrm{Di}} + 2.9\mathrm{i}\Delta_1 \right) X_\perp - \frac{2}{3} \kappa_B Y_\perp + \frac{16}{15} \omega_{\mathrm{Di}} Z_\perp \right] \qquad (10.53)$$

and, finally,

the inertial equation

$$\left(Q_1^{(0)} - k_z^2 c_{\mathrm{A}}^2 \right) \frac{\tilde{B}_z}{B_0} + \omega_{\mathrm{Di}} \left[k_z V_z - \left(\omega - 3\omega_{\mathrm{Di}} + \mathrm{i} \frac{6}{5} \Delta_1 - \frac{k_z^2 c_{\mathrm{A}}^2}{\omega_{\mathrm{Ti}}} \right) X_\perp \right.$$
$$\left. - \frac{\kappa_B}{\omega_{\mathrm{Di}}} \left(\omega - \frac{3}{2} \omega_{\mathrm{Di}} \right) Y_\perp - \left(\frac{3}{2} \omega_{\mathrm{Di}} - \frac{k_z^2 c_{\mathrm{A}}^2}{\omega_{\mathrm{Ti}}} \right) Z_\perp \right] = 0. \qquad (10.54)$$

Here $\Delta_1 = (2/3) 3.9 k_z^2 T_0 / M_{\mathrm{i}} \nu_{\mathrm{i}}$, the quantity $Q_1^{(0)}$ was introduced in section 10.2.2.

The equation set (10.49)–(10.54) will be used in section 12.3.3.

10.3.4 *Perturbations in a strongly collisional plasma*

In this section we shall obtain the dispersion relation for the *inertialess perturbations* allowing for the *heat transfer*, the finiteness of the *longitudinal electron heat conductivity* and the *ion longitudinal motion.* Such a dispersion relation will contain the results of sections 10.2.3, 10.3.1 and 10.3.2 as particular cases. In addition, it will be suitable for examining *oblique perturbations* in the *strongly collisional plasma regime.*

We start with the following linearized hydrodynamic equations [10.3]:

the electron continuity equation

$$\omega\frac{\tilde{n}}{n_0} - \omega_{De}\frac{\tilde{p}_e}{p_{0e}} - k_z V_z + i\frac{cE_y}{B_0}(\kappa_n - \kappa_B) + (\omega - \omega_{pe}^*)\frac{\tilde{B}_z}{B_0} = 0 \quad (10.55)$$

the electron heat-balance equation

$$(-\gamma_0\omega_{De} + i\nu_{ie})\frac{\tilde{n}}{n_0} - [\omega - 2\gamma_0\omega_{De} + i(\nu_{ie} + \Delta)]\frac{\tilde{p}_e}{p_{0e}} + i\nu_{ie}\frac{\tilde{p}_i}{p_{0i}}$$

$$+ \gamma_0 k_z V_z - i\frac{cE_y}{B_0}(\kappa_p - \gamma_0\kappa_B)$$

$$+ \gamma_0(\omega - 2\omega_{pe}^* + \omega_{ne})\frac{\tilde{B}_z}{B_0} - \Delta\frac{\kappa_T}{k_z}\frac{\tilde{B}_x}{B_0} = 0$$

$$(10.56)$$

the ion heat-balance equation

$$\gamma_0\omega_{Di}\frac{\tilde{n}}{n_0} - i\nu_{ie}\frac{\tilde{p}_e}{p_{0e}} + (\omega - 2\gamma_0\omega_{Di} + i\nu_{ie})\frac{\tilde{p}_i}{p_{0i}} - \gamma_0 k_z V_z$$

$$+ \frac{icE_y}{B_0}(\kappa_p - \gamma_0\kappa_B) - \gamma_0(\omega - 2\omega_{pi}^* + \omega_{ni})\frac{\tilde{B}_z}{B_0} = 0$$

$$(10.57)$$

the electron longitudinal motion equation

$$k_z\left(s\frac{\tilde{n}}{n_0} - (1+s)\frac{\tilde{p}_e}{p_{0e}} - i\frac{\kappa_p}{\omega_{pe}^*}\frac{cE_y}{B_0}\right)$$

$$- i[\omega + s\omega_{ne} - (1+s)\omega_{pe}^*]\frac{\kappa_T}{\omega_{Te}}\frac{\tilde{B}_x}{B_0} = 0$$

$$(10.58)$$

and the ion longitudinal motion equation written with an allowance for the electron longitudinal motion equation

$$\frac{M_i V_z}{T}(\omega - \omega_{Di}) + 2i\kappa_p\left(\frac{\tilde{B}_x}{B_0} + \frac{ik_z}{\kappa_p}\frac{\tilde{p}_i + \tilde{p}_e}{2p_{0i}}\right) = 0. \quad (10.59)$$

Here it is assumed that $T_{0i} = T_{0e} \equiv T$ so that $p_{0e} = p_{0i}$, $\omega_{pe}^* = -\omega_{pi}^*$, $\omega_{ne} = -\omega_{ni}$, $\omega_{De} = -\omega_{Di}$. The quantity Δ was defined in section 10.3.1.

We express V_z and $\tilde{B}_x$ in terms of the remaining variables by means of equations (10.58) and (10.59). Then, as usual, we introduce the variables X, Y, Z defined by equations (10.4). As a result, we obtain three equations for these variables. Equating the determinant of these equations to zero, we find the following dispersion relation [10.3]:

$$(\omega - \omega_{\mathrm{Di}})Q(\omega) - k_z^2 T R(\omega)/M_{\mathrm{i}} = 0. \qquad (10.60)$$

Here

$$
\begin{aligned}
Q(\omega) = &\left\{ \left(1 + \frac{5}{6}\beta\right)\omega^2 + \frac{5}{12}\beta^2\omega_{\mathrm{pi}}^*\left[\left(1 - \frac{5}{6}\beta\right)\omega_{\mathrm{pi}}^* - \frac{10}{3}\omega_{\mathrm{ni}}\right] \right. \\
&\left. + 2i\nu_{\mathrm{ie}}\left(1 + \frac{5}{6}\beta\right)\omega\right\}[\omega - s\omega_{\mathrm{ni}} + (1+s)\omega_{\mathrm{pi}}^*] \\
&+ i\Delta\left\{ \left(1 + \frac{2}{3}\beta\right)\omega^2 + \left[\left(1 + \frac{5}{6}\beta\right)\omega_{\mathrm{ni}} + \frac{5}{6}\beta\left(1 + \frac{1}{2}\beta\right)\omega_{\mathrm{pi}}^*\right]\omega \right. \\
&\left. + \frac{\beta}{4}\left[\frac{5}{3}(\omega_{\mathrm{ni}} + 2\omega_{\mathrm{pi}}^*)\omega_{\mathrm{ni}} + \left(\frac{5}{3}\beta - 1\right)\omega_{\mathrm{pi}}^{*2}\right]\right\} \qquad (10.61)
\end{aligned}
$$

$$
\begin{aligned}
R(\omega) = &\frac{10}{3}\omega^2 + 2s\left(\omega_{\mathrm{pi}}^* - \frac{5}{3}\omega_{\mathrm{ni}}\right)\omega + \frac{25}{18}\beta^2\omega_{\mathrm{pi}}^{*2} \\
&+ \frac{5}{3}(1+s)\left(\omega_{\mathrm{pi}}^* - \frac{5}{3}\omega_{\mathrm{ni}}\right)\beta\omega_{\mathrm{pi}}^* - 2\nu_{\mathrm{ie}}\Delta + i\Delta \\
&\times \left[\frac{8}{3}\omega - \left(\omega_{\mathrm{pi}}^* - \frac{5}{3}\omega_{\mathrm{ni}}\right) - \frac{5}{3}\beta\omega_{\mathrm{pi}}^*\right] + 2i\nu_{\mathrm{ie}} \\
&\times \left[\frac{10}{3}\omega + 2s\left(\omega_{\mathrm{pi}}^* - \frac{5}{3}\omega_{\mathrm{ni}}\right)\right] + \frac{\nu_{\mathrm{ie}}\Delta M_{\mathrm{i}}}{k_z^2 T} \\
&\times (\omega - \omega_{\mathrm{Di}})\left[\left(1 + \frac{\beta}{2}\right)\omega + \frac{\beta}{2}\omega_{\mathrm{pi}}^* + \omega_{\mathrm{ni}}\right]. \qquad (10.62)
\end{aligned}
$$

When $k_z = 0$, from (10.60) we obtain (10.26) as well as (10.33) with $k_\perp^2 \to 0$ and the equation for the *transverse ion magneto-drift oscillation branch*

$$\omega = \omega_{\mathrm{Di}}. \qquad (10.63)$$

When $k_z \neq 0$ but $k_z^2 v_{T\mathrm{i}}^2/\omega^2 \to 0$ and the heat transfer is neglected, $\nu_{\mathrm{ie}} \to 0$, then from equation (10.60) we find (10.38) and (10.63).

When $\Delta \to \infty$ and $\nu_{\mathrm{ie}} \to 0$, (10.60) reduces to (10.48) corresponding to the perturbations with $k_z \simeq \omega/v_{T\mathrm{i}}$ in the *weakly collisional plasma regime*.

Finally, when one takes $\nu_{ie} \to \infty$, $\Delta \to 0$ in (10.60), we arrive at the following dispersion relation for the *strongly collisional plasma*

$$\omega(\omega - \omega_{\mathrm{Di}})[\omega + s\omega_{ne} - (1+s)\omega_{pe}^*]$$
$$- \frac{k_z^2 v_{Te}^2}{1 + 5\beta/6}\left[\frac{5}{3}\omega + s\left(\omega_{pi}^* - \frac{5}{3}\omega_{ni}\right)\right] = 0. \qquad (10.64)$$

This equation has been found in [10.4] for $\nabla T = 0$ and in [10.3] for arbitrary ∇T. We shall use it in sections 11.4 and 12.4.

10.4 Dispersion relations for inertialess perturbations in a plasma flow with an inhomogeneous velocity profile

10.4.1 Starting equations

As before, we describe each particle species (ions and electrons) by their density n, velocity $\boldsymbol{V}$ and pressure p. In addition, the viscosity tensor $\boldsymbol{\pi}$ and the heat flux $\boldsymbol{q}$ are taken into account for the ions. The densities and the velocities of the particles are determined by the equations of continuity and motion (9.19) and (9.20) with $\boldsymbol{R} = 0$.

We assume the electron heat conductivity to be infinite so that the electron heat-balance equation has the form of (10.39). The ion heat-balance equation is written in the form (cf. (9.21))

$$\frac{dp_i}{dt} - \gamma_0 T_i \frac{dn}{dt} + (\gamma_0 - 1)\mathrm{div}\,\boldsymbol{q}_{\|} + \gamma_0 \mathrm{div}\,\boldsymbol{q}_{\perp} = 0 \qquad (10.65)$$

where the ion adiabatic exponent γ_0 is unspecified as yet. For such a description, the vector $\boldsymbol{q}_{\perp}$ means (cf. (9.33))

$$\boldsymbol{q}_{\perp} = \frac{p_i}{M_i \Omega_i}\boldsymbol{e}_0 \times \nabla T_i \qquad (10.66)$$

whilst $\boldsymbol{q}_{\|} = \boldsymbol{e}_0 q_{\|}$ is the longitudinal heat flux which in the case of a collisionless plasma is defined by the equation

$$\frac{dq_{\|}}{dt} = -\frac{\boldsymbol{e}_0}{Mn} \cdot (p\nabla \cdot \boldsymbol{\pi} - \boldsymbol{\pi} \cdot \nabla p). \qquad (10.67)$$

This equation can be constructed using the first equation of (9.26) augmented by the term with dq/dt.

We use (10.67) when the gradient of the equilibrium pressure is non-essential. Otherwise, (10.67) should be replaced by a more complex relation.

In section 9.3 we have given the expressions for the viscosity tensor components $\pi_{\alpha\beta}$ in the reference frame for which one of the axes is directed along the magnetic field (see (9.30)). However, for subsequent calculations we require $\pi_{\alpha\beta}$ in an arbitrary reference frame. According to [9.3, 9.4], neglecting the dissipative terms, the required form of $\pi_{\alpha\beta}$ is

$$\pi_{\alpha\beta} = \frac{p}{2\Omega}(W_{3\alpha\beta} + 2W_{4\alpha\beta}) \qquad (10.68)$$

where

$$\begin{aligned}
W_{3\alpha\beta} &= \frac{1}{2}(\delta^\perp_{\alpha\mu}\epsilon_{\beta\gamma\nu} + \delta^\perp_{\beta\nu}\epsilon_{\alpha\gamma\mu})h_\gamma W^{(1)}_{\mu\nu} \\
W_{4\alpha\beta} &= (h_\alpha h_\mu \epsilon_{\beta\gamma\nu} + h_\beta h_\nu \epsilon_{\alpha\gamma\mu})h_\gamma W^{(1)}_{\mu\nu}.
\end{aligned} \qquad (10.69)$$

Here $W^{(1)}_{\alpha\beta}$ is defined by (9.25), $\delta^\perp_{\alpha\beta} = \delta_{\alpha\beta} - h_\alpha h_\beta$, and $\boldsymbol{h} \equiv \boldsymbol{e}_0$ is, as above, the unit vector along the magnetic field.

10.4.2 *Inertialess perturbations in a plasma flow with* $\nabla n_0 = \nabla T_0 = 0$

In this section we assume $\nabla n_0 = \nabla T_0 = 0$. Therefore, according to (1.42), $\nabla B_0 = 0$. With these assumptions the equilibrium velocity of the plasma components (ions and electrons) across the magnetic field vanishes, so $\boldsymbol{V}_0 = \boldsymbol{e}_z U$. The velocity U is assumed to be dependent upon x, $dU/dx \neq 0$.

It follows from (10.68) that at $dU/dx \neq 0$ the equilibrium values of the viscosity-tensor components, π_{yz} and π_{zy}, differ from zero

$$\pi_{0zy} = \pi_{0yz} = \frac{p_{0i}}{\Omega_i}\frac{dU}{dx}. \qquad (10.70)$$

Linearizing the equations of motion (9.20), for each particle species we obtain (cf. (9.32))

$$\begin{aligned}
\tilde{V}_x &= \frac{c}{B_0}\left(E_y + \frac{U}{c}\tilde{B}_x - \frac{ik_y\tilde{p}}{en_0}\right) \\
\tilde{V}_y &= \frac{c}{B_0}\left(-E_x + \frac{U}{c}\tilde{B}_y + \frac{1}{en_0}\frac{\partial\tilde{p}}{\partial x}\right).
\end{aligned} \qquad (10.71)$$

When $\nabla T_0 = 0$, linearization of (10.66) yields simply

$$\tilde{q}_\perp = 0. \qquad (10.72)$$

Allowing for (10.71) and (10.72), the linearized ion equations of continuity (9.19) and heat balance (10.65) take the form (cf. (10.1),

(10.2))

$$-\tilde{\omega}\frac{\tilde{n}}{n_0} + \tilde{\omega}\frac{\tilde{B}_z}{B_0} - \frac{i\tilde{B}_x}{B_0}\frac{dU}{dx} + k_z\tilde{V}_z = 0$$

$$-\tilde{\omega}\left(\frac{\tilde{p}_i}{p_{0i}} - \gamma_0\frac{\tilde{n}}{n_0}\right) + k_z(\gamma_0 - 1)\frac{\tilde{q}_z}{p_{0i}} = 0 \qquad (10.73)$$

where $\tilde{\omega} = \omega - k_z U$. According to (9.20) and (10.67), $\tilde{V}_z$ and $\tilde{q}_z$ satisfy the equations

$$M_i n_0\left(-i\tilde{\omega}\tilde{V}_z + \tilde{V}_x\frac{dU}{dx}\right) + \boldsymbol{e}_z \cdot \nabla \cdot \tilde{\pi} = -ik_z\tilde{p}_i + en_0 E_z$$

$$-i\tilde{\omega}\tilde{q}_z = -\frac{1}{M_i n_0}\left(p_{0i}\boldsymbol{e}_z \cdot \nabla \cdot \tilde{\pi} - ik_y\pi_{0zy}\tilde{p}_i\right). \qquad (10.74)$$

As follows from (10.68), the perturbed viscosity-tensor components contained in these equations are equal to (note—only the essential terms are written)

$$\tilde{\pi}_{zx} = -\frac{ip_{0i}}{\Omega_i}k_y\tilde{V}_z$$

$$\tilde{\pi}_{zy} = \frac{p_{0i}}{\Omega_i}\left[\frac{\partial\tilde{V}_z}{\partial x} + \frac{dU}{dx}\left(\frac{\tilde{p}_i}{p_{0i}} - \frac{\tilde{B}_z}{B_0}\right)\right] \qquad (10.75)$$

and $\tilde{\pi}_{zz} = 0$, while π_{0zy} is determined by (10.70). Using these expressions, we obtain the relations

$$M_i n_0\tilde{V}_x\frac{dU}{dx} + \boldsymbol{e}_z \cdot \nabla \cdot \tilde{\pi} = \frac{M_i n_0}{B_0}\frac{dU}{dx}\left(cE_y + U\tilde{B}_x - ik_y\frac{T_{0i}}{M_i\Omega_i}\tilde{B}_z\right)$$

$$\boldsymbol{e}_z \cdot \nabla \cdot \tilde{\pi} - ik_y\pi_{0zy}\frac{\tilde{p}_i}{p_{0i}} = -\frac{ik_y p_{0i}\tilde{B}_z}{\Omega_i B_0}\frac{dU}{dx}.$$

(10.76)

Note that in the first equation of (10.76) the terms with $\tilde{p}_i dU/dx$, contained in the product $\tilde{V}_x dU/dx$ and in the divergence of the viscosity tensor, are *mutually compensated*.

Allowing for (10.76), it follows from (10.74) that

$$\tilde{V}_z = \frac{i}{k_z}\frac{\tilde{B}_x}{B_0}\frac{dU}{dx} + \frac{1}{M_i\tilde{\omega}}\left(ieE_z(1-\alpha_i) + \frac{k_z\tilde{p}_i}{n_0} - \alpha_i k_z T_{0i}\frac{\tilde{B}_z}{B_0}\right)$$

$$\tilde{q}_z = -\alpha_i\frac{k_z T_{0i}p_{0i}}{M_i\tilde{\omega}}\frac{\tilde{B}_z}{B_0}$$

(10.77)

where

$$\alpha_i = \frac{k_y}{k_z \Omega_i} \frac{dU}{dx}. \tag{10.78}$$

Substitution of (10.77) into (10.73) yields

$$-\frac{\tilde{n}}{n_0} + \frac{k_z^2 T_{0i}}{M_i \tilde{\omega}^2} \frac{\tilde{p}_i}{p_{0i}} + \left(1 - \alpha_i \frac{k_z^2 T_{0i}}{M_i \tilde{\omega}^2}\right) \frac{\tilde{B}_z}{B_0} + \frac{i e k_z E_z}{M_i \tilde{\omega}^2}(1 - \alpha_i) = 0$$

$$\frac{\tilde{p}_i}{p_{0i}} - \gamma_0 \frac{\tilde{n}}{n_0} + (\gamma_0 - 1)\alpha_i \frac{k_z^2 T_{0i}}{M_i \tilde{\omega}^2} \frac{\tilde{B}_z}{B_0} = 0. $$

$$\tag{10.79}$$

Note that the term with $\tilde{B}_z$ in the second equation of (10.79) is due to the longitudinal heat flux $\tilde{q}_z$. Neglecting $\tilde{q}_z$, from the second equation of (10.79) the trivial result $\tilde{p}_i = \gamma_0 T_0 \tilde{n}$ should follow.

We supplement (10.79) by the pressure-balance equation (10.3) and by the relations

$$\tilde{T}_e = 0$$
$$-i k_z T_{0e} \tilde{n} + e_e n_0 E_z = 0 \tag{10.80}$$

following from (10.39) and (9.20).

For the case where $\tilde{\omega}^2 \gg k_z^2 T_{0i}/M_i$, the dispersion relation follows from the above equations:

$$1 + \frac{\beta_i \gamma_0 + \beta_e}{2} + \alpha_i \frac{k_z^2 T_{0i}}{M_i \tilde{\omega}^2}\left(\frac{T_{0e}}{T_{0i}} - \gamma_0 \beta_i + \frac{\beta_i - \beta_e}{2}\right)$$

$$- \frac{\beta_e}{2}(\gamma_0 - 1)\left(\alpha_i \frac{k_z^2 T_{0i}}{M_i \tilde{\omega}^2}\right)^2 = 0. \tag{10.81}$$

Here, as usual, $\beta_{i,e} = 8\pi n_0 T_{0i,e}/B_0^2$.

We shall use equation (10.81) in section 13.5.1.

10.4.3 Allowance for the density gradient

Let us supplement the results of section 10.4.2 by the terms with ∇n_0. For finite β, according to (1.42), it is also necessary to take into account ∇B_0 in addition to ∇n_0. Describing the ions, we shall follow the isothermal approximation, $\gamma_0 = 1$, neglecting the ion temperature perturbations, $\tilde{T}_i = 0$, and the ion heat flux, $q_i = 0$.

By analogy with (10.73), we reduce the continuity equation to the form

$$-(\tilde{\omega} - \omega_{Di})\hat{n} + (\tilde{\omega} - \omega_{ni})\hat{B} - i e E_z(\omega_{ni} - \omega_{Di})/k_z T_{0i} + k_z \hat{V}_z = 0. \tag{10.82}$$

Here new variables are introduced:

$$\hat{n} = \frac{\tilde{n}}{n_0} - \frac{i\kappa_n}{k_z}\frac{\tilde{B}_x}{B_0}$$

$$\hat{B} = \frac{\tilde{B}_z}{B_0} - \frac{i\kappa_B}{k_z}\frac{\tilde{B}_x}{B_0} \tag{10.83}$$

$$\hat{V}_z = \tilde{V}_z - \frac{i}{k_z}\frac{dU}{dx}\frac{\tilde{B}_x}{B_0}.$$

The ion longitudinal motion equation now yields

$$(\tilde{\omega} - \omega_{\mathrm{Di}})\hat{V}_z = \frac{ie}{M_{\mathrm{i}}}E_z(1 - \alpha_{\mathrm{i}}) + \frac{k_z^2 T_{0\mathrm{i}}}{M_{\mathrm{i}}}\hat{n} - \frac{\alpha_{\mathrm{i}}k_z T_{0\mathrm{i}}}{M_{\mathrm{i}}}\hat{B}. \tag{10.84}$$

We supplement (10.82) and (10.84) by the equations

$$\hat{B} + \beta\hat{n}/2 = 0$$
$$e_{\mathrm{e}}E_z - ik_z T_{0\mathrm{e}}\hat{n} = 0 \tag{10.85}$$

which are similar to (10.3) and (10.80).

By means of (10.82)–(10.85) we obtain the dispersion relation

$$\left(1 + \frac{\beta}{2}\right)(\tilde{\omega} - \omega_{n\mathrm{e}})\left(\tilde{\omega} - \frac{\beta}{2}\frac{T_{0\mathrm{i}}}{T_{0\mathrm{e}}}\omega_{n\mathrm{e}}\right)$$
$$- \frac{k_z^2 T_{0\mathrm{e}}}{M_{\mathrm{i}}}\left[1 + \frac{T_{0\mathrm{i}}}{T_{0\mathrm{e}}} - \alpha_{\mathrm{i}}\left(1 - \frac{\beta}{2}\frac{T_{0\mathrm{i}}}{T_{0\mathrm{e}}}\right)\right] = 0. \tag{10.86}$$

We shall consider equation (10.86) in section 13.5.3.

10.5 Description of Alfven perturbations in a plasma flow with an inhomogeneous velocity profile

To obtain equations for the *Alfven perturbations* in a *plasma flow with an inhomogeneous velocity profile* it is necessary to add one more independent equation to the equations of section 10.4. We take the inertial equation (9.58) as such an equation.

10.5.1 Alfven perturbations when neglecting drift effects
Let us neglect the viscosity in (9.58). Calculating the perturbed velocities (see (10.71)), we neglect the ion perturbed pressure so that

$$\boldsymbol{V}_\perp = c\boldsymbol{E} \times \boldsymbol{e}_z/B_0 + U\tilde{\boldsymbol{B}}_\perp/B_0. \tag{10.87}$$

We neglect the last term in the derivative $\mathrm{d}/\mathrm{d}t \equiv \partial/\partial t + U\partial/\partial z + \boldsymbol{V}_\perp \cdot \nabla$ as being of the order of ω^*/ω. In addition, we neglect the perturbation $\tilde{B}_z$ so that $\tilde{\boldsymbol{B}} = [\tilde{B}_x, (\mathrm{i}/k_y)\partial\tilde{B}_x/\partial x, 0]$. The equilibrium magnetic field is assumed to be uniform, $\nabla B_0 = 0$. Finally, assuming $E_z = 0$, we express $\boldsymbol{E}_\perp$ in terms of $\tilde{B}_x$:

$$E_x = \mathrm{i}\frac{\omega}{ck_y k_z}\frac{\partial\tilde{B}_z}{\partial x}$$
$$E_y = -\frac{\omega}{ck_z}\tilde{B}_x. \tag{10.88}$$

As a result of this, we obtain that (9.58) reduces to

$$\nabla_\perp \cdot [(\tilde{\omega}^2 - c_A^2 k_z^2)\nabla_\perp\tilde{B}_x] = 0. \tag{10.89}$$

This is the starting equation for the *ordinary Kelvin–Helmholtz instability*. We shall examine (10.89) in section 13.1.

10.5.2 Magnetic viscosity tensor for $\mathrm{d}U/\mathrm{d}x \neq 0$, with $\tilde{B}_x \neq 0$

To add the drift effects to (10.89) it is necessary to have expressions for $\pi_{xx}, \pi_{yy}, \pi_{xy} = \pi_{yx}, \pi_{zx}, \pi_{yz}$. The two last quantities are given by (10.75) and (10.70). We find the remaining components of the viscosity tensor by linearizing (10.68) with respect to h_x and h_y:

$$\pi_{xx} = -\frac{p}{2\Omega}\left(\frac{\partial\tilde{V}_x}{\partial y} + \frac{\partial\tilde{V}_y}{\partial x} - \frac{\tilde{B}_y}{B_0}\frac{\mathrm{d}U}{\mathrm{d}x}\right)$$
$$\pi_{yy} = -\pi_{xx} + 2\frac{p}{\Omega}\frac{\tilde{B}_y}{B_0}\frac{\mathrm{d}U}{\mathrm{d}x} \tag{10.90}$$
$$\pi_{xy} = \pi_{yx} = \frac{p}{2\Omega}\left(\frac{\partial\tilde{V}_x}{\partial x} + \frac{\tilde{B}_x}{B_0}\frac{\mathrm{d}U}{\mathrm{d}x} - \frac{\partial\tilde{V}_y}{\partial y}\right).$$

Note that to obtain $\tilde{\boldsymbol{B}}_\perp\,\mathrm{d}U/\mathrm{d}x$ it is necessary to use (10.68) and (10.69) but not (9.57). In [10.8], equation (9.57) has been used incorrectly.

10.5.3 Allowance for the drift effects for $\nabla n_0 = \nabla T_0 = 0$

Let us reduce (9.58) using the fact that the variables $\hat{n}$, $\hat{B}$, and $\hat{V}_z$ defined by (10.83) vanish in the case of Alfven perturbations. As in section 10.4.3, we assume $\nabla T_0 = 0$. In addition, we take $\nabla n_0 = 0$. Then, according to (10.83)

$$\tilde{n} = \tilde{B}_z = 0$$
$$\tilde{V}_z = \frac{\mathrm{i}}{k_z}\frac{\tilde{B}_x}{B_0}\frac{\mathrm{d}U}{\mathrm{d}x}. \tag{10.91}$$

In this case p and Ω in (10.90) should be considered to be equal to their equilibrium values independent of the coordinates. The terms with $\tilde{p}_i$ and $\tilde{B}_z$ should now be omitted in (10.75). In (10.71), as in section 10.5.1, we omit the term with $\tilde{p}$ and express $\tilde{E}_x$, $\tilde{E}_y$, and $\tilde{B}_y$ in terms of $\tilde{B}_x$ by means of equations (10.88) and of the relation $\tilde{B}_y = (\mathrm{i}/k_y)\partial\tilde{B}_x/\partial x$. Then, as in section 10.5.1, we have

$$\tilde{V}_x = -\frac{\tilde{\omega}}{k_z}\frac{\tilde{B}_x}{B_0}$$
$$\tilde{V}_y = -\frac{\mathrm{i}\tilde{\omega}}{k_y k_z B_0}\frac{\partial\tilde{B}_x}{\partial x}. \tag{10.92}$$

Using (10.75) and (10.90), we find

$$\mathrm{curl}_z \nabla \cdot \boldsymbol{\pi} = \frac{p_{0\mathrm{i}}}{2\Omega_\mathrm{i}}\left[\nabla_\perp^2 \nabla \cdot \tilde{\boldsymbol{V}}_\perp + \frac{1}{B_0}\frac{\partial}{\partial x}\left(\tilde{B}_x \frac{\mathrm{d}^2 U}{\mathrm{d}x^2} - \frac{\partial \tilde{B}_x}{\partial x}\frac{\mathrm{d}U}{\mathrm{d}x}\right)\right.$$
$$\left. + k_y^2 \frac{\tilde{B}_x}{B_0}\frac{\mathrm{d}U}{\mathrm{d}x}\right] + \frac{\mathrm{i}k_z p_{0\mathrm{i}}}{\Omega_\mathrm{i}}\nabla_\perp^2 \tilde{V}_z. \tag{10.93}$$

Further transformation of (9.58) using the above relations leads to

$$\nabla_\perp \cdot [(\tilde{\omega}^2 - c_\mathrm{A}^2 k_z^2 + \alpha_\mathrm{i} k_z^2 T_{0\mathrm{i}}/M_\mathrm{i})\nabla_\perp \tilde{B}_x] = 0 \tag{10.94}$$

where α_i is defined by (10.78).

Equation (10.94) will be used in the analysis of section 13.3.

10.5.4 *Dispersion relation of small-scale Alfven perturbations for* $\nabla n_0 = \nabla T_0 = 0$

When $|\partial\ln\tilde{B}_x/\partial x| \gg |\mathrm{d}\ln U/\mathrm{d}x|$, the dispersion relation follows from (10.94):

$$\tilde{\omega}^2 - c_\mathrm{A}^2 k_z^2 + \alpha_\mathrm{i} k_z^2 T_{0\mathrm{i}}/M_\mathrm{i} = 0. \tag{10.95}$$

Exactly the same dispersion relation is also obtained in the *kinetic approach* (see (2.5), (2.7) and the first equation of (2.26)).

The consequences following from (10.95) will be considered in section 13.4.1.

10.5.5 *Dispersion relation of small-scale Alfven perturbations for* $\nabla n_0 \neq 0$, *with* $\nabla T_0 = 0$

Restricting ourselves to the case of *small-scale perturbations*, we reduce (9.58) to (10.42). In contrast to section 10.5.4, we now consider $\nabla n_0 \neq 0$ and $\nabla B_0 \neq 0$. By analogy with section 10.5.3, we find that the right-hand side of (10.93) is augmented by the term $\delta\{\mathrm{curl}_z \nabla \cdot \boldsymbol{\pi}\}$, where

$$\delta\{\mathrm{curl}_z \nabla \cdot \boldsymbol{\pi}\} = \frac{\mathrm{d}}{\mathrm{d}x}\left(\frac{p_{0i}}{\Omega_i}\right)\nabla_\perp^2 \tilde{V}_x. \qquad (10.96)$$

The procedure for obtaining the dispersion relation by means of (10.42) reduces to expressing all the perturbed quantities in terms of $\tilde{B}_x$ making allowance for the quantities (10.83) which vanish, i.e., (cf. (10.10), (10.91))

$$\frac{\tilde{n}}{n_0} = \frac{\mathrm{i}\kappa_n}{k_z}\frac{\tilde{B}_x}{B_0}$$

$$\frac{\tilde{B}_z}{B_0} = \frac{\mathrm{i}\kappa_B}{k_z}\frac{\tilde{B}_x}{B_0} \qquad (10.97)$$

$$\tilde{V}_z = \frac{\mathrm{i}}{k_z}\frac{\mathrm{d}U}{\mathrm{d}x}\frac{\tilde{B}_x}{B_0}.$$

Then the part of $\mathrm{curl}_z \nabla \cdot \boldsymbol{\pi}$ given by the right-hand side of (10.93) is an unchanged function of $\tilde{B}_x$ when $\nabla n_0 \neq 0, \nabla B_0 \neq 0$, which can be proved by means of the expression for the perturbed velocity

$$\tilde{\boldsymbol{V}}_\perp = c\left(\frac{\boldsymbol{E}}{B_0} - \frac{T_0 \nabla n}{enB}\right) \times \boldsymbol{e}_z + U\frac{\tilde{\boldsymbol{B}}_\perp}{B_0} - \boldsymbol{e}_y V_n \qquad (10.98)$$

and by equations (10.97). (Here, as above, $V_n = cT_0\kappa_n/eB_0$.) The right-hand side of (10.96) is expressed in terms of $\tilde{B}_x$ using (10.97), (10.98) and (10.88). Then

$$\tilde{V}_x = -(\tilde{\omega} - \omega_{ni})\tilde{B}_x/k_z B_0. \qquad (10.99)$$

Finally, the quantity $\tilde{V}_y$ contained in (10.42) under the operator $\mathrm{d}/\mathrm{d}t = -\mathrm{i}(\tilde{\omega} - \omega_{ni})$ can be replaced by $(\mathrm{i}/k_y)\partial\tilde{V}_x/\partial x$ with the help of the approximate condition $\mathrm{div}\,\tilde{\boldsymbol{V}} = 0$. As a result, from (10.42) we find the dispersion relation

$$(\tilde{\omega} - \omega_{ni})(\tilde{\omega} - \omega_{Di}) - c_A^2 k_z^2 + \alpha_i k_z^2 T_{0i}/M_i = 0. \qquad (10.100)$$

This equation coincides with that obtained by the *kinetic method* (cf. (2.5), (2.7), (2.9), and (2.26)). We shall use (10.100) in section 13.4.2.

11 Instabilities in a Collisional Finite-β Plasma

Using the results of Chapters 9 and 10 as starting equations, we shall now analyse the low-frequency long-wavelength instabilities in a collisional finite-β plasma. This analysis will also be continued in Chapter 12 where the material related to a plasma with $\beta \gg 1$ and $|\,\partial \ln p/\partial \ln n\,| \ll 1$ will be presented, and in Chapter 13 where a plasma flow with an inhomogeneous velocity profile will be discussed.

According to Chapters 9 and 10, the instabilities which are of interest to us, as in the case of a collisionless plasma, are associated with two types of perturbations, namely, inertialess and inertial ones. The inertial (Alfven) perturbations are less sensitive to the degree of plasma collisionality than inertialess ones. Therefore, the analysis of *inertialess instabilities* occupies the majority of this chapter (sections 11.1–11.4). In section 11.1 we assume $k_z = 0$, corresponding to the case of *magneto-drift entropy waves*. In section 11.2 we study perturbations with $k_z \simeq (\nu_e \omega)^{1/2}/v_{Te}$, i.e., *oblique inertialess perturbations* under conditions where the finiteness of the *electron heat conductivity* is important. Sections 11.3 and 11.4 are devoted to inertialess instabilities with $k_z \simeq \omega/v_{Ti}$, i.e., to those for which the *ion longitudinal motion* is important. The *plasma* is considered to be *weakly collisional* in section 11.3 and *strongly collisional* in section 11.4, in the meaning given in Chapters 9 and 10. The *Alfven (inertial) instabilities in the collisional regime* are discussed in section 11.5. There we deal with only the *dissipative effects* of order $k_\perp^2 \rho_i^2 \nu_i/\omega$ since, as mentioned in Chapter 10, the terms describing the remaining dissipative effects are not contained in the dispersion relations for these instabilities.

Regarding the classification of the types of perturbations discussed, sections 11.1–11.4 are similar to Chapter 3 where the inertialess perturbations in a collisionless plasma have been considered, whereas section 11.5, similar to Chapter 4, is devoted to inertial perturbations in such a plasma.

Comparing the collisionless and collisional regimes, one can find what properties the perturbations in these regimes have in common. It is clear that the collisionless regime should first of all be compared with the *weakly collisional regime*. This is just the aspect which we shall consider below. It is also of specific interest to compare the results of the weakly and strongly collisional regime as discussed below.

According to section 11.1.1, in a weakly collisional plasma there are two branches of the inertialess magneto-drift entropy oscillations which grow when the *temperature gradient* is compared with the density gradient. In this respect we have an obvious analogy with *Tserkovnikov's magneto-drift instabilities* in a collisionless plasma discussed in section 3.1. As for the *dissipative effects* of order ω/ν_i and $(k_\perp \rho_i)^2 \nu_i/\omega$ discussed in sections 11.1.2 and 11.1.3, they cause the *entropy-wave instability* even for $\nabla T = 0$. This is evidence of the fact that a collisional plasma is more unstable than a collisionless one. At the same time, according to section 11.1.4, the entropy waves, as well as Tserkovnikov's waves, exist only for not too a large collision frequency, $\nu_{ie} \lesssim \omega$ (*the weakly collisional regime*). In the opposite case, when $\nu_{ie} > \omega$, we have *two essentially dissipative oscillation branches* instead of the entropy waves. One of those branches is excited under the same temperature gradients as the entropy waves, but its growth rate becomes small as the ratio ω/ν_{ie}. Note that such a sensitivity of the entropy waves to the degree of collisionality is explained by the fact that, when $\nu_{ie} > \omega$, the collisional heat transfer prevents differences in the perturbed temperatures of different particle species as required for these waves.

Recall that in section 3.2 we examined the *transverse drift waves* in a *collisionless plasma*. However, as follows from section 10.1.4., this oscillation branch exists also in a collisional plasma. In section 3.2 we allowed for the small terms of order M_e/M_i and observed the transverse drift-wave instability with the growth rate of the above-mentioned order for $\nabla T = 0$. That instability has been of interest since all the remaining oscillation branches become stable for $\nabla T = 0$. However, as is clear from the above, in a collisional plasma there are oscillation branches which are unstable due to dissipative effects even for $\nabla T = 0$. Therefore, in such a plasma the effects of order M_e/M_i are not of interest, and we do not consider them.

As mentioned in section 3.3, when k_z increases to magnitudes of order ω/v_{Te} and higher, in a *collisionless plasma*, first, the inertialess magneto-drift branches become coupled with the transverse drift waves, and, second, one of the magneto-drift branches disappears due to the *broadening* of the *electron magneto-drift resonance* by the electron longitudinal thermal motion. As a result, in that case we had two branches of inertialess perturbations: the *electron drift branch* and the single *magneto-drift branch*. Those branches can be excited only in the presence of a temperature gradient. The analysis of section 11.2 shows that similar modification of the inertial perturbation branches also holds in a *collisional plasma*. However, some differences are now revealed. One of them concerns the *characteristic longitudinal wave number* for which the above modification occurs. This wave number becomes of order $k_z \simeq (\omega \nu_e)^{1/2}/v_{Te}$ which is essentially larger than that in the case of a collisionless plasma. The given k_z increases with increasing collision frequency. When $\nu_{ie} \simeq \omega$, i.e., just as under the condition when the *heat transfer* is important (see above), this k_z becomes of order ω/v_{Ti}. In this case the *ion longitudinal motion* is important. At the same time, the condition $\nu_{ie} \simeq \omega$ corresponds to the *strongly collisional plasma regime*. Therefore, the above-mentioned analogy in the behaviour of the oscillation branches concerns only the *weakly collisional plasma regime*.

Another essential difference in the behaviour of the inertialess branches with $k_z \neq 0$ in the collisional and collisionless regimes is associated with difference of the temperature gradient ranges for which these branches are unstable. In particular, the perturbations with such k_z in the collisional plasma become unstable even for $\nabla T = 0$. Such an instability is due to the *finite electron longitudinal heat conductivity*. Therefore, although the effect of the electron longitudinal heat conductivity is similar to that of the electron longitudinal thermal motion, there is no complete analogy here.

The analogy between properties of the inertialess perturbations in the collisionless and collisional plasma is re-established in the limit of *infinitely large electron heat conductivity*. This approximation can be used, in particular, in the case where $k_z \simeq \omega/v_{Ti}$ discussed in section 11.3. It then appears that, if $\nabla T = 0$, both a weakly collisional plasma and a collisionless one are stable against the perturbations considered.

Summarizing, we conclude that *inertialess perturbations in a weakly collisional plasma* (when neglecting dissipative effects) are very similar to those in a collisionless plasma. Besides the rest, this analogy is also of methodological interest since, if necessary, a more complicated kinetic description can be replaced by a hydrodynamic one without an essential loss of information. However, there is no complete analogy between dissipative effects in a collisional plasma and kinetic resonant effects (collisionless dissipation) in a collisionless one.

The *inertialess perturbations* in a *strongly collisional plasma* have essentially different properties. According to section 11.4, they are excited for all β even in the case where $\nabla T = 0$. This can be considered as one more evidence of the fact that the collisions as a whole favour a destabilization of the inertialess perturbations.

The role of collisions is more passive in the *inertial perturbations.* According to section 11.5, when $\nabla T = 0$, the dissipative effects do not result in their excitation for all β. It should also be remembered that, as shown in Chapter 4, when $\nabla T = 0$, the inertial perturbations are stable also in the absence of collisions.

The above remarks do not concern force-free or almost force-free high-β plasmas which will be discussed in Chapter 12.

The results presented in this chapter have been obtained in [2.1, 2.2, 3.1, 9.1, 9.9, 10.3–10.5, 11.1–11.3]. The specific contributions of these papers will be mentioned when considering the respective problems.

Non-linear effects related to the corresponding types of perturbations were considered in [10.7, 11.4–11.6].

As a rule, in the present chapter and in Chapters 12 and 13 we omit the subscript zero in the equilibrium number density and magnetic field, i.e., $n_0 \rightarrow n, B_0 \rightarrow B$.

11.1   Magneto-drift entropy instabilities

11.1.1 Entropy perturbations when neglecting dissipative effects
According to section 10.1.1, the *entropy perturbations* when neglecting dissipative effects are described by dispersion relation (10.6). From (10.6), for $T_e = T_i \equiv T$ the *instability criterion* follows [3.1, 2.1]

$$\frac{\partial \ln T}{\partial \ln p} > 1 - \frac{1}{2\gamma_0} + \frac{\beta}{4} \equiv \frac{7}{10} + \frac{\beta}{4}. \qquad (11.1)$$

In particular, when $\beta \ll 1$, this means that a *plasma is unstable* if

$$\eta > \frac{7}{3} \qquad \text{or} \qquad \eta < -1. \qquad (11.2)$$

The growth rate is of order

$$\gamma \simeq \omega_{\mathrm{D}}. \qquad (11.3)$$

It can be seen that a collisional plasma, as in the case of a collisionless one (cf. section 3.1), is unstable when the temperature gradient is greater than the density gradient.

It also follows from (10.6) that the most unstable plasma is that with $T_e = T_i$. When the difference between the electron and ion temperatures is sufficiently large, the plasma is stable.

When condition (11.1) is not satisfied, the frequencies of the perturbations of type (10.6) are real. For $\beta \ll 1$ they are defined by the relation

$$\omega^{(\pm)} = \pm\omega_{\mathrm{Di}}\left[\gamma_0\left(2\gamma_0\frac{\kappa_n}{\kappa_p} - 1\right)\right]^{1/2} \tag{11.4}$$

Below we shall discuss the possibility of exciting these oscillation branches by *dissipative effects* of order ω/ν_i (section 11.1.2) and of order $(k_\perp\rho_i)^2\nu_i/\omega$ (section 11.1.3). In section 11.1.4 the *heat transfer* is taken into account.

11.1.2 The role of dissipative effects of order ω/ν_i

The dissipative effects of order ω/ν_i (including the *gyro-relaxation effect*) in the perturbations of type (11.4), are described by dispersion relation (10.19). Using (10.19), we find that the growth rates of the branches (11.4) are defined by the expression [9.9]

$$\gamma^{(\pm)} = -\frac{0.96}{20}\frac{\omega_{\mathrm{Di}}^2}{\nu_i}\left\{3.78\frac{\kappa_n}{\kappa_p} + 17.89 \pm \left[\gamma_0\left(2\gamma_0\frac{\kappa_n}{\kappa_p} - 1\right)\right]^{1/2}\right.$$

$$\left. \times \left[2.69\frac{\kappa_n}{\kappa_p} + 16.71 + \left(22.55\frac{\kappa_n}{\kappa_p} - 0.59\right)\left(2\gamma_0\frac{\kappa_n}{\kappa_p} - 1\right)^{-1}\right]\right\}. \tag{11.5}$$

In particular, when $\nabla T = 0(\kappa_n/\kappa_p = 1)$, it follows from (11.5) that the $\omega^{(-)}$ branch is excited with growth rate

$$\gamma^{(-)} \approx \frac{1}{20}\frac{\omega_{\mathrm{Di}}^2}{\nu_i}. \tag{11.6}$$

Instability takes place also in any range of positive values of κ_n/κ_p. In particular, when $\kappa_n/\kappa_p \gg 1, (\eta \approx -1)$, we obtain from (11.5) that

$$\gamma^{(\pm)} = \mp 0.3(\kappa_n/\kappa_p)^{3/2}\omega_{\mathrm{Di}}^2/\nu_i. \tag{11.7}$$

As in the case where $\nabla T = 0$, the branch $\omega = \omega^{(-)}$ is *unstable*.

11.1.3 The role of dissipative effects of order $(k_\perp\rho_i)^2\nu_i/\omega$

In finding out the role of dissipative effects of order $(k_\perp\rho_i)^2\nu_i/\omega$ in the *entropy perturbations*, we start with (10.22). As a result, we obtain that, when $\beta \ll 1$ and $\nabla T = 0$, the growth rates of these perturbations are [9.1]

$$\gamma^{(\pm)} = \mp(k_\perp\rho_i)^2\nu_i(12 + 2\sqrt{35})/3\sqrt{35}. \tag{11.8}$$

It can be seen that, as in the case of the effects of order ω/ν_i (see section 11.1.2), the branch $\omega = \omega^{(-)}$ is *growing*.

11.1.4 *The role of the heat transfer*
It follows from (10.26) that the heat transfer results in a reduction of the growth rate of the *entropy waves* when $\nu_{ie} \gtrsim \omega_D$. In particular, when $\nu_{ie} \gg \omega_D$, we find from (10.26), [3.1]:

$$\omega = i\frac{\gamma_0 \omega_{Di}^2}{\nu_{ie}}\left(1 + 2\gamma_0 \frac{\kappa_B - \kappa_n}{\kappa_p - \gamma_0 \kappa_B}\right) \tag{11.9}$$

so that

$$\gamma \simeq \omega_D^2/\nu_{ie}. \tag{11.10}$$

Then instability criterion (11.1) holds.

11.2 Inertialess instabilities with $k_z \simeq (\nu_e\omega)^{1/2}/v_{Te}$

The inertial perturbations with finite $k_z v_{Te}/(\nu_e\omega)^{1/2}$ are described by equation (10.38). When $\lambda = 0$ (i.e., $k_z = 0$), this equation *splits into two equations*. One of them is (10.6) (for $T_e = T_i$) which corresponds to the perturbations with $E \perp B_0$. The second has the form of (10.33). It describes the perturbations with $E \parallel B_0$, i.e., the transverse drift waves. Following [10.1], we now consider equation (10.38) for finite λ assuming $\nabla T = 0$. In this case it also splits into two dispersion relations. One of them coincides with (3.29) (see also (3.39)) while the second is of the form [10.1]

$$\left(1 + \frac{\gamma_0\beta}{2}\right)\left\{\omega^2 + \gamma_0\omega_{Di}^2\left[1 - 2\gamma_0\left(1 + \frac{\beta}{2}\right)\left(1 + \frac{\gamma_0\beta}{2}\right)\right]\right\}$$
$$+ i\Delta\left[\omega\left(1 + \frac{2\beta}{3}\right) - 2\omega_{Di}\left(1 + \frac{5\beta}{12}\right)\right] = 0. \tag{11.11}$$

Evidently, (3.29) corresponds to stable oscillations of the *electron drift branch* (cf. section 3.3.1). It can be seen that this branch has the same dispersion law in both collisionless and collisional plasma.
When $\Delta/\omega_{Di} \gg 1$, (11.11) has the solution

$$\omega = 2\omega_{Di}\frac{1 + 5\beta/12}{1 + 2\beta/3}. \tag{11.12}$$

This corresponds to the *magneto-drift branch* similar to that discussed in section 3.3.2.

As a collisional analogue of (3.28) describing the connection between the electron drift and magneto-drift branches for $\nabla T \neq 0$, the following dispersion relation is obtained from (10.38) for $\lambda \to \infty$:

$$\omega^2 \left(1 + \frac{2}{5}\gamma_0\beta\right) - \omega\left[\omega_{ne}\left(1 + \frac{\gamma_0\beta}{2}\right) + \gamma_0\omega_{Di}\left(1 + \frac{\beta}{2}\right)\right]$$
$$+ \gamma_0\omega_{ne}\omega_{Di}\left(\frac{\kappa_n}{2\kappa_p} + 1 - \frac{\kappa_B}{\kappa_n} - \frac{\kappa_p}{2\gamma_0\kappa_B}\right) = 0. \tag{11.13}$$

It follows hence that (3.29) is valid also for $\nabla T \neq 0$ if $\beta \ll 1$; this is in accordance with section 3.3.1. Note also that, according to (11.13), for $\beta \ll 1$ and $\nabla T \neq 0$ the magneto-drift branch has the frequency (cf. (11.12), (3.42), (3.48))

$$\omega = 2\omega_{Di}[1 - \eta/6(1 + \eta)]. \tag{11.14}$$

In section 11.2.1 we consider the instabilities predicted by (10.38) for the case where $\nabla T = 0$, which reduces to the analysis of (11.11). In addition, as in section 3.3.1, we allow for the temperature gradient in the perturbations of type (3.29) and investigate the possibility of their excitation, section 11.2.2.

11.2.1 Magneto-drift instability in a plasma with $\nabla T = 0$
In the case where $\beta \ll 1$ equation (11.11) reduces to [10.1]

$$\omega^2 - \frac{35}{9}\omega_{Di}^2 + i\Delta(\omega - 2\omega_{Di}) = 0. \tag{11.15}$$

For small k_z when $\Delta/\omega_{Di} \ll 1$, it follows from (11.15) that (cf. (11.4)–(11.6))

$$\mathrm{Re}\,\omega = \omega^{(\pm)} = \pm(35)^{1/2}\omega_{Di}/3$$
$$\gamma = \gamma^{(\pm)} \equiv -\frac{\Delta}{2}\left(1 \pm \frac{6}{(35)^{1/2}}\right). \tag{11.16}$$

It can be seen that the root $\omega = \omega^{(-)}$ corresponds to an instability. The growth rate is *numerically small*, $\gamma \approx \Delta/144$.

For large k_z when $\Delta \gg \omega_{Di}$, it follows from (11.15) that unstable perturbations have the frequency and the growth rate defined by (cf. (8.12))

$$\mathrm{Re}\,\omega = 2\omega_{Di}$$
$$\gamma = \omega_{Di}^2/9\Delta. \tag{11.17}$$

The instability described by (11.11) does not hold for too large β. Equating the real and imaginary parts of (11.11) to zero at $\gamma = 0$, we find the *instability boundary*

$$\beta_0 = 2/\gamma_0 \equiv 6/5. \tag{11.18}$$

When $\beta > \beta_0$, the plasma is stable.

It follows from the foregoing analysis that a plasma with $\nabla T = 0$ and finite β is unstable. The instability is due to the fact that in plasma with finite, albeit small, β there is a branch of sufficiently slow oscillations, $\omega/k_y \simeq V_D$, i.e., the *magneto-drift oscillations*, which are excited as a result of the *negative dissipation* related to the *finite electron heat conductivity*. The longitudinal wave number of this instability is of order

$$k_z \simeq (\nu_e \omega_{ne} \beta)^{1/2}/v_{Te}. \tag{11.19}$$

The maximum growth rate is numerically small compared with the frequency

$$\gamma \simeq 0.1 \text{Re}\,\omega \simeq 0.1 \beta (ck_y T/eB) \partial \ln n/\partial x. \tag{11.20}$$

Note that the nature of the excitation is essentially associated with the *temperature perturbations*, whereas the presence of an equilibrium temperature gradient is not required.

11.2.2 Instability of electron drift waves in a plasma with $\nabla T \neq 0$
By analogy with section 3.3.1, we now consider the influence of a small *temperature gradient* on the branch (3.29). Assuming $|\eta| \ll 1$ and keeping the small terms of order η in (10.38), we find the growth rate of this branch

$$\gamma = -\eta\Delta\frac{[s(1+\beta/2)+1](1+\gamma_0\beta+\gamma_0\beta^2/2)}{(1+\gamma_0\beta+\gamma_0\beta^2/2)^2+\lambda^2(1+\beta/2)^2}. \tag{11.21}$$

Here, according to section 10.3.1, $\lambda = \Delta/\omega_{ni}$.

It can be seen that the electron drift waves are excited in a plasma with $\eta < 0$. This result is in accordance with the similar prediction of the collisionless plasma theory (compare (11.21) with (3.30)).

When $\beta \to 0$, (11.21) describes the instability due to the *finite electron heat conductivity* initially examined in [3.23] (see also [11.7]). This instability was discussed in [2], section 5.3.

11.3   Inertialess instabilities with $k_z \simeq \omega/v_{Ti}$ in a weakly collisional plasma

Let us consider the instabilities described by (10.48). At first, following [2.2], we shall present an analytical examination of equation (10.48) and then the results of the numerical calculations of [2.2] refined in [11.1, 11.3] will be given.

When $\beta \to 0$, (10.48) means [11.8, 11.9]

$$x^3 + \frac{x}{1+\eta} - \alpha\left(\frac{8}{3}x + \frac{2/3 - \eta}{1+\eta}\right) = 0. \qquad (11.22)$$

The consequences of (11.22) were discussed in [2], section 5.3. This equation describes *unstable perturbations* when

$$\eta > 2/3 \qquad \text{or} \qquad \eta < -2. \qquad (11.23)$$

For $\beta \gg 1$ and $|1+\eta| \gg 1/\beta$, in the leading order in $1/\beta$, all the roots of (10.48) are real. Two roots coincide for $\alpha = (5/4)(\beta/4)^3$. In the next order in $1/\beta$, we find that the coinciding roots are complex when

$$\eta > 1. \qquad (11.24)$$

Then

$$\mathrm{Re}\,\omega = -\frac{5}{4}\omega_{Di}$$
$$\gamma = \frac{|\,\mathrm{Re}\,\omega\,|}{2(3\beta)^{1/2}}\left(\frac{\eta-1}{\eta+1}\right)^{1/2}. \qquad (11.25)$$

Thus, it can be seen that a plasma with a sufficiently large positive temperature gradient is *unstable* for both small and large β. However, increasing β from a small values up to large values, the instability range of a plasma with $\eta > 0$ is not continues. This range has a gap for $\beta \simeq 1$ (see figure 11.1), [2.2].

The instability region for a plasma with $\eta < 0$ is also *non-trivial* (see figure 11.1). It can be seen that the instability described by the second inequality of (11.23) disappears for finite $\beta < 1$. Instead, for $\beta = 1.2$ a new instability appears. Its boundary is displaced, with increasing β, from $\eta = 0$ (for $\beta = 1.2$) up to $\eta = -1$ (for $\beta = \infty$), [2.2], (see figure 11.1). In addition, as seen from figure 11.1, there is one more stability gap in the region of negative η [11.1, 11.3].

As for a plasma with $\nabla T = 0$, it is stable against the discussed perturbations for all β.

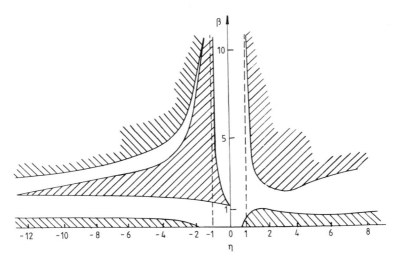

Figure 11.1 Instability ranges (shaded) of collisional plasma.

11.4   Inertialess instabilities in a strongly collisional plasma

The inertialess perturbations in a strongly collisional plasma are described by dispersion relation (10.64). We restrict ourselves to analysing it for $\nabla T = 0$ [10.3, 10.4]. An analysis of (10.64) for arbitrary ∇T was carried out in [10.3, 11.2].

When $\nabla T = 0$, (10.64) takes the form

$$x\left(x + \frac{\beta}{2}\right)(x + 1) - 4\alpha\left(x - \frac{2s}{5}\right) \bigg/ \left(\beta + \frac{6}{5}\right) = 0. \qquad (11.26)$$

Here $x = \omega/\omega_{ni}, \alpha = k_z^2 T/M_i\omega_{ni}^2$. This equation describes the so-called *thermal-force instability* which occurs for both small and large β. For small β this instability was initially studied in [11.10] and then in [11.8, 11.9]. The corresponding results of these papers are presented in [2], section 5.4.

When $\beta \gg 1$, one of the roots of (11.26) is large (it corresponds to stable oscillations) while the two other roots satisfy the equation

$$x(x + 1) - \frac{8\alpha}{\beta^2}\left(x - \frac{2s}{5}\right) = 0. \qquad (11.27)$$

Hence it follows that there is an instability if

$$0.2 < \alpha^{1/2}/\beta < 0.6. \qquad (11.28)$$

This means that the characteristic k_z for the unstable perturbations increases with increasing β, $k_z \sim \beta$.

The oscillation frequency and the growth rate of the *thermal-force instability* are of order ω_{ni}.

11.5 Inertial instabilities

In analysing the inertial instabilities, we restrict ourselves to the case of perturbations with $k_z = 0$. According to section 10.1.2, in neglecting dissipative effects, the dispersion relation of the inertial perturbations with $k_z = 0$ reduces to (4.1) as has been examined in section 4.1.1. Following from section 10.2.1, dissipative effects of order ω/ν_i do not make a contribution in the dispersion relation for such perturbations. The same concerns the heat transfer (see section 10.2.3). Therefore, we should consider only the effects of order $(k_\perp \rho_i)^2 \nu_i/\omega$.

For this, we use dispersion relation (10.22). We analyse it in the case where $\beta \ll 1, \nabla T = 0$. (A general analysis of the role of the effects of order $(k_\perp \rho_i)^2 \nu_i/\omega$ was carried out in [10.5].).

Then, according to section 4.1.1, we have two oscillation branches with frequencies

$$\begin{aligned} \operatorname{Re}\omega_1 &= \omega_{ni} \\ \operatorname{Re}\omega_2 &= \omega_{Di}. \end{aligned} \tag{11.29}$$

From (10.22) we find the imaginary additions to these frequencies

$$\begin{aligned} \gamma_1 &= 0 \\ \gamma_2 &= -0.3(k_\perp \rho_i)^2 \nu_i. \end{aligned} \tag{11.30}$$

This means that the *perturbations* corresponding to the ω_1 *branch neither grow nor are damped* while those corresponding to the ω_2 *branch are damped.*

12 Instabilities in Force-free and Almost Force-free High-β Plasmas

The study of instabilities in inhomogeneous *force-free* or *almost force-free high-β plasmas* (i.e., in those with vanishing or almost vanishing pressure gradients, $\nabla p \approx 0$, and with $\beta \gg 1$) is of interest, in particular, in connection with the problem of hot-plasma confinement due to the pressure of a dense *blanket of neutral gas*. This problem was first discussed in [12.1] and then in [12.2–12.5] and in a number of other papers. As noted in [12.6], similar conditions can be realized in a *laser-produced plasma*. In [2.6, 12.6] it has been mentioned that the case where $\beta \gg 1$ and $\nabla p \approx 0$ is typical for many space situations, in particular, for certain *solar wind* regions.

Interest in force-free high-β plasma has also increased in connection with the problem of plasma compression by *imploding liner* (see the literature on this problem, e.g., in [12.7, 12.8]). It is clear that plasma with $\beta \gg 1$ and a small pressure gradient can be realized in the central region of *theta-pinches*, in usual *mirror machines* as well as in the *tandem mirror experiments* (TMX) [6.14, 6.15], and in *reversed field configurations* (see, e.g., [12.9, 12.10]). The above situation is typical also for the problem of *magnetically insulated inertial confinement fusion* discussed in [12.11–12.13].

Using the basic equations concerning force-free or almost force-free high-β plasmas given in Chapters 2 and 10, we shall examine *inertialess instabilities* in such plasmas in sections 12.1–12.4 and the *inertial (Alfven) instabilities* in section 12.5.

In sections 12.1 and 12.2 the plasma is assumed to be *collisionless*. Then in section 12.1 it is considered to be *strictly force-free*, $\nabla B = 0$, while in section 12.2 it is *almost force-free*, $\nabla p \approx 0, \nabla B \neq 0$. In contrast to sections 12.1 and 12.2, in sections 12.3 and 12.4 a *collisional plasma* is studied. We deal with a *weakly collisional plasma* in section 12.3 and with a *strongly collisional one* in section 12.4 (within the meaning explained in sections 10.3.2 and 10.3.4)

In section 12.5, in analysing the *Alfven instabilities* we allow for both collisionless and collisional effects.

When the problem of plasma confinement by a *neutral gas blanket* is kept in mind, in studying instabilities it is necessary to distinguish the *central region*, where the plasma has maximum temperature and minimum density, and the *peripheral region*, where the temperature is low and the density is high. Recall that the characteristic frequency of gradient (drift) perturbations increases with temperature ($\omega^* \sim T$) and the binary collision frequency increases with density, but decreases with increasing temperature. Then it is clear that the theory of gradient instabilities in the *peripheral region* must clearly allow for *collisions* while the *central region* can be described in some cases by the *collisionless* approximation.

In accordance with Chapters 1 and 2 (see sections 1.2.3 and 2.1.4), the force-free plasma ($\nabla p = \nabla B = 0$) is of interest from the physical viewpoint since in such a plasma there are no particle magnetic drift and Larmor (diamagnetic) current. That is why at first sight it can seem that there is no reason for instabilities. However, on more careful consideration of the equilibria of force-free plasma it can be seen that, due to the presence of temperature and density gradients (when $\nabla p = 0$, these gradients are related by the expression $\nabla(nT) = 0$) in such a plasma, there are *higher-order fluxes*, e.g., *heat flux*. It is these higher-order fluxes which are the reason for instabilities of force-free plasma. Hence it follows, in particular, that in a hydrodynamic description of instabilities of force-free (and almost force-free) plasmas, generally speaking, it is necessary to use the equations correctly allowing for the higher-order moments of the distribution function. Then instabilities in weakly collisional and collisionless force-free plasmas are similar (cf. section 12.1 with section 12.3, and also section 12.4.1 with section 12.4.3). A common property for both cases is an instability of the oblique ($k_z \neq 0$) inertialess perturbations for $\beta \to \infty$ and $T_e = T_i$ and the stabilization of these perturbations for finite $1/\beta$ and $T_e \neq T_i$.

It is necessary to keep in mind that when $\beta \gg 1$, the presence of even a small equilibrium plasma pressure gradient causes a significant equilibrium *magnetic-field gradient* (this follows from the equilibrium condition (1.42)). Therefore, instabilities in an almost force-free plasma must be investigated making allowance for the magnetic-field gradient and the particle *magnetic drift* associated with it. We follow this procedure in section 12.2 and in certain parts of sections 12.3–12.5. Then the picture of instabilities depends very essentially on the presence of a magnetic-field gradient and on the direction of this gradient. As a whole, we find that collisionless and weakly collisional plasmas with $\partial \ln B / \partial \ln n < 0$ are more stable than those with magnetic-field and density gradients in the same direction.

The condition $\partial \ln B / \partial \ln n < 0$ means, for the problem of hot-plasma confinement by the pressure of a *dense blanket of neutral gas*, that the magnetic field is greater at the centre than at the periphery. It is precisely this type of magnetic system which is most appropriate to the case where the field is used *to reduce thermal conductivity and diffusion* but not to contain the plasma. In principle, the presence of a magnetic field is essential only at the centre and not at the periphery where, due to frequent collisions, the magnetization of the plasma is of no importance. Hence it is on the whole a fair conclusion that a plasma with $\partial \ln B / \partial \ln n < 0$ has enhanced stability, though we must not forget that even in this most favourable case the plasma is still not entirely free of instabilities, especially, if the role of the collisional dissipative effects is not negligible. As for a strongly collisional plasma, according to section 12.4, it is most unstable just for $\partial \ln B / \partial \ln n < 0$, although the growth rate of the perturbations is small as $\beta^{-1/2}$ in this case (it is assumed that $\beta \gg 1$).

In contrast to the case where $\nabla T = 0$ dealt with in Chapter 11, the *Alfven instabilities* can play a more active role for the case $\nabla p \approx 0$ discussed in the present chapter. According to section 12.5, such instabilities can be excited even if resonant and dissipative effects are neglected, if $-4 < \partial \ln B / \partial \ln n < 0$. The above-mentioned effects enhance the instability region (see, for more detail, section 12.5).

On the whole, force-free and almost force-free plasmas are *more unstable* than a plasma with $\nabla T = 0$. However, it should be noted that in the most unstable case the coefficients of heat conductivity and diffusion caused by the instabilities discussed cannot essentially exceed the Bohm values.

The results presented in this chapter have been found in [2.6, 3.5, 3.12, 9.9, 10.1–10.3, 12.14]. In addition, papers [11.1–11.3] are associated with the topic of this chapter, where the boundaries of instabilities for various values of β and η in various collisionality regimes were calculated. In [12.15] the short-wavelength perturbations ($k_\perp \rho_i \gg 1$) were studied.

A part of the results of [2.6, 3.12] has been reproduced in [12.6]. Paper [12.6] was critiqued in [12.16]. However, this critique as well as the results of [12.16] are wrong.

Instabilities in a collisional high-β plasma with $k_z = 0$ were also studied in [12.17]. However, the author of [12.17] did not consider a difference in the perturbed temperatures of electrons and ions. Therefore, the growth rate of perturbations considered in [12.17] wrongly depends on the difference $z - 1$ where z is the ion charge number.

Non-linear effects related to the above instabilities were discussed in [12.8].

12.1 Inertialess instabilities in a force-free collisionless high-β plasma

The *low-frequency perturbations* ($\omega \ll \Omega_i$) of a collisionless plasma with $\nabla p = \nabla B = 0$ are described by dispersion relation (1.1) with $\epsilon_{\alpha\beta}$ of the form of (2.20). In the case of *long-wavelength perturbations* ($k_\perp \rho_i \ll 1$) and a *high plasma pressure* ($\beta \gg 1$) equation (1.1) reduces to

$$(\omega^2 - k_z^2 c_A^2)(\epsilon_{22}\epsilon_{33} + \epsilon_{23}^2) - (9/8)z_i\beta_i k_\perp^2 c^2 (1 - \omega_{Ti}/\omega)^2 \epsilon_{33} = 0. \quad (12.1)$$

In the zeroth approximation with respect to z_i, equation (12.1) is separated into two dispersion relations one of which has the form of (4.11) and the second is

$$\epsilon_{22}\epsilon_{33} + \epsilon_{23}^2 = 0. \quad (12.2)$$

Equation (4.11) describes the *Alfven waves* whilst (12.2) corresponds to *magnetoacoustic perturbations*. It is clear that in the above approximation the problem of instabilities consists of analysing (12.2). We shall examine (12.2) in the case where $T_e = T_i$ in section 12.1.1 and in the case where $T_e \neq T_i$ in section 12.1.2.

When β is finite and $z_i \to 0$, we have from (1.1) that (12.1) is replaced by (see (2.6) and (2.8))

$$(\epsilon_{22} - N^2)\epsilon_{33} + \epsilon_{23}^2 = 0. \quad (12.3)$$

We shall analyse (12.3) in section 12.1.3.

12.1.1 Plasma with $T_e = T_i$ and $\beta \to \infty$
When $\omega \ll k_z v_{Te}$, we find from (12.2) and (2.20) (cf. (3.52))

$$\frac{1}{T}\hat{l}_i^0(TxW) + \frac{i\pi^{1/2}}{4}\left[2\hat{l}_i^0\left(\frac{xW}{T}\right)\hat{l}_i^0(TxW) - \left(\hat{l}_i^0(xW)\right)^2\right] = 0. \quad (12.4)$$

Here $x = \omega/|k_z| v_{Ti}$, $W = W(x)$, and the operator $\hat{l}_i^0$ is determined by (2.21).

When $x \gg 1$, from (12.4) a quadratic equation in ω is obtained [3.12]

$$(\omega - \omega_{Ti})^2 + \omega^2/2 = 0 \tag{12.5}$$

with complex solutions

$$\omega = \frac{2}{3}\left(1 \pm \frac{i}{2^{1/2}}\right)\omega_{Ti}. \tag{12.6}$$

It is clear that an instability occurs. This is the *hydrodynamic instability* of a *force-free plasma* (it is not associated with the resonant particles).

It follows from (12.4) that an instability also occurs for $x \ll 1$ when we have [3.12]

$$\mathrm{Re}\,\omega = \omega_{Ti}/2$$
$$\gamma = \frac{\pi - 2}{2^{1/2}4}\frac{\omega_{Ti}^2}{|k_z|\,v_{Ti}}. \tag{12.7}$$

This is the *kinetic instability of a force-free plasma*.

The collisionless instabilities described by (12.6) and (12.7) have collisional analogues which will be considered in section 12.3.

12.1.2 Plasma with $T_e \neq T_i$ and $\beta \to \infty$

According to section 12.1.1, the perturbations in a plasma with $\beta \to \infty$ and $T_e = T_i$ with $v_{Ti} \ll \omega/k_z \ll v_{Te}$ are *hydrodynamically unstable* while those with $\omega \ll k_z v_{Ti}$ are *kinetically unstable*. We shall now show that all the perturbations are damped when T_e is a few times smaller than T_i.

(a) Perturbations with $v_{Ti} \ll \omega/k_z \ll v_{Te}$

For $T_i \neq T_e$ instead of (12.5) we find from (2.20) and (12.2) that

$$(1 - \omega_{Ti}/\omega)(1 - \tau\omega_{Ti}/\omega) + \tau/2 = 0 \tag{12.8}$$

where $\tau \equiv T_e/T_i$. The solutions of (12.8) are

$$\omega = \omega_{Ti}[1 + \tau \pm (1 - 2\tau - \tau^2)^{1/2}]/(2 + \tau). \tag{12.9}$$

The frequencies of (12.9) are real if [2.6]

$$T_e/T_i < 2^{1/2} - 1 = 0.41. \tag{12.10}$$

This is the *condition of stabilization of hydrodynamically unstable perturbations* of type (12.6).

(b) Perturbations with $\omega \ll k_z v_{Ti}$

For $T_i = T_e$ we find from (2.20) and (12.2) the following generalization of (12.7)

$$\omega = \frac{\omega_{Ti}}{2} + i\left(\frac{\pi}{1+1/\tau} - 1\right)\frac{\omega_{Ti}^2}{2\pi^{1/2}\,|\,k_z\,|\,v_{Ti}}. \tag{12.11}$$

Hence we obtain that the *kinetic instability* described by (12.7) *is suppressed* if [2.6]

$$T_e/T_i < (\pi - 1)^{-1} \approx 0.45. \tag{12.12}$$

As a whole, the force-free plasma with $T_e/T_i < 1$ is more stable than that with $T_e/T_i > 1$.

12.1.3 Instability boundaries for finite β

The instabilities discussed in sections 12.1.1 and 12.1.2 are specific for $\beta \gg 1$ and are suppressed if β is not too large. We now give the critical values of β corresponding to these instabilities.

(a) Boundary of hydrodynamic instability

For *finite β* we find from (12.3) the following generalization of (12.5)

$$\left(1 - \frac{\omega_{Ti}}{\omega} + \frac{1}{\beta_i}\right)\left(1 - \frac{\omega_{Ti}}{\omega}\right) + \frac{1}{2} = 0. \tag{12.13}$$

Hence we obtain that ω is complex (i.e., the instability takes place) only if [2.6]

$$\beta_i > 2^{-1/2}. \tag{12.14}$$

For $T_e = T_i$ this means $\beta > 2^{1/2}$.

(b) Boundary of kinetic instability

Instead of (12.7) for *finite β* we have from (12.3)

$$\omega = \frac{\omega_{Ti}}{2} + \frac{2i\omega^2}{\pi^{1/2}\,|\,k_z\,|\,v_{Ti}}\left(\frac{\pi}{2} - 1 - \frac{k_z^2 v_{Ti}^2}{2\omega^2\beta}\right). \tag{12.15}$$

For instability it is necessary that [2.6]

$$\beta > \frac{k_z^2 v_{Ti}^2}{2\omega^2(\pi/2 - 1)}. \tag{12.16}$$

Since $k_z v_{Ti} \gtrsim \omega$, we find that (12.16) cannot be satisfied for $\beta \simeq 1$.

12.2 Inertialess instabilities in an almost force-free collisionless high-β plasma

As in section 12.1, we now examine the *inertialess perturbations* described by (12.3) (i.e., by (2.8)), however, in contrast to that section, it is now assumed that $\nabla p \neq 0$. We suppose $\beta \gg 1$ and restrict ourselves to the study of situations where $\mid \partial \ln p / \partial \ln n \mid \lesssim 1/\beta$, i.e., $\mid 1 + \eta \mid \lesssim 1/\beta$. Then $\mid \kappa_B / \kappa_n \mid \lesssim 1$. Therefore, it is convenient to use a dimensionless parameter $b \equiv \kappa_B / \kappa_n$ characterizing the *magnetic-field inhomogeneity* since the range of variation of this parameter in the given situations is of the order of unity, $\mid b \mid \lesssim 1$.

We begin our analysis by considering the perturbations with $k_z < \omega / v_{Te}$ (section 12.2.1). In this approximation we deal with *Tserkovnikov's perturbations* (see section 3.1) and the *transverse drift waves* (see section 3.2). Then we consider perturbations with $v_{Ti} \ll \omega / k_z \ll v_{Te}$ (section 12.2.2) and with $k_z \ll \omega / v_{Ti}$ (section 12.2.3). The subsequent consideration follows [3.6].

12.2.1 Perturbations with small k_z
The *inertialess perturbations* with $k_z = 0$ were also examined in sections 3.1 and 3.2. However, the case where $\beta \lesssim 1$ and $\eta \gtrsim 1$ was emphasized there. In contrast, we are now interested in the case where $\beta \gg 1$ and $\eta \approx -1$. However, in both cases we deal with the same types of waves, being perturbations with $\boldsymbol{E} \perp \boldsymbol{B}_0$, described by (2.14) with ϵ_{22} of the form of (2.17), and those with $\boldsymbol{E} \parallel \boldsymbol{B}_0$, described by (2.15) with ϵ_{33} of the form of (2.18). We shall consider separately both types of waves in the above case for $\beta \gg 1$ and $b \lesssim 1$.

(a) Perturbations with $\boldsymbol{E} \perp \boldsymbol{B}_0$
An *instability criterion* following from (2.14) (see also (3.1)) for arbitrary β and b is determined by (3.4) and (3.7). Let us allow for the identity

$$\eta = -1 - 2b/\beta. \tag{12.17}$$

Then, (3.4) and (3.7) mean that a plasma is unstable if

$$b > 0 \qquad \text{or} \qquad b < -\beta. \tag{12.18}$$

Since the second inequality of (12.18) is not realized when $\beta \gg 1$ and $b \lesssim 1$, we conclude that the instability takes place if the magnetic-field gradient is in the same direction as the density gradient.

(b) Perturbations with $\boldsymbol{E} \parallel \boldsymbol{B}_0$
Let us assume that ∇p is sufficiently small so that $\mid b \mid \ll 1$. Then the solution of equation (2.15) with ϵ_{33} of the form of (2.18) satisfies

the condition $\omega_{\mathrm{De}} \ll \omega \ll \omega_{ne}$ and is given by

$$\omega^2 = -b\omega_{ne}^2. \tag{12.19}$$

Hence it follows that the perturbations with $E \parallel B_0$ are unstable when

$$b > 0 \tag{12.20}$$

i.e., when the perturbations with $E \perp B_0$ are also unstable (cf. (12.20) with the first inequality of (12.18)).

As in section 3.2, let us investigate the role of finite k_z in perturbations with $E \parallel B_0$ assuming $k_z \ll \omega/v_{T\mathrm{e}}$. Then we allow for the influence of the *resonant ions* on the pure electron branches of *transverse drift waves* described by (12.19) in the case where $b < 0$ and $\omega^2 > 0$. Then we obtain the dispersion relation (cf. (3.24))

$$\epsilon_{33} + \mathrm{i}\frac{|\epsilon_{23}|^2}{|\epsilon_{22} - N^2|^2}\mathrm{Im}\,\epsilon_{22}^{(\mathrm{i})} = 0 \tag{12.21}$$

where

$$\epsilon_{23} = \mathrm{i}\frac{\omega_{pe}^2 T k_z k_\perp}{M_e \Omega_e \omega^3}\left(1 - \frac{\omega_{Te}}{\omega}\right) \tag{12.22}$$

and ϵ_{22} and ϵ_{33} are determined by (2.17) and (2.18).

We find by means of (12.21) the imaginary part of the oscillation frequencies of (12.19) which proves to be positive. Consequently, when $b < 0$, an instability takes place. The perturbations considered are of *negative energy* so that the energy of the resonant ions increases under their interaction with the waves (cf. the similar situation in section 3.2).

12.2.2 Perturbations with $v_{Ti} \ll \omega/k_z \ll v_{Te}$
According to (3.28), in the approximation $\beta \gg 1, \kappa_p/\kappa_n \ll 1$ and $v_{Ti} \ll \omega/k_z \ll v_{Te}$ we have the dispersion relation

$$1 + \frac{\omega_{Te}}{\omega} - \left[\left(1 - \frac{\omega_{Ti}}{\omega}\frac{\partial}{\partial \ln T}\right)\left\langle\frac{\omega\lambda}{\omega - \omega_{Di}\lambda}\right\rangle\right]\left(1 + \frac{\omega_{Te}}{\omega} - \frac{\omega_{De}}{\omega}\right) = 0. \tag{12.23}$$

where $\lambda = M_i v_\perp^2/2T$.

When $b = 0$, (12.23) has the solution (12.5) corresponding to growing perturbations. If b increases up to values of order of and larger than unity, the instability associated with these perturbations is suppressed, regardless of the sign of $\partial \ln p/\partial \ln n$.

12.2.3 Perturbations with $\omega \ll k_z v_{Ti}$
Considering the perturbations with $\omega \ll k_z v_{Ti}$, we start with equation (3.52) assuming the argument of W-functions to be small. When $\beta \gg 1$ and $\kappa_p/\kappa_n \ll 1$, we obtain

$$\mathrm{Re}\,\omega = \omega_{Ti}/2$$

$$\gamma = \frac{\omega_{Ti}^2}{4\pi^{1/2}\,|\,k_z\,|\,v_{Ti}}(\pi - 2 + 12b). \qquad (12.24)$$

For $\nabla p = 0(b = 0)$ this result coincides with (12.7) and describes the *kinetic instability* discussed in section 12.1.1. It is clear from (12.24) that the magnetic-field gradient leads to *destabilization* when $b > 0$, which is similar to the case of perturbations with $k_z = 0$. However, the *instability is suppressed* when

$$b < -1/12. \qquad (12.25)$$

For $\partial \ln p/\partial \ln n < 0$, the presence of a pressure gradient makes instability possible.

12.3   Inertialess instabilities in a weakly collisional plasma

12.3.1 Perturbations with $k_z = 0$
In neglecting dissipative effects, the inertialess perturbations in a collisional plasma with $k_z = 0$ are described by dispersion relation (10.6). The instability criterion following from (10.6) for arbitrary β and η is of the form of (11.1). It follows from (11.1) that, when $\beta \gg 1$ and $\eta \approx -1$, a plasma is unstable if b is in the range of (12.18). In other words, in the case of a high-β plasma with low pressure gradient, the *instability criteria* of the inertialess perturbations with $k_z = 0$ prove to be *the same* in both collisionless and collisional plasmas.

We now assume that b is not in the range of (12.18) and allow for *dissipative effects* of order ω/ν_i. Starting with equation set (10.17), we obtain the dispersion relation [9.9]

$$\omega^2 - \gamma_0\omega_{Di}^2\left(1 - \frac{2}{b}\right) + i\frac{0.96}{10}\frac{\omega^3}{\nu_i}F\left(\frac{\omega}{\omega_{Di}}\right) = 0. \qquad (12.26)$$

Here

$$F(x) = x(x - 0.59)(x + 1 - 1/b)$$
$$+ \frac{42}{41}[(8/3)x - 2 - (3/5b)(x + 5)](x + 22.02). \qquad (12.27)$$

Using (12.26), we find two additional ranges of *instabilities of inertialess perturbations*:

$$2 < b < 2.6 \tag{12.28}$$
$$-0.1 < b < 0. \tag{12.29}$$

The growth rate of these instabilities is of order

$$\gamma \simeq \omega_{Di}^2 / \nu_i. \tag{12.30}$$

It becomes comparable to $\mathrm{Re}\,\omega$ when $\nu_i \simeq \omega_{Di}$, i.e., at the validity limit of the collisional approximation.

12.3.2 Perturbations with finite k_z

When $k_z \gtrsim (\omega\nu_e)^{1/2}/v_{Te}$, the *electron longitudinal heat conductivity* is important. According to (10.38), in this case we have the following dispersion relation replacing (12.26) [10.1]

$$
\begin{aligned}
&[\omega^2 - \gamma_0\omega_{Di}^2(1 - 2/b)](\omega - 0.71\omega_{ni}) \\
&+ i\Delta_e[(1 + \gamma_0^{-1})\omega^2 + 2(1 - b)\omega\omega_{ni} + \omega_{ni}^2] = 0
\end{aligned}
\tag{12.31}
$$

where $\Delta_e = (3.16/3)k_z^2 T/M_e\nu_e$.

In the limit $\Delta_e \to \infty$, from (12.31) we obtain the quadratic equation

$$(1 + \gamma_0^{-1})\omega^2 + 2(1 - b)\omega\omega_{ni} + \omega_{ni}^2 = 0. \tag{12.32}$$

This equation describes *unstable perturbations* when

$$-0.3 < b < 2.3. \tag{12.33}$$

In particular, this range of b includes the value $b = 0$ corresponding to the case of a homogeneous magnetic field. Then equation (12.32) has solutions

$$\omega = \omega_{Ti}\frac{1 \pm i\gamma_0^{-1/2}}{1 + \gamma_0^{-1}}. \tag{12.34}$$

This expression has been obtained for $\gamma_0 = 5/3$, however, it is valid for arbitrary γ_0. In particular, when $\gamma_0 = 2$, (12.34) means the same as (12.6).

When b is not in the range (12.33), the roots of (12.31) are real either for $\Delta_e \to \infty$ or for $\Delta_e \to 0$. In these cases, damping or excitation of perturbations is determined by the imaginary terms of the dispersion relation corresponding to the effects of finite electron heat conductivity. Then a *dissipative instability* is revealed in the range

$$-\infty < b < -0.3. \tag{12.35}$$

In the particular case where $|b| \gg 1$, the solution of (12.31) has the form

$$\omega = 0.71 \frac{\omega_{ni} - 0.3i\Delta_e/b}{1 + (0.3\Delta_e/\omega_{Di})^2}. \tag{12.36}$$

When $k_z \simeq (\omega_{Di}\nu_e)^{1/2}/v_{Te}$ and $b \approx -1$, the growth rate is of order ω_{ni}. It decreases if $|b|$ and k_z increase.

12.3.3 Perturbations with $k_z \gtrsim (\omega\nu_i)^{1/2}/v_{Ti}$

We will now consider perturbations with $k_z \gg (\omega\nu_i)^{1/2}/v_{Ti}$, i.e., those with the shortest longitudinal wavelength. As the starting equation set, we use (10.49)–(10.54).

Since κ_p is small compared with κ_B and κ_T, of order $1/\beta$ for $\beta \gg 1$, the terms with κ_p in (10.49)–(10.54) can be neglected. Assuming $\Delta_1 \gg \omega$, we conclude that, in the zeroth approximation in the small parameters ω/Δ_1 and $1/\beta$, the equation of ion longitudinal motion (10.52) means simply that the *ion longitudinal viscosity* vanishes

$$\pi_{zz} = 0. \tag{12.37}$$

The following relationship for the perturbed quantities $X_\perp, Y_\perp, Z_\perp$ and V_z can be determined from (12.37) and (10.53):

$$\frac{4}{3}k_z V_z - 2.9i\Delta_1 X_\perp - \frac{2}{3}\kappa_B Y_\perp + \frac{16}{15}\omega_{Di} Z_\perp = 0. \tag{12.38}$$

Recall that the term with $\Delta_1 X_\perp$ is due to the dependence of the *ion longitudinal viscosity* on the *ion longitudinal heat conductivity* (i.e., on the *second derivatives of temperature*). It is of the same order as the other terms in (12.38); to check this, π_{zz} should be expressed through $X_\perp, Y_\perp, Z_\perp$ using the *ion heat-balance equation* (10.51).

From (12.38) and equations (10.49)–(10.51), in which the small terms of order ω/Δ_1 and $1/\beta$ are neglected, we find the dispersion relation [10.2]

$$\omega = 1.7\omega_{Ti}. \tag{12.39}$$

Before proceeding to take account of the effects of order ω/Δ_1 which can excite the branch (12.39), let us consider the matter of the limits of validity of (12.37) and explain the consequences of neglecting the pressure gradient with respect to the viscosity tensor gradient in the equation of ion longitudinal motion written in the form of (10.52). Within an order of magnitude we find from (10.53)

$$\pi_{zz} \simeq \frac{p_{0i}\omega}{\nu_i} \frac{\tilde{B}_z}{B_0}. \tag{12.40}$$

Therefore, using (10.14), we obtain the estimate

$$(\tilde{p}_e + \tilde{p}_{\perp i})/\pi_{zz} \simeq \nu_i/\omega\beta. \tag{12.41}$$

It can be seen that the viscosity is more important than the pressure if β is sufficiently large:

$$\beta > \nu_i/\omega. \tag{12.42}$$

Since $\nu_i \gg \omega$ (condition of the *collisional approximation*), it also follows from (12.42) that for $\beta \lesssim 1$ the longitudinal viscosity can lead only to small corrections, of order ω/ν_i, to the terms related to the pressure gradient. Such correctional effects causing dissipative instabilities were studied in Chapter 11.

Now we return to (12.39) and augment it by the terms of order ω/Δ_1. Then we obtain the following expression for the oscillation frequency [10.2]

$$\omega = 1.7\omega_{Ti} + i1.6(1 + 0.9b)\omega_{Ti}^2 M_i\nu_i/k_z^2 T. \tag{12.43}$$

According to (12.43), the perturbations are excited if

$$b > -1.1. \tag{12.44}$$

In particular, the plasma in a *homogeneous magnetic field* ($b = 0$, i.e., $\nabla B_0 = 0$) appears to be unstable.

Let us compare (12.43) with the corresponding result for collisionless plasma, i.e., with (12.7) and (12.24). It can be seen that the collisional and collisionless dispersion relations have qualitatively the same structure. In the limit of their validity, i.e., for $\nu_i \simeq k_z v_{Ti}$, they give results of the same order of magnitude.

When the terms of order $1/\beta$ are taken into account, we find instead of (12.43) [10.2]

$$\omega = 1.7\omega_{Ti} + i1.6(1 + 0.9b)\omega_{Ti}^2 M_i\nu_i/k_z^2 T - i\nu_i/0.7\beta. \tag{12.45}$$

It can be seen that a stabilizing effect related to the terms of order $1/\beta$ takes place for sufficiently large k_z (cf. (12.15)). The order of magnitude of the upper limit of k_z of growing perturbations is given by (cf. (12.16))

$$k_z \lesssim \beta^{1/2}\omega_{Ti}/v_{Ti}. \tag{12.46}$$

For simplicity we have considered the limiting case of perturbations with $\omega \ll \Delta_1$. The growth rate of such perturbations is small compared with their frequency, $\gamma \ll \mathrm{Re}\,\omega$. The maximum growth rate corresponds to $\omega \simeq \Delta_1$. Then $\gamma \simeq \mathrm{Re}\,\omega$ and the optimum k_z is of order

$$k_z \simeq (\omega_{Ti}\nu_i)^{1/2}/v_{Ti}. \tag{12.47}$$

Since we assume $\nu_i \gg \omega_{Ti}$, condition (12.47) means that $k_z \gg \omega_{Ti}/v_{Ti}$.

12.4   Inertialess instabilities in a strongly collisional plasma

Let us consider dispersion relation (10.64) for $\beta \gg 1, \nabla p \approx 0$. When the pressure gradient is completely neglected, i.e., when $\nabla p = \nabla B = 0$, the roots of (10.64) are real and are defined by

$$\omega^2_{1,2} = k_z^2 c_A^2 \tag{12.48}$$

$$\omega_3 = s\omega_{ne}. \tag{12.49}$$

Thus, a strongly collisional plasma with $\nabla p = 0$ is stable against inertial perturbations.

Let us explain that equation (12.48) coincides formally with the dispersion law of Alfven waves (cf. (4.11)), however, as a matter of fact it describes *slow magnetoacoustic waves* for $\beta \to \infty$. The difference between the dispersion law of slow magnetoacoustic waves and that of Alfven waves is revealed for finite β when, according to (10.64), equation (12.48) is replaced by

$$\omega^2_{1,2} = \frac{k_z^2 c_A^2}{1 + 6/5\beta}. \tag{12.50}$$

It can be seen that for $\beta \to 0$ equation (12.50) becomes

$$\omega^2 = (10/3)k_z^2 T/M_i \tag{12.51}$$

which is the well-known dispersion law for *ion-acoustic waves* (cf. [2], section 5.4).

Allowing for the magnetic-field gradient in (10.64) and neglecting the pressure gradient which is small compared with the magnetic-field gradient as $1/\beta$, we obtain instead of (12.48)

$$\omega(\omega - \omega_{\mathrm{Di}}) - k_z^2 c_A^2 = 0 \qquad (12.52)$$

while (12.49) remains valid. Consequently, in the zeroth approximation in $1/\beta$, a strongly collisional plasma with $\nabla B \neq 0$ as well as with $\nabla B = 0$, is stable against inertial perturbations.

Let us now allow for the terms of order $1/\beta$ assuming $\nabla B \neq 0$. Such terms are substantial when one of the solutions of (12.52) coincides with (12.49). Such an intersection of the roots takes place if

$$(k_z c_A/\omega_{\mathrm{ni}})^2 = s(s - b). \qquad (12.53)$$

Allowing for the terms of order $1/\beta$, we find that the frequencies of intersecting roots are shifted by a magnitude $\delta\omega$, where, according to (10.64),

$$(\delta\omega)^2 = \frac{2}{\beta} \frac{b}{2s - b} k_z^2 c_A^2. \qquad (12.54)$$

Hence we conclude that an instability takes place if [10.3]

$$b < 0. \qquad (12.55)$$

The ratio of the growth rate to the frequency is of order

$$\gamma/\mathrm{Re}\,\omega \sim \beta^{-1/2}. \qquad (12.56)$$

Comparing the results obtained here with those of section 12.3, we arrive at the conclusion that *strong collisions suppress instabilities* in a plasma with $\nabla B = 0$ and, at the same time, contribute to the *destabilization* of plasma with $\partial \ln B/\partial \ln n < 0$.

12.5   Alfven instabilities in a high-β plasma

In the approximation $k_\perp^2 \rho_i^2 \to 0$, Alfven perturbations in both collisionless and collisional plasmas are described by dispersion relation (4.10). When $\beta \gg 1$ and $|\,b\,| \lesssim 1$, it becomes

$$\omega^2 - \omega_{\mathrm{Di}}\omega - \omega_{\mathrm{Di}}^2/b - k_z^2 c_A^2 = 0. \qquad (12.57)$$

Hence it follows that *Alfven perturbations are excited* if

$$-4 < b < 0. \qquad (12.58)$$

The growth rate of these perturbations is of order $|\omega_{Di}|$. The instability criterion (12.58) concerns the case where $k_z = 0$. Instability proves to be suppressed as k_z increases (cf. section 4.1.2).

Following [12.14], we now assume criterion (12.58) not to be satisfied and consider the excitation of the oscillation branches by *resonant particles* in a collisionless plasma and by *dissipative effects* in a collisional plasma. An analysis will be made using, in the main, the expression for the growth rate similar to (2.33)

$$\gamma = -\frac{\operatorname{Im} D_1^{(1)}}{\partial D_1^{(0)}/\partial \omega} \tag{12.59}$$

the dispersion relation being written in the form (cf. (2.32))

$$D_1^{(0)} + D_1^{(1)} = 0. \tag{12.60}$$

Here $D_1^{(0)}$ is defined by (2.5) and $D_1^{(1)}$ is an addition to the dispersion relation due to resonant particles or dissipative effects.

To elucidate the physical meaning of instabilities considered below, we shall treat (12.59), similarly to Chapter 4, in the energy terms. Then $\omega \partial D_1^{(0)}/\partial \omega$ is taken to be the *dimensionless oscillation energy* and $\omega \operatorname{Im} D_1^{(1)}$ is the *rate of oscillation energy dissipation*.

In the case considered, when $\beta \gg 1$ and $|b| \lesssim 1$, and when resonant and dissipative effects are neglected, such that equation (12.57) holds, we conclude that (as is clear from (2.5) and (2.9)) the function $D_1^{(0)}$ is of the form

$$D_1^{(0)} = \frac{c^2}{c_A^2 \omega^2}\left(\omega^2 - \omega \omega_{Di} - \frac{\omega_{Di}^2}{b} - k_z^2 c_A^2\right) \tag{12.61}$$

while the dimensionless oscillation energy is

$$W \equiv \omega \frac{\partial D_1^{(0)}}{\partial \omega} = \frac{c^2}{c_A^2}\left(2 - \frac{\omega_{Di}}{\omega}\right). \tag{12.62}$$

It follows hence that the energy of the Alfven perturbations is negative, $W < 0$, if

$$\omega_{Di}/\omega > 2. \tag{12.63}$$

Using (12.61) and (12.63), we find that waves with $W < 0$ exist if

$$b < -(k_z c_A/\omega_{ni})^2 \tag{12.64}$$

and if, in addition, the inequality

$$(b + 2)^2 > 4[1 - (k_z c_A/\omega_{ni})^2] \tag{12.65}$$

is satisfied, which should be taken into account for $| k_z c_A | < | \omega_{ni} |$.

However, $W > 0$ even if one of the inequalities fails to be true. In particular, both Alfven branches are of *positive energy* in the case of a homogeneous magnetic field, i.e., when $b = 0$. It is for the case where $b = 0$ that we begin our analysis assuming the plasma to be collisionless (section 12.5.1). We will then deal with instabilities in a collisionless plasma with $b \neq 0$, section 12.5.2, and with those in a collisional plasma, section 12.5.3.

12.5.1 Instabilities in a force-free collisionless plasma

If $\beta \gg 1$ and $\omega \simeq k_z c_A$, we have that $\omega / k_z v_{Ti}$ is small. Then we find from (12.1) that

$$D_1 \equiv \frac{c^2}{c_A^2 \omega^2} \left(\omega^2 - c_A^2 k_z^2 + i \frac{9}{8\pi^{1/2}} z_i \omega \mid k_z \mid v_{Ti} \frac{(1 - \omega_{Ti}/\omega)^2}{1 - \omega_{Ti}/2\omega} \right) = 0.$$
(12.66)

Note that the electron temperature does not appear here in contrast to (12.8) and (12.11).

Assuming z_i to be sufficiently small, we find the real part of the oscillation frequency to be determined by (4.11) while, according to (12.59), the growth rate is

$$\gamma = -\frac{9 z_i \beta_i^{1/2}}{16\pi^{1/2}} \mid k_z \mid c_A \frac{(1 - \omega_{Ti}/\omega)^2}{1 - \omega_{Ti}/2\omega}.$$
(12.67)

It is evident that the *growth rate is positive* for one of the two oscillation branches if $k_z < \omega_{Ti}/2c_A$.

A more precise interpretation of (12.67) is required if the frequencies of the Alfven perturbations (4.11) coincide with the leading term in the right-hand side of (12.11); this occurs when

$$k_z = \omega_{Ti}/2c_A.$$
(12.68)

From (12.66) we find that in such an oscillation branch intersection the growth rate reaches a maximum of order

$$\gamma \simeq z_i^{1/2} \beta_i^{1/4} \omega_{Ti}.$$
(12.69)

The assumption that the terms of order z_i are small is justified if

$$z_i < \beta_i^{-1/2}.$$
(12.70)

Under the condition opposite to (12.70) and if $k_z \simeq \omega_{Ti}/c_A$, the term with z_i in (12.66) becomes the largest so that (12.66) can be solved by a series expansion in inverse powers of z_i. We then obtain

$$\omega = \omega_{Ti} \left\{ 1 \pm \frac{2}{3} \left[\frac{i\pi^{1/2}}{z_i} \frac{\omega_{Ti}}{\mid k_z \mid v_{Ti}} \left(\frac{k_z^2 c_A^2}{\omega_{Ti}^2} - 1 \right) \right]^{1/2} \right\}.$$
(12.71)

Obviously, the instability takes place for both $k_z < \omega_{Ti}/c_A$ and $k_z > \omega_{Ti}/c_A$. The growth rate increases for both k_z decreasing and increasing, i.e., there are *two ranges of instability*. The growth rate becomes of order of the frequency in the region of small k_z if

$$k_z \simeq \omega_{Ti}/c_A z_i \beta_i^{1/2} \tag{12.72}$$

and in the region of large k_z if

$$k_z \simeq \omega_{Ti} z_i \beta_i^{1/2}/c_A. \tag{12.73}$$

These results are valid up to $z_i \simeq 1$.

12.5.2 Kinetic instabilities in a collisionless plasma for $\nabla B \neq 0$

As in section 12.5.1, we now assume $k_z v_{Ti} \gg (\omega, \omega_{Di})$. Allowing for $\beta \gg 1$, the expression for $D_i^{(0)}$ is now given by (12.61) while $\operatorname{Im} D_i^{(1)}$ is given by the terms of order z_i of (12.66). Although (12.66) has been derived in the assumption $b = 0$, the expression for $\operatorname{Im} D_i^{(1)}$ given by (12.66) remains valid for $b \neq 0$.

Since the case $b = 0$ has already been examined (section 12.5.1) we restrict ourselves to the analysis of the case where $|b| \gg 1$. The roots of (12.57) are then real and are defined by

$$\omega_1 = \omega_{Di} \tag{12.74}$$
$$\omega_2 = -\omega_{ni} - k_z^2 c_A^2/\omega_{Di} \tag{12.75}$$

so that $|\omega_1| \gg |\omega_2|$.

In the case where $\omega = \omega_1$ we have

$$\omega \operatorname{Im} D_i^{(1)} > 0 \tag{12.76}$$

$$W > 0. \tag{12.77}$$

Consequently, according to (12.59), the perturbations of type (12.74) *are damped*. If $\omega = \omega_2$ and $(k_z c_A)^2 < \omega_{ni}^2 |b|$, for both signs of b we have inequality (12.76) corresponding to the *positive* (ordinary) *dissipation*. However, the sign of the oscillation energy W depends on the sign of b, and for $b < 0$ it follows that

$$W < 0 \tag{12.78}$$

i.e., the *wave energy is negative*. Therefore the oscillation branch $\omega = \omega_2$ for $b < 0$ is unstable due to the negative wave energy. The growth rate of such an instability is of order

$$\gamma \simeq z_i k_z c_A. \tag{12.79}$$

For $k_z \simeq |b|^{1/2} \omega_{ni}/c_A$ it reaches a maximum of order

$$\gamma_{max} \simeq z_i (|b|\beta)^{1/2} \omega_{ni}. \tag{12.80}$$

The instabilities examined in this section are due to the interaction of resonant ions with waves, i.e. they are *essentially kinetic*. However, as will be shown in section 12.5.3, these instabilities are similar to those in a *collisional plasma* where *kinetic effects* are replaced by *dissipative ones*.

12.5.3 Dissipative instabilities in a collisional plasma
As in section 12.3.3, we will now consider perturbations with $k_z \gtrsim (\omega \nu_i)^{1/2}/v_{Ti}$, but now, as in sections 12.5.1 and 12.5.2, we are interested in Alfven branches rather than inertialess ones. The dispersion relation for Alfven waves with the given k_z can be derived from the general hydrodynamic equations with the expressions for the viscosity tensor and heat flux given in section 9.3. This dispersion relation can be written in the form of (12.59) where $D_i^{(0)}$ is determined by (12.61) while the function $\mathrm{Im}\, D_i^{(1)}$ has the form

$$\mathrm{Im}\, D_i^{(1)} = \frac{\nu_i z_i c^2}{3\omega c_A^2} \frac{0.9 + 2.4\omega_{ni}/\omega + 0.1(\omega_{ni}/\omega)^2}{1 + 1.7\omega_{ni}/\omega}. \tag{12.81}$$

When $b = 0$ and $k_z c_A \ll \omega_{ni}$, it follows from (12.59), (12.61) and (12.81) that the plasma is unstable against the perturbations considered. Their frequency is defined by (4.11), while the growth rate is equal to

$$\gamma = \frac{z_i}{204} \frac{\nu_i |\omega_{ni}|}{|k_z| v_{Ti}} (2\beta)^{1/2}. \tag{12.82}$$

It is evident that for $k_z v_{Ti} \simeq \nu_i$ expressions (12.82) and (12.67) are qualitatively similar. Like (12.67), equation (12.82) describes the excitation of positive energy waves, $W > 0$, due to the effect of *negative dissipation*, $\omega \mathrm{Im}\, D_i^{(1)} < 0$. Here, however, the dissipation is due to binary collisions and not to the interaction of resonant particles with the waves.

For $|b| \gg 1$ and $(k_z c_A)^2 \ll |b| \omega_{ni}^2$ we obtain, instead of (12.82)

$$\gamma = -\frac{2}{3b} z_i \nu_i. \tag{12.83}$$

As for the real part of the frequency, it is given by $\mathrm{Re}\,\omega = \omega_2$ (see (12.75)).

It can be seen from (12.83) that, when $b < 0$, an instability will occur. This result is similar to (12.79) and (12.80).

13 Electromagnetic Kelvin–Helmholtz Instabilities

In the present chapter we consider a plasma moving along the magnetic field with a velocity U dependent on the transverse coordinate x so that $dU/dx \equiv U' \neq 0$. Within the collisional approximation, the permittivity tensor of such a plasma has been calculated in section 2.1.5. This enables one to examine small-scale instabilities in a collisionless plasma caused by the *velocity profile inhomogeneity* U'. As is evident from sections 2.1.2 and 2.1.5, such instabilities can be associated with both the *inertialess (magnetoacoustic)* and *Alfven (inertial) oscillation branches*. In sections 10.4 and 10.5 we have given the hydrodynamic derivation of the dispersion relations for the inertialess perturbations in a plasma flow with $U' \neq 0$ and have obtained starting equations for large-scale and small-scale Alfven perturbations.

Using all these results, we shall now examine various specific types of instabilities in a plasma flow with $U' \neq 0$—called *Kelvin–Helmholtz instabilities*. As an abbreviation we shall write simply KH instead of "Kelvin–Helmholtz".

The main topic of this chapter is the theory of the so-called *drift KH instabilities*. However, such instabilities are important only when the so-called *hydromagnetic*, or *ordinary*, KH *instability* does not appear. Therefore, at first we shall consider the ordinary KH instability in sections 13.1 and 13.2. Then we shall consider the *drift-Alfven KH instabilities*.

Large-scale and small-scale varieties of these instabilities are discussed in sections 13.3 and 13.4, respectively. Finally, in section 13.5 we present the theory of *drift magnetoacoustic KH instabilities*.

The notion of the ordinary KH instability can be obtained in the simplest manner by considering a plasma with a *step-function velocity profile*. We follow such an approach in section 13.1. Then it becomes clear that, for this instability to be excited, a sufficiently large jump in velocity is required. In the simplest case of a plasma with homogeneous density, that jump, ΔU, must exceed twice the Alfven velocity c_A, i.e., $|\Delta U| > 2c_A$. If this condition is satisfied, the excitation of *large-scale Alfven perturbations* with a growth rate of order $|\Delta U|/a$ can be expected, where a is the transition layer thickness. The condition for excitation of the ordinary KH instability becomes more rigorous if the plasma density is varied together with the velocity—this is associated with increasing the mean Alfven velocity.

The above results of the ordinary KH instability theory are presented in section 13.1 within the scope of the *two-fluid approach* that enables one to take into account the magnetic viscosity and other non-standard effects and thereby to consider a wider class of instabilities. At the same time, in section 13.1 we elucidate the standard, i.e., *single-fluid approach* to the ordinary KH instability theory. In addition, we develop there the *electrodynamic approach* to the theory of this instability and a similar treatment of the electromagnetic instability of *interpenetrating plasmas*.

It is also of interest to investigate the relationship between the ordinary KH instability and the *Rayleigh–Taylor instability*. This problem is discussed in section 13.1.

Section 13.2 is devoted to the kinetic theory of the ordinary KH instability. The *kinetic approach* enables one to consider the *plasma pressure anisotropy effect* on this instability. Thereby, we find out the connection between the ordinary KH instability and the *fire-hose instability*.

When k_z is sufficiently small, the ordinary KH instability becomes sensitive to *drift effects* which can be described within the scope of the single-fluid approach if the magnetic viscosity is taken into account. In section 13.3 we obtain an instability criterion for the large-scale Alfven perturbations whilst allowing for the above drift effects. As a particular case, it reduces to the criterion for ordinary KH instability, and when the latter one fails, it reduces to a criterion of the *large-scale drift-Alfven KH instability*. Thereby, we demonstrate the transition from the ordinary KH instability to the drift-Alfven instability. Note that such a smooth transition holds only in the case of *large-scale perturbations* covering regions with essentially different velocities. In the case of *small-scale perturbations*, the physical reason for the ordinary KH instability, i.e., the electromagnetic interaction between regions of different velocities, disappears. As for the drift-Alfven KH instability, its physical basis lies in the drift convection of the longitudinal momentum which holds for both large-scale and small-scale perturbations. Thus, there is no small-scale ordinary KH instability, but there exists *a small-scale drift-Alfven* кн *instability.* The latter is considered in section 13.4.

In contrast to the ordinary кн instability, the drift-Alfven кн instabilities, both large-scale and small-scale, may be considered as *microinstabilities.* A specific feature of these microinstabilities, as well as of all drift instabilities, is the fact that their wave number k_y is not equal to zero (recall that y is the direction perpendicular to the equilibrium magnetic field and to the velocity gradient). As usual for Alfven waves, the *Alfven* кн *microinstabilities* lead to a bending of the magnetic field lines. Since $k_y \neq 0$, development of these microinstabilities should excite, in particular, a magnetic field in the direction of the velocity gradient.

The growth rate of the Alfven кн microinstabilities is small compared with that of ordinary кн instabilities, roughly speaking, of order ρ_i/a. However, it increases with $k_\perp$. At the validity limits of the notion of Alfven waves, i.e., when $k_\perp \rho_i \simeq 1$, it proves to be of the same order as the growth rate of the ordinary кн instability.

The Alfven microinstabilities are sensitive to the *ion-pressure gradient* which leads to their suppression. In the case of a plasma with homogeneous temperature, when a pressure gradient is due to a density gradient, such a stabilizing effect may be considered as a density-gradient effect. However, one should keep in mind that a density gradient does not lead to suppression of the Alfven кн microinstabilities if, in addition to it, there is a temperature gradient in the opposite direction so that $\nabla \ln T = -\nabla \ln n$. This corresponds to the case of a force-free plasma, $\nabla p = 0$, discussed in Chapter 12. Thus, a *force-free plasma* with inhomogeneous density and temperature is just as liable to Alfven кн microinstabilities as that with $\nabla n = \nabla T = 0$.

Unlike the Alfven кн microinstabilities, the *magnetoacoustic* кн *instabilities* are sensitive to more specific effects, such as *resonant kinetic effects* in the case of a collisionless plasma, and *dissipative effects* in a collisional plasma. In section 13.5, which is devoted to these instabilities, we restrict ourselves to considering collisionless plasma. It is then found that their character depends essentially on the value of the longitudinal wave number k_z and to be more precise, on the ratio $\tilde{\omega}/k_z v_{Ti}$ (see the designations in section 2.1). It is also noteworthy that, unlike the Alfven кн instabilities when *one oscillation branch* is excited, the magnetoacoustic кн instabilities are associated with the excitation of *two oscillation branches* differing in the sign of k_y/k_z.

When $\beta \ll 1$, the growth rate of one of the unstable magnetoacoustic branches does not depend on β, while that of the other is small, of order $\beta^{1/2}$. The magnetoacoustic branch with the larger growth rate corresponds to the *electrostatic perturbations* discussed in [2], section 3.5. This branch is associated with the excitation of a longitudinal electric field, $E_z \neq 0$, and is a modification of the ion-acoustic branch. The branch with the smaller growth rate is essentially electromagnetic. Its excitation is accompanied by a build up of the perturbed magnetic field $\tilde{B}_z$, corresponding to compressible perturbations. However, when $\beta \gtrsim 1$, both branches are characterized by growth rates of the same order. Similarly to Alfven KH instabilities, magnetoacoustic instabilities are suppressed by a sufficiently large *density gradient* (when $\nabla T = 0$).

On the whole, the picture of *microinstabilities* due to a longitudinal-velocity gradient is similar to that of microinstabilities driven by a *temperature gradient* (see the preceding chapters). Hence it may be considered that, like the *anomalous transverse heat conductivity effect* which, according to non-linear theory, should occur during the development of instabilities in a plasma with $\nabla T \neq 0$, the analogous effect of instabilities in a plasma with $\nabla U \neq 0$ should be an *anomalous viscosity*.

The *history of KH instabilities* in a plasma moving along the magnetic field goes back to [13.1]. There the above-mentioned ordinary KH instability has been demonstrated. The results of [13.1] have been described in books [13.2, 13.3]. The electromagnetic treatment of the ordinary KH instability and its resemblance to the *Alfven instability* in *interpenetrating plasmas* were discussed in [1.1], section 7.

The role of perturbations with $k_z a \gtrsim 1$ in the ordinary KH instability was studied in [13.4]. According to [13.4], perturbations with $k_z a \gtrsim 1$ are stable, and the growth rate of unstable perturbations is reached when $k_z a \simeq 1$. This is in agreement with the qualitative estimates of section 13.1. The finiteness of the transition layer thickness was also taken into account in [13.5–13.7].

The role of the *compressibility* in the ordinary KH instability within the scope of the *single-fluid approach* was discussed, in particular, in [13.4, 13.8, 13.9].

The problem of a *kinetic description* of the ordinary KH instability was initially discussed in [13.10] (see also [13.11, 13.12]). The author of [13.10] has noted that a hydrodynamic description can be invalid, firstly, due to the fact that the Larmor radius of particles can be comparable with other characteristic lengths of the problem (e.g., the thickness of the transition layer and the wavelength of perturbations), and, secondly, due to possible *wave-particle resonance*.

Plasma pressure anisotropy was allowed for in the theory of the ordinary KH instability, in particular, in [13.13, 13.14], where the *Chew–Goldberger–Low (CGL) approximation* was used.

We also note an interesting problem—*hydromagnetic wave amplification* at a *tangential velocity discontinuity*, which is associated with the ordinary KH instability, see [13.14–13.19].

Drift KH instabilities were initially studied in the *electrostatic approximation*, primarily for plasma with $\beta = 0$. A corresponding group of papers was mentioned in [2], Chapter 3. An important step forward in the theory of such instabilities in a finite-β plasma was made by [10.8]. The author of [10.8] has demonstrated that, apart from the *magnetoacoustic KH instabilities* (including electrostatic ones when $\beta \to 0$), in a finite-β plasma, the *Alfven instabilities* can also be excited. Furthermore, he has shown that, when the parameter β is finite or large, the class of magnetoacoustic KH instabilities in a plasma with $\nabla U \neq 0$ is not limited to electrostatic instabilities and has to be described by relationships taking into account the *perturbed magnetic field*. The author of [10.8] has based his work on the equations of *two-fluid hydrodynamics* with the *magnetic viscosity tensor*. However, as shown in [10.6], linearizing the viscosity tensor, the author of [10.8] has made a mistake. Therefore, some of the results of [10.8] are not correct.

The two-fluid theory of drift KH instabilities in a finite-β plasma was further developed in [10.6], while the *kinetic theory* of drift KH instabilities in such a plasma was presented in [2.7]. The results of [10.6] are presented below.

We mention also [13.20, 13.21] where drift-Alfven KH instabilities in a low-β plasma were discussed.

Space and astrophysical problems are traditional topics of application of the theory of KH instabilities. Usually, however, only the concepts of the ordinary KH instability are used, as in the following problems:

(1) stability of *Earth's magnetopause* (i.e., the boundary between the *magnetosphere* and the *solar wind plasma* overflowing it); numerous papers on this problem were mentioned in [1.1];

(2) interaction between *adjacent streams of different velocities in the solar wind* [13.22];

(3) *filamentary structure of the solar corona* [13.23, 13.13];

(4) *helical structure of ionized comet tails* [13.24];

(5) interaction of the *solar wind* with *Venus* [13.25].

The allowance for drift KH instabilities in space and astrophysical problems was initiated by [13.26] which was concerned with *ionized comet tails*, and by [13.27] which was devoted to instabilities in the *solar wind streams*. Among the space applications of the theory of drift KH instabilities we also note [13.28] in which the instability of the *polar cusp boundary* was considered in the electrostatic approximation. In addition, we mention [13.29–13.32] in which, side by side with specific space problems, various aspects of the theory of drift

KH instabilities in a *two-energy component* plasma were studied.

Drift KH instabilities can be of interest also for *controlled fusion problems* and, in particular, for the problem of a *tokamak with fast-neutral injection.* In this connection, we should mention [13.33, 13.34] in which these instabilities were analysed under tokamak conditions in the electrostatic approximation, and [13.20] which was devoted to Alfven instabilities. Note also paper [13.35] which dealt with experiments at the Princeton tokamak PLT with neutral injection. Here, in the absence of microinstabilities, a flow with inhomogeneous velocity profile along the torus should have been established. But it has been experimentally ascertained that the plasma actually moves along the torus with uniform velocity. In the opinion of [13.35], such a smoothing of the velocity profile demonstrates the *anomalous plasma viscosity* caused by microinstabilities. Later on, the problem of *anomalous viscosity in tokamaks* was discussed, in particular, in [13.36, 13.37].

13.1   The ordinary Kelvin–Helmholtz instability

Let us consider (10.89) for the simplest case of a *step-function velocity profile.* Following the *surface-wave method,* [1], section 1.7, we find the dispersion relation for perturbations localized in the vicinity of the jump:

$$(\omega - k_z U_1)^2 + (\omega - k_z U_2)^2 - 2k_z^2 c_A^2 = 0. \tag{13.1}$$

Here U_1, U_2 are velocities on the two sides of the jump. The density and magnetic field are assumed to be the same on both sides of the jump so that $c_A = \text{const}.$

It follows from (13.1) that an instability takes place if

$$|\Delta U| > 2c_A \tag{13.2}$$

where $\Delta U \equiv U_2 - U_1$. This is the *ordinary* KH *instability.*

According to (13.1), under condition (13.2), the growth rate of the perturbations is

$$\gamma = |k_z| [(\Delta U)^2/4 - c_A^2]^{1/2}. \tag{13.3}$$

Since the only significant restriction on k_z is $k_z a \ll 1$, where a is the transition layer thickness, we have, in order of magnitude,

$$\gamma_{\max} \simeq |\Delta U|/a. \tag{13.4}$$

When $\nabla n \neq 0$ and $\nabla B \neq 0$, equation (10.89) is modified as follows

$$\nabla_\perp \cdot [(\tilde{\omega}^2/c_A^2 - k_z^2)\nabla_\perp \tilde{B}_x] = 0. \tag{13.5}$$

Then (13.1) is replaced by

$$(\omega - k_z U_1)^2/c_{A1}^2 + (\omega - k_z U_2)^2/c_{A2}^2 - 2k_z^2 = 0. \tag{13.6}$$

where subscripts 1, 2 in c_A^2 denote the values on both sides of the jump. Instability condition (13.2) is replaced by

$$|\Delta U|^2 > 2(c_{A1}^2 + c_{A2}^2). \tag{13.7}$$

This instability condition becomes substantially more rigorous than (13.2) when the plasma density on one side of the jump is small since the corresponding value of c_A^2 is large.

We shall consider how the above results can be obtained by means of the single-fluid approximation, section 13.1.1, and by means of the electrodynamic one, section 13.1.2. In addition, we shall discuss a resemblance of the instability studied, firstly, with the Alfven instability of two interpenetrating plasmas, section 13.1.3, and, secondly, with the Rayleigh–Taylor instability, section 13.1.4.

13.1.1 Single-fluid approach
We introduce the *plasma displacement* $\boldsymbol{\xi}$ defined by $d\boldsymbol{\xi}/dt = \tilde{\boldsymbol{V}}$. The perturbed magnetic field $\tilde{\boldsymbol{B}}$ is related to $\boldsymbol{\xi}$ by

$$\tilde{\boldsymbol{B}} = \nabla \times (\boldsymbol{\xi} \times \boldsymbol{B}_0). \tag{13.8}$$

When $\boldsymbol{B}_0 = \text{const}$, (13.8) means

$$\tilde{\boldsymbol{B}} = \mathrm{i}k_z B_0 \boldsymbol{\xi} - \boldsymbol{B}_0 \nabla \cdot \boldsymbol{\xi}. \tag{13.9}$$

In terms of $\boldsymbol{\xi}$, the single-fluid equation of motion has the form

$$-\rho_0 \tilde{\omega}^2 \boldsymbol{\xi} = (\nabla\nabla \cdot \boldsymbol{\xi} - k_z^2 \boldsymbol{\xi})B_0^2/4\pi \tag{13.10}$$

where ρ_0 is the equilibrium *mass density* and $\tilde{\omega} = \omega - k_z U$. Assuming the terms with $\tilde{\omega}^2$ and k_z^2 to be small, from (13.10) we find, in the zeroth approximation in these terms, the *incompressibility condition*

$$\nabla \cdot \boldsymbol{\xi} = 0. \tag{13.11}$$

This condition yields a relation between the components ξ_x and ξ_y, so that

$$\xi_y = (\mathrm{i}/k_y)\partial \xi_x/\partial x. \tag{13.12}$$

One more relation between ξ_x and ξ_y can be obtained by taking curl_z of both sides of (13.10)

$$\text{curl}_z[(\rho_0\tilde{\omega}^2 - k_z^2 B_0^2/4\pi)\boldsymbol{\xi}] = 0. \tag{13.13}$$

Using (13.12) to eliminate ξ_y, we arrive at the equation

$$\frac{\partial}{\partial x}\left[\left(\rho_0\tilde{\omega}^2 - \frac{k_z^2 B_0^2}{4\pi}\right)\frac{\partial \xi_x}{\partial x}\right] - k_y^2\left(\rho_0\tilde{\omega}^2 - \frac{k_z^2 B_0^2}{4\pi}\right)\xi_x = 0. \tag{13.14}$$

According to (13.9), $\xi_x = -\mathrm{i}\tilde{B}_x/k_z B_0$. Then (13.14) reduces to (13.5).

13.1.2 *Electrodynamic approach*
This is based on the fact that, in the *ideal longitudinal conductivity approximation*, the low-frequency long-wavelength perturbations are described by the electrodynamic equations

$$(\nabla \times \nabla \times \boldsymbol{E})_\alpha = (\omega/c)^2 \epsilon_\perp E_\alpha \tag{13.15}$$

where $\alpha = 1, 2$ and $E_z = 0$, and

$$\epsilon_\perp = [1 - k_z U(x)/\omega]^2 c^2/c_A^2(x). \tag{13.16}$$

Eliminating E_x from (13.15), we find

$$\frac{\partial}{\partial x}\left(\frac{k_z^2 - \omega^2\epsilon_\perp/c^2}{k_y^2 + k_z^2 - \omega^2\epsilon_\perp/c^2}\frac{\partial E_y}{\partial x}\right) - \left(k_z^2 - \frac{\omega^2}{c^2}\epsilon_\perp\right)E_y = 0. \tag{13.17}$$

When k_z/k_y is small and $(\omega/ck_y)^2 \ll 1/\epsilon_\perp$, it follows that

$$\frac{\partial}{\partial x}\left[\left(k_z^2 - \frac{\omega^2}{c^2}\epsilon_\perp\right)\frac{\partial E_y}{\partial x}\right] - k_y^2\left(k_z^2 - \frac{\omega^2}{c^2}\epsilon_\perp\right)E_y = 0. \tag{13.18}$$

Since, according to (10.88), $E_y = -(\omega/ck_z)\tilde{B}_x$, we conclude that (13.18) is the same as (13.5).

13.1.3 *Resemblance of the ordinary Kelvin–Helmholtz instability to the Alfven instability in interpenetrating plasmas*
The electrodynamic approach enables one to reveal a resemblance between the *ordinary* KH *instability* and the *Alfven instability in interpenetrating plasmas*, [1.1], section 7.

Using (2.3), (2.5) and (2.7), we find the following dispersion relation for the case of interpenetrating plasmas:

$$\epsilon_{11} - c^2 k_z^2 / \omega^2 = 0 \qquad (13.19)$$

where

$$\epsilon_{11} = \sum_{\alpha=1,2} \frac{c^2}{c_{A\alpha}^2} \left(1 - \frac{k_z U_\alpha}{\omega} \right)^2. \qquad (13.20)$$

Here $c_{A\alpha}$ is the Alfven velocity of each plasma and U_α is its corresponding longitudinal velocity.

It follows from (13.19) and (13.20) that

$$(\omega - k_z U_1)^2 / c_{A1}^2 + (\omega - k_z U_2)^2 / c_{A2}^2 - k_z^2 = 0. \qquad (13.21)$$

This dispersion relation coincides with (13.6) even to the coefficient of k_z^2. Such a coincidence means that, under *large-scale* (with respect to the transition layer thickness) *perturbations*, two adjacent plasmas with different velocities behave qualitatively like two interpenetrating plasmas (cf. [1], section 1.7).

13.1.4 Correspondence between Kelvin–Helmholtz and Rayleigh–Taylor instabilities

Let us include in our analysis a *gravity force* $\boldsymbol{g} = (g,0,0)$ (cf. section 4.5). Using the results of section 6.1 of [2], we then find that (13.18) is replaced by

$$\nabla_\perp \cdot \left[\left(k_z^2 - \frac{\omega^2}{c^2} \epsilon_\perp \right) \nabla_\perp E_y \right] - k_y^2 g E_y \frac{\partial}{\partial x} \left(\frac{1}{c_A^2} \right) = 0. \qquad (13.22)$$

Hence the following generalization of the dispersion relation (13.6) is obtained

$$(\omega - k_z U_1)^2 / c_{A1}^2 + (\omega - k_z U_2)^2 / c_{A2}^2 - 2k_z^2 + g \mid k_y \mid (1/c_{A1}^2 - 1/c_{A2}^2) = 0. \qquad (13.23)$$

It describes growing perturbations when

$$k_z^2 \left(2 - \frac{(U_1 - U_2)^2}{c_{A1}^2 + c_{A2}^2} \right) - g \mid k_y \mid \left(\frac{1}{c_{A1}^2} - \frac{1}{c_{A2}^2} \right) < 0. \qquad (13.24)$$

If $g = 0$, (13.24) is the criterion of the KH instability (13.7). If $U_1 = U_2$ and $k_z = 0$, it reduces to the known condition of the *Rayleigh–Taylor* (RT) *instability*

$$g(n_2 - n_1) < 0 \qquad (13.25)$$

thus perturbations are unstable when the gravity force is directed against the density gradient. When $U_1 = U_2$, the terms with k_z^2 in (13.24) describe suppression of the RT instability due to bending of the magnetic field lines in perturbations with finite longitudinal wavelength.

It can be seen from (13.23) and (13.24) that KH and RT instabilities are related to the same (Alfven) oscillation branches. By means of (13.24) we find that the KH instability is more important than the RT instability if

$$k_z^2(\Delta U)^2 n_1 n_2 \gtrsim \mid k_y g(n_1^2 - n_2^2) \mid. \tag{13.26}$$

Let us also note that, unlike the KH instability, the RT instability reveals itself not only in the *jump transition layer approximation* but also in the WKB approximation (see [2], section 6.1).

13.2 Kinetic theory of the ordinary Kelvin–Helmholtz instability

In the *ideal longitudinal conductivity approximation* ($\epsilon_{33} \to \infty$, $E_z \to 0$), the *large-scale perturbations* are described by the electrodynamic equations

$$(k_y^2 + k_z^2)E_x + ik_y\frac{\partial E_y}{\partial x} = \frac{\omega^2}{c^2}(\hat{\epsilon}_{xx}E_x + \hat{\epsilon}_{xy}E_y)$$
$$ik_y\frac{\partial E_x}{\partial x} - \frac{\partial^2 E_y}{\partial x^2} + k_z^2 E_y = \frac{\omega^2}{c^2}(\hat{\epsilon}_{yx}E_x + \hat{\epsilon}_{yy}E_y). \tag{13.27}$$

Here $\hat{\epsilon}_{\alpha\beta}$ ($\alpha, \beta = x, y$) are operators for which the explicit form can be obtained by using expressions (2.3) for ϵ_{11} and ϵ_{22}.

Neglecting the compressibility, we find

$$\begin{aligned}\hat{\epsilon}_{xx} &= \hat{\epsilon}_{yy} = \epsilon_\perp \\ \hat{\epsilon}_{xy} &= \hat{\epsilon}_{yx} = 0\end{aligned} \tag{13.28}$$

where

$$\epsilon_\perp = -\frac{4\pi e^2}{M\Omega^2}\left\langle \frac{v_\perp^2}{2}(1 - \frac{k_z v_z}{\omega})G \right\rangle. \tag{13.29}$$

The function G is assumed to be defined by (1.54) with $\partial F/\partial x \to 0$. In particular, for a *moving plasma* with a *bi-Maxwellian velocity distribution*, i.e., for

$$F \sim \exp\left(-\frac{Mv_\perp^2}{2T_\perp} - \frac{M(v_z - U)^2}{2T_\parallel}\right) \tag{13.30}$$

expression (13.29) means

$$\epsilon_\perp = \frac{c^2}{c_A^2}\left[\left(1 - \frac{k_z U}{\omega}\right)^2 + \frac{(T_\parallel - T_\perp)k_z^2}{M_i\omega^2}\right]. \qquad (13.31)$$

From (13.27) we find equation (13.17) with $\epsilon_\perp$ of the form of (13.29) or (13.31). Assuming $(\omega/ck_y)^2 \ll 1/\epsilon_\perp$, we arrive at (13.18). As a result, in the case of a step-function velocity profile we obtain the following dispersion relation replacing (13.1) and (13.6) (cf. (13.19))

$$\epsilon_{\perp 1} + \epsilon_{\perp 2} - 2c^2 k_z^2/\omega^2 = 0 \qquad (13.32)$$

where the meaning of the subscripts 1, 2 is obvious.

When F is of the form of (13.30) and $\epsilon_\perp$ is defined by (13.31), instability condition (13.2) is replaced by

$$\frac{(\Delta U)^2}{4} + \frac{T_\parallel - T_\perp}{M_i} > c_A^2. \qquad (13.33)$$

For $\Delta U = 0$ this inequality reduces to the condition of the *fire-hose instability* (see, e.g., [1.1], section 6.7)

$$(T_\parallel - T_\perp)/M_i > c_A^2. \qquad (13.34)$$

Therefore, when $\nabla U \neq 0$, we deal with a hybrid of the fire-hose and KH instabilities.

13.3   The large-scale drift-Alfven Kelvin–Helmholtz instability

Let us assume that instability criterion (13.2) fails. Augmenting (10.89) by the *drift effects*, we obtain (10.94). In the case of a *step-function velocity profile*, from (10.94) we find the dispersion relation replacing (13.1) to be

$$(\omega - k_z U_1)^2 + (\omega - k_z U_2)^2 - 2k_z^2 c_A^2 + k_y^2 k_z T_i(U_2 - U_1)\mathrm{sgn}\,k_y/M_i\Omega_i = 0. \qquad (13.35)$$

Now we have, instead of (13.2), the instability criterion

$$(\Delta U)^2 - 4c_A^2 + 2k_y^2 T_i \Delta U \,\mathrm{sgn}\,k_y/k_z M_i\Omega_i > 0. \qquad (13.36)$$

It is evident that an instability occurs even though criterion (13.2) fails, when k_z is sufficiently small. We call such an instability the *large-scale drift-Alfven KH instability*.

Remembering that the *surface-wave approximation* holds only when the transition layer thickness a is small compared to k_y^{-1} ($ak_y < 1$), we find that the term in (13.36) which is additional to (13.2) is essential only if

$$k_z a \lesssim \rho_i v_{Ti}/a\Delta U. \tag{13.37}$$

When $v_{Ti} \simeq \Delta U$, it follows that, at the limit of validity,

$$k_z a \simeq \rho_i/a. \tag{13.38}$$

This corresponds to the characteristic k_z in the usual theory of drift instabilities (see, e.g., [2], Chapter 3, and also the preceding chapters of the present book).

When $\beta \simeq 1$, $v_{Ti} \simeq \Delta U$ and $k_y a \simeq 1$, the characteristic growth rate of the instability considered is of order

$$\gamma \simeq \rho_i v_{Ti}/a^2 \tag{13.39}$$

i.e., of order of the drift frequencies. This growth rate is small compared with the growth rate of the ordinary KH instability (13.4) as ρ_i/a. Therefore, it is evident that the drift-Alfven KH instability, as well as other instabilities in a flow with an inhomogeneous velocity profile considered below, can be important only if the criterion of type (13.2) fails.

When $|\Delta U| \ll c_A$, from (13.35) it follows that the maximum growth rate is

$$\gamma_{\max} = \beta_i^{1/2} k_y^2 \rho_i \,|\Delta U|\,/8. \tag{13.40}$$

It is attained when

$$k_z = k_{z0} = k_y^2 \beta_i \Delta U \, \mathrm{sgn} k_y /8\Omega_i. \tag{13.41}$$

Hence it can be seen that the instability considered depends essentially on β and is only of interest when β is not too small.

13.4 The small-scale drift-Alfven Kelvin–Helmholtz instability

13.4.1 Plasma with $\nabla n = \nabla T = 0$
From (10.95) (or from (2.5), (2.7), (2.26)) it follows that *small-scale Alfven perturbations* in a plasma with $\nabla U \neq 0$ are unstable if

$$\alpha_i > 2/\beta_i. \tag{13.42}$$

The maximum growth rate is attained when

$$k_z = k_{z0} \equiv \beta_\mathrm{i} k_y U'/4\Omega_\mathrm{i}. \tag{13.43}$$

It is equal to

$$\gamma_{\max} = \beta_\mathrm{i}^{1/2} k_y \rho_\mathrm{i} U'/4. \tag{13.44}$$

At the limits of validity, i.e., at $k_y \simeq 1/a$, equations (13.40) and (13.44) are qualitatively the same.

13.4.2 The role of the density gradient
When $\nabla n \neq 0$ and $\nabla T = 0$, we have (10.100) instead of (10.95). Equilibrium equation (1.42) yields a relationship between $\omega_{\mathrm{D}i}$ and $\omega_{\mathrm{n}i}$ of the form of $\omega_{\mathrm{D}i} = -\beta\omega_{\mathrm{n}i}/2$. Allowing for this result, from (10.100) the criterion for the *suppression* of the *small-scale drift-Alfven* KH *instability by the density gradient* follows:

$$\left| \frac{\partial \ln n}{\partial \ln U} \right| \gtrsim \frac{\Delta U}{c_\mathrm{A}(1 + \beta/2)}. \tag{13.45}$$

Under condition (13.45), the perturbations for all k_z are stabilized. As for perturbations with a particular k_z, their stability condition can be written in the form

$$\frac{k_y}{k_z}\xi < \frac{2}{\beta_\mathrm{i}}\left[1 + \frac{1}{8}\left(1 + \frac{\beta}{2}\right)^2 \beta_\mathrm{i}\frac{k_y^2}{k_z^2}\frac{\rho_\mathrm{i}^2}{L_n^2}\right]. \tag{13.46}$$

Here $\xi = U'/\Omega_\mathrm{i} \equiv (k_z/k_y)\alpha_\mathrm{i}$ and $L_n = (\partial \ln n/\partial x)^{-1}$ is the density gradient length. The left-hand side of inequality (13.46) is assumed to be positive, otherwise condition (13.42) fails. Unlike (13.45), the stability condition (13.46) takes into account the difference between T_e and T_i.

13.4.3 The role of the temperature gradient
It is evident from (2.5), (2.7), (2.9) and (2.26) that, when $\nabla n \neq 0$ and $\nabla T \neq 0$, the dispersion relation for *small-scale Alfven perturbations* in a *plasma flow* with $\nabla U \neq 0$ is of the form

$$\tilde{\omega}^2 - k_z^2 c_\mathrm{A}^2 - \tilde{\omega}\omega_{\mathrm{p}i}^*\left(1 - \frac{\beta}{2}\right) - \frac{\beta}{2}\omega_{\mathrm{p}i}^{*2}\left(1 + \frac{\kappa_T}{\kappa_p}\right) + \alpha_\mathrm{i}\frac{k_z^2 T_\mathrm{i}}{M_\mathrm{i}} = 0. \tag{13.47}$$

Using (13.47), we find the stabilization criterion

$$\left| \frac{\partial \ln p}{\partial \ln U} \right| \left[\left(1 + \frac{\beta}{2}\right)^2 + 2\beta\frac{\partial \ln T}{\partial \ln p}\right]^{1/2} \gtrsim \frac{|\Delta U|}{c_\mathrm{A}}. \tag{13.48}$$

The expression in the square brackets is assumed to be positive since otherwise the instability considered in section 4.1 takes place.

When $\nabla T = 0$, (13.48) reduces to (13.45). When $\nabla n = 0$ and $\nabla T \neq 0$, (13.48) means

$$\left| \frac{\partial \ln T}{\partial \ln U} \right| \gtrsim \frac{|\Delta U|}{c_A} \left[\left(1 + \frac{\beta}{2} \right)^2 + 2\beta \right]^{-1/2}. \tag{13.49}$$

Thus, as well as the density gradient, the *temperature gradient suppresses* the *instability* considered. Such a suppression occurs also in the presence of both gradients if $\partial \ln T / \partial \ln n \neq -1$, see section 13.4.4.

13.4.4 Force-free and almost force-free high-β plasma

When there are density and temperature gradients, $\nabla n \neq 0$ and $\nabla T \neq 0$, satisfying the relationship $\partial \ln T / \partial \ln n = -1$, we have $\nabla p = \nabla B = 0$. Then, according to (13.47), equation (10.95) holds, thus demonstrating the instability considered in section 13.4.1.

For $\beta \gg 1$ the case of an *almost force-free plasma* with $\nabla p \approx 0$ and $\nabla B \neq 0$ is of interest. For this case, (13.48) reduces to the following stabilization criterion

$$\left| \frac{\partial \ln B}{\partial \ln U} \right| \gtrsim \frac{|\Delta U|}{c_A} \left(1 + \frac{2}{b} \right)^{-1/2} \tag{13.50}$$

where, as in section 12.2, $b = \partial \ln B / \partial \ln n$.

At first sight it may seem that, in the presence of a magnetic-field gradient comparable with the density gradient, the stabilization criterion in a plasma with $\beta \gg 1$, equation (13.50), looks approximately like that in the case of a plasma with $\beta \lesssim 1$ and $\nabla T = 0$ (cf., e.g., (13.45)). However, one should keep in mind that, when $\beta \gg 1$, the Alfven velocity is small compared with the ion thermal velocity, $c_A / v_{Ti} \simeq \beta^{-1/2}$. Therefore, stabilization criterion (13.50) is really more rigorous than, e.g., (13.45).

13.5 Drift magnetoacoustic Kelvin–Helmholtz instabilities

Under the kinetic approach, *drift magnetoacoustic* KH *instabilities* are described by dispersion relation (2.8) with D_2 of the form of (2.6) and with ϵ_{22}, ϵ_{23}, ϵ_{33} of the form of (2.26). Within the scope of the two-fluid approach we have obtained dispersion relations (10.81) and (10.86) describing the same type of instabilities. Now we will analyse these dispersion relations. In sections 13.5.1 and 13.5.2 we assume $\nabla n = \nabla T = 0$. The perturbations with $\tilde{\omega} \gg k_z v_{Ti}$ will be considered in section 13.5.1, and those with $\tilde{\omega} \ll k_z v_{Ti}$ will be dealt with in section 13.5.2. In section 13.5.3 the density gradient will be taken into account.

13.5.1 Perturbations with $v_{Ti} \ll \tilde{\omega}/k_z \ll v_{Te}$
When $v_{Ti} \ll \tilde{\omega}/k_z \ll v_{Te}$, equations (2.6), (2.8) and (2.26) yield the dispersion relation

$$1 + \alpha_i \frac{k_z^2 T}{M_i \tilde{\omega}^2} + \frac{\beta}{4}\left(1 - \alpha_i \frac{k_z^2 T}{M_i \tilde{\omega}^2}\right)\left(3 + \alpha_i \frac{k_z^2 T}{M_i \tilde{\omega}^2}\right) = 0. \qquad (13.51)$$

For simplicity we assume here $T_e = T_i \equiv T$.

Note that (13.51) can also be obtained from (10.81) if we assume $\gamma_0 = 2$. Such a coincidence of the kinetic dispersion relation with the hydrodynamic one holds also when $T_i \neq T_e$.

Let us consider separately the consequences of equation (13.51) in the limiting cases where $\beta \ll 1$ and $\beta \gg 1$.

(a) Low-β plasma
When $\beta \ll 1$, equation (13.51) yields first of all the result of the *electrostatic approximation* (see [2], section 3.5)

$$\tilde{\omega}^2 = -\alpha_i k_z^2 T/M_i \qquad (13.52)$$

demonstrating the *ion-acoustic KH instability* for $\alpha_i > 0$. The growth rate of this instability is

$$\gamma = \alpha_i^{1/2} k_z (T/M_i)^{1/2}. \qquad (13.53)$$

In addition, when $\beta \ll 1$, equation (13.51) describes one more pair of oscillation branches with frequencies

$$\tilde{\omega}^2 = \alpha_i \beta k_z^2 T/4 M_i. \qquad (13.54)$$

These frequencies satisfy the assumption adopted above; $\tilde{\omega} \gg k_z v_{Ti}$ if $|\alpha_i| \gg 1/\beta$. They prove to be purely imaginary when $\alpha_i < 0$. In this case

$$\gamma = (-\alpha_i \beta/4)^{1/2} k_z (T/M_i)^{1/2}. \qquad (13.55)$$

As β increases to values of the order of unity, the frequencies in (13.52) and (13.54) become of the same order.

(b) High-β plasma
When $\beta \gg 1$, instead of (13.52) and (13.54) from (13.51) we find the following *two pairs* of oscillation branches

$$\tilde{\omega}^2 = -\alpha_i k_z^2 T/3 M_i \qquad (13.56)$$
$$\tilde{\omega}^2 = \alpha_i k_z^2 T/M_i. \qquad (13.57)$$

Clearly, unlike the case where $\beta \ll 1$, for $\beta \gg 1$ perturbations with $\alpha_i > 0$ and $\alpha_i < 0$ have growth rates of the same order.

13.5.2 Perturbations with $\tilde{\omega} \ll k_z v_{Ti}$

When $\tilde{\omega} \ll k_z v_{Ti}$, equations (2.6), (2.8) and (2.26) yield the dispersion relation

$$y^2 + 4y(1 - 1/\beta) - 8/\beta = 0 \qquad (13.58)$$

where

$$y = -\alpha_i + i\pi^{1/2}(1 - \alpha_i)\tilde{\omega}/ \mid k_z \mid v_{Ti}. \qquad (13.59)$$

Equation (13.58) has solution $\tilde{\omega} = 0$ provided α_i satisfies the relationship

$$\alpha_i = \alpha_i^{\pm} \equiv 2\left[1 - \frac{1}{\beta} \pm \left(1 + \frac{1}{\beta^2}\right)^{1/2}\right]. \qquad (13.60)$$

In the vicinity of $\alpha_i = \alpha_i^{\pm}$ the solution of (13.58) has the form

$$\tilde{\omega} = -\frac{i}{\pi^{1/2}} \frac{\mid k_z \mid v_{Ti}}{1 - \alpha_i} \frac{(\alpha_i - \alpha_i^{+})(\alpha_i - \alpha_i^{-})}{2(\alpha_i - 2 + 2/\beta)}. \qquad (13.61)$$

Equation (13.60) plays the role of the instability boundary. Two signs in (13.60) mean that there are *two types of instabilities* similar to those considered in section 13.5.1 for $\tilde{\omega} \gg k_z v_{Ti}$. To ascertain the properties of these instabilities, let us analyse the limiting cases of low- and high-β plasmas.

(a) Low-β plasma
It follows from (13.60) for $\beta \ll 1$ that

$$\alpha_i^{+} = 2 \qquad \alpha_i^{-} = -4/\beta. \qquad (13.62)$$

In the vicinity of $\alpha_i = 2$ the solution (13.61) reduces to

$$\tilde{\omega} = i\pi^{-1/2}(\alpha_i - 2) \mid k_z \mid v_{Ti}. \qquad (13.63)$$

Instability occurs if

$$\alpha_i > 2. \qquad (13.64)$$

The instability described by (13.63) may be treated as an extension of the instability of type (13.52) into the range of smaller values of α_i. Note that this equation as well as equation (13.52) can be obtained if dispersion relation (2.8) is replaced by the electrostatic dispersion relation $\epsilon_{33} = 0$.

When α_i is in the vicinity of $\alpha_i^- \equiv -4/\beta$, it follows from (13.61) that

$$\tilde{\omega} = -i\frac{\beta}{4\pi^{1/2}} \mid k_z \mid v_{Ti}\left(\alpha_i + \frac{4}{\beta}\right). \qquad (13.65)$$

The range

$$\alpha_i < -4/\beta \qquad (13.66)$$

corresponds to instability. It is evident that instability conditions in the case where $\alpha_i \approx \alpha_i^-$ are similar to those in the case of equation (13.54) (in both cases it is necessary that $\alpha_i < 0$). Therefore, we conclude that in both cases we deal with *one and the same instability* but for different values of α_i.

(b) High-β plasma
For $\beta \gg 1$ it follows from (13.60) that

$$\alpha_i^+ = 4 \qquad \alpha_i^- = -2/\beta. \qquad (13.67)$$

According to (13.61), in the vicinity of these values of α_i the oscillation frequencies are

$$\tilde{\omega} = i\pi^{-1/2} \mid k_z \mid v_{Ti}(\alpha_i - 4)/3 \qquad (13.68)$$
$$\tilde{\omega} = -i\pi^{-1/2} \mid k_z \mid v_{Ti}(\alpha_i + 2/\beta). \qquad (13.69)$$

Comparing (13.68) with (13.63), one can see that, as β increases, the boundary value of $\alpha_i = \alpha_i^+$ is displaced from $\alpha_i^+ = 2$ to $\alpha_i^+ = 4$, i.e., the instability condition becomes somewhat more rigorous, although the growth rate of the instability is of the same order of magnitude.

As for $\alpha_i = \alpha_i^-$, when β increases, it is displaced from a very large negative value $\alpha_i = -4/\beta$ at small β to an almost zero value $\alpha_i = -2/\beta$ at large β, so that the conditions of the second instability prove to be substantially less rigorous. The growth rate of this instability increases substantially as β increases (cf. (13.65) with (13.69)). Note also that (13.69) holds even far from the instability boundary, including $\beta \gg 2/\mid\alpha_i\mid$ (only the inequality $\mid \alpha_i \mid \gg 1$ is required). In this case it follows from (13.69), as well as from (13.57), that, for large β, the growth rate of the second instability does not depend on β.

13.5.3 The role of the density gradient
The density gradient in the problem of *drift magnetoacoustic* KH *instabilities* is taken into account by dispersion relation (10.86). From

it, we obtain the *stabilization criterion* similar to (13.46)

$$\left(1 - \frac{\beta}{2}\frac{T_e}{T_i}\right)\alpha_i < 1 + \frac{T_i}{T_e} + \frac{1}{4}\left(1 + \frac{\beta}{2}\right)\left(1 + \frac{\beta}{2}\frac{T_i}{T_e}\right)^2 \frac{T_e}{T_i}\frac{k_y^2\rho_i^2}{k_z^2 L_n^2}. \quad (13.70)$$

Hence it follows that, as well as the Alfven KH instabilities, magnetoacoustic instabilities can be *suppressed by a density gradient.*

14 Electromagnetic Drift Instabilities in a Low-β Plasma

As is well known (see, e.g., [1.3]), *drift instabilities in a low-β plasma* are subdivided into *electrostatic* and *electromagnetic ones*. The theory of electrostatic drift instabilities was presented in [2]. In this chapter we shall study *electromagnetic low-β plasma instabilities*. Within this study, we shall also deal with some *electrostatic drift instabilities* related to the electromagnetic instabilities under consideration.

Note that the term '*drift instabilities*' means the same as '*gradient instabilities*' (or '*gradient-driven instabilities*') implying that the instabilities are due to *density* and *temperature gradients*.

Note also that, according to section 2.4, specifying the low-β range is related to the fact that the longitudinal perturbed magnetic field $\tilde{B}_z$ and the magnetic drift of particles, caused by the gradient of the equilibrium magnetic field, are neglected, while the transverse perturbed magnetic field $\tilde{B}_\perp$ is allowed for (when $\tilde{B}_\perp$ is also neglected, we deal with the case of *zero-β plasma instabilities*).

In section 14.1, based on the *collisionless approach* of section 2.4, some results of the theory of the well-known *kinetic drift-Alfven instability* will be recalled. This instability is an electromagnetic modification of the *electrostatic drift (universal) instability* discussed in [2], section 3.1.

In section 14.2 we shall consider the *two-fluid approach* to the *resistive drift-Alfven instability* and calculate the growth rate of this instability. Similarly to the above kinetic instabilities, such an instability is an electromagnetic modification of the *electrostatic resistive drift instability* discussed in [2], section 5.2.

As is known, the transition from kinetic instabilities to the corresponding resistive ones, by increasing the degree of plasma collisionality, can be qualitatively studied by means of the *kinetic equation with a model collisional term*, see [2], section 4.7. In section 14.3 we shall use such an approach to obtain a unified picture of the *drift-Alfven instabilities*, taking into account both kinetic and resistive effects.

Then we shall consider the *suppression* of the above drift-Alfven instabilities *by finite-β effects* (section 14.4) and study the *electromagnetic short-wavelength temperature-gradient instabilities* (section 14.5) associated with the *electrostatic short-wavelength temperature-gradient instabilities* (see [2], section 3.2). Finally, we shall comment on the *hydrodynamic drift-Alfven instability* in a *highly rarefied plasma*, $\beta \ll M_e/M_i$ (section 14.6).

The *kinetic drift-Alfven instability* was initially pointed out in [2.10]. The dispersion relations for the *drift-Alfven oscillation branches* and the growth rates of these branches for arbitrary $k_\perp^2 \rho_i^2$ and $\partial \ln T/\partial \ln n$ were obtained in that paper. The characteristic value $\beta \simeq M_e/M_i$, above which it becomes necessary to allow for the *electromagnetic effects* in the *kinetic density-gradient instabilities*, was found there. In addition to the electromagnetic approach, in [2.10] the *hydrodynamic interpretation* of the *drift term* in the dispersion relation for the *drift-Alfven waves* is also presented. According to [2.10], this term is due to the *ion magnetic viscosity*. A more detailed consideration of the electrodynamic approach in the theory of this instability was given in [1.3], see also [1.2]. One more approach for describing the drift-Alfven waves was presented in the book of [14.1].

As mentioned in Chapter 5, the kinetic drift-Alfven instability was discussed in connection with *magnetospheric problems*. In this respect, paper [14.2] has been important, see also [14.3]. Then the role of collisionless drift-Alfven instabilities, studied in the *straight field-line approximation*, was considered in magnetospheric problems in a great number of other papers, in particular, in [14.4–14.11]. It is possible that some aspects of observations in the magnetospheric plasma can be interpreted by means of these instabilities. At the same time, it is clear that in such interpretations one should also take into account other instabilities, in particular, those discussed in the foregoing chapters and those instabilities which are dependent on the *magnetic-field geometry*, see Chapters 15 and 16.

The *resistive drift-Alfven instability* was initially investigated in [14.12]. The authors of [14.12] used an *electrodynamical approach* different from that presented in Chapters 1 and 2, augmenting the *Vlasov kinetic equation* by the *model collisional term* of type [14.13]. The *two-fluid approach* for this instability was developed in [14.14]. We shall use the results of [14.14] in section 14.2 whilst those of [14.12] will be used in section 14.3.

Laboratory observation of the resistive drift-Alfven instability was reported in [14.15], while the role of the drift-Alfven instabilities in *tokamaks* was initially discussed in [14.16].

The *non-linear theory* of the kinetic drift-Alfven instability was originally developed in [14.17, 14.18]. For further investigations on this topic see, in particular, [14.19–14.21].

The problem of the *suppression* of drift-Alfven instabilities by *finite-β effects* was posed in [1.7]. Other references on this problem will be considered in section 14.4.

The *electromagnetic short-wavelength temperature-gradient instabilities* in a collisionless low-β plasma were initially studied in [6.39]. According to [6.39], these instabilities are due to an *electron-temperature gradient* so that they can be of importance for a *hot-electron plasma*. We follow [6.39] in section 14.5.

The *hydrodynamic drift-Alfven instability* in a *highly rarefied plasma* was pointed out in [1.3, 2.10]. Applications of this instability will be considered in section 14.6.

14.1   Kinetic drift-Alfven instability in a low-β plasma

As a whole, the kinetic drift-Alfven instability is well-known. It was discussed in particular, in section 3.3 of [2] and Chapter 2 of [1.3]. Therefore, the main goal of the present section is a brief summary of the corresponding results. Consideration of the *suppression of the kinetic drift-Alfven instability* by *ion longitudinal motion* and *cold electron admixture* is new to [1, 1.3].

14.1.1   The simplest results
The basic dispersion relation for *drift-Alfven instabilities* can be obtained by means of (2.75) and (2.76). For the low-frequency ($\omega \ll \Omega_i$) long-wavelength ($k_\perp \rho_i \ll 1$) perturbations with $v_{Ti} \ll \omega/k_z \ll v_{Te}$, and for $\nabla T = 0$, we then have (cf. equation (3.43) of [2])

$$\left(1 - \frac{\omega_{ne}}{\omega}\right)\left(1 + i\pi^{1/2}\frac{\omega}{|k_z| v_{Te}}\right)\left[1 - \frac{\omega^2}{c_A^2 k_z^2}\left(1 - \frac{\omega_{ni}}{\omega}\right)\right]$$
$$+ z_i \frac{T_e}{T_i}\left(1 - \frac{\omega_{ni}}{\omega}\right) = 0. \tag{14.1}$$

In the $z_i \to 0$ approximation, equation (14.1) yields

$$\omega_{1,2} = -|\omega_{ni}|/2 \pm [(\omega_{ni}/2)^2 + (k_z c_A)^2]^{1/2} \tag{14.2}$$
$$\omega_3 = \omega_{ne}. \tag{14.3}$$

For definiteness we assume $\omega_{ne} > 0$ so that $\omega_{ni} < 0$. We adopt below that $T_i = T_e$. Then $\omega_{ni} = -\omega_{ne}$.

Solutions (14.2) correspond to the *drift-Alfven oscillation branches (drift-Alfven waves)* whilst (14.3) describes the *electron drift waves* (or, simply, *drift waves*). Note that (14.3) is simply (3.29). As for (14.2), it is a limiting case of (4.10) when $\beta \ll 1$, $\eta = 0$.

Following section 3.3 of [2] or section 2 of [1.3], one can find that, in allowance for the terms of order z_i in (14.1), the roots $\omega_{1,2,3}$ become complex. The maximum growth rate is attained for oscillations whose frequency lies near the point of intersection of the branches ω_1 and ω_3, i.e. near

$$k_z = k_z^{(0)} \equiv 2^{1/2}\omega_{ne}/c_A. \tag{14.4}$$

Then

$$\gamma_{max} = \left(\frac{\pi}{6}\frac{M_e}{M_i\beta}z_i\right)^{1/2}\omega_{ne}. \tag{14.5}$$

The instability is related to the electron-drift oscillation branch $\omega = \omega_3$ when $k_z > k_z^{(0)}$, and to the drift-Alfven branch when $k_z < k_z^{(0)}$ (see in detail [1.3]).

It can be seen that the growth rate increases with increasing $k_\perp$ as $k_\perp^2$. At the validity limits of the long-wavelength approximation, i.e., when $z_i \simeq 1$, we have

$$\mathrm{Re}\,\omega \simeq v_{Ti}/a$$
$$\gamma_{max} \simeq (M_e/M_i\beta)^{1/2}v_{Ti}/a \tag{14.6}$$
$$k_z \simeq \beta^{1/2}/a$$

where the value a is the characteristic length of the density gradient.

When $z_i \gtrsim 1$, equation (14.1) is replaced by

$$\left(1 - \frac{\omega_{ne}}{\omega}\right)\left(1 + i\pi^{1/2}\frac{\omega}{|k_z|v_{Te}}\right)$$
$$\times \left[1 - \frac{\omega^2}{c_A^2 k_z^2}\left(1 - \frac{\omega_{ni}}{\omega}\right)\frac{1 - I_0\exp(-z_i)}{z_i}\right] \tag{14.7}$$
$$+ \left(1 - \frac{\omega_{ni}}{\omega}\right)\left[1 - I_0\exp(-z_i)\right] = 0.$$

If $z_i \gg 1$, the unstable root satisfies the condition $\omega \ll \omega_{ne}$. Under this condition (14.7) reduces to

$$2 + \left(\frac{\omega_{ne}}{k_\perp\lambda_D ck_z}\right)^2 - \frac{\omega_{ne}}{\omega(2\pi z_i)^{1/2}} - \frac{i\pi^{1/2}\omega_{ne}}{|k_z|v_{Te}} = 0. \tag{14.8}$$

Assuming $\gamma/\text{Re}\,\omega \ll 1$, it follows that

$$\text{Re}\,\omega = \frac{\omega_{ne}}{(2\pi z_i)^{1/2}} \frac{1}{2 + \beta \kappa_n^2 k_y^2 / k_z^2 k_\perp^2} \tag{14.9}$$

$$\gamma = \pi^{1/2} \omega_{ne} \text{Re}\,\omega / \mid k_z \mid v_{Te}. \tag{14.10}$$

As a function of k_z, the growth rate has a maximum

$$\gamma_{\text{max}} = 2^{-3/2} (M_e/M_i\beta)^{1/2} \omega_{ne} \tag{14.11}$$

attained at

$$k_z = k_{z\text{opt}} = (\beta/8)^{1/2} \kappa_n. \tag{14.12}$$

The assumption $\gamma \ll \text{Re}\,\omega$ is justified when $z_i \ll \beta M_i/M_e$.

Let us discuss equations (14.6) for the oscillation frequency and the growth rate concerning the perturbations with $k_\perp \rho_i \simeq 1$. It can be seen that the ratio of the growth rate to the oscillation frequency, under the above assumption $\beta > M_e/M_i$, is a small value of order $(M_i/M_e\beta)^{1/2}$. This ratio can also been written as c_A/v_{Te}, i.e., it is the same as the ratio of the phase velocity of the Alfven waves to the electron thermal velocity. Physically, this ratio characterizes the relative number of resonance electrons.

14.1.2 The role of the temperature gradient

The *resonant electrons* can excite the *drift-Alfven waves* not only for $\nabla T = 0$ but also for a certain range of $\eta \neq 0$. In order to be convinced of this, let us consider (2.76) for small z_i. We then obtain

$$\omega^2 + \omega \omega_{ne}(1 + \eta) - c_A^2 k_z^2 = -i\pi^{1/2}\frac{\omega}{v_{Te}}$$

$$\times c_A^2 \mid k_z \mid (k_\perp \rho_i)^2 \frac{[1 + \omega_{ne}(1 + \eta)/\omega][1 - \omega_{ne}(1 - \eta/2)/\omega]}{(1 - \omega_{ne}/\omega)^2}. \tag{14.13}$$

Hence we find that the kinetic drift-Alfven instability takes place if

$$-4 < \eta < 2. \tag{14.14}$$

We shall consider other electromagnetic instabilities in a collision-less low-β plasma with $\nabla T \neq 0$ in section 14.5.

14.1.3 The role of the ion longitudinal motion

In the simplest case $T_e = T_i$, $\nabla T = 0$, in allowance for the *ion longitudinal motion* (for the terms of order $k_z^2 v_{Ti}^2/\omega^2$), it follows from

(2.75) and (2.76) that (cf. (14.1))

$$\left[\left(1-\frac{\omega_{ne}}{\omega}\right)\left(1+i\pi^{1/2}\frac{\omega}{|k_z|v_{Te}}\right)-\frac{k_z^2 T}{M_i\omega^2}\left(1+\frac{\omega_{ne}}{\omega}\right)\right]$$
$$\times\left[1-\frac{\omega^2}{c_A^2 k_z^2}\left(1+\frac{\omega_{ne}}{\omega}\right)\right]+z_i\frac{T_e}{T_i}\left(1+\frac{\omega_{ne}}{\omega}\right)=0. \tag{14.15}$$

We find the *instability boundary* by equating the real and imaginary terms of (14.15) to zero. We then conclude that the instability takes place only if

$$z_i > z_{i\,min} = \frac{k_z^2 T}{M_i\omega_{ne}^2}-\frac{\beta}{2}\equiv\frac{T}{M_i\omega_{ne}^2}(k_z^2-k_z^{(0)2}) \tag{14.16}$$

where $k_z^{(0)}$ is defined by (14.4). It is clear that such a restriction on z_i concerns only the perturbations with

$$k_z > k_z^{(0)}. \tag{14.17}$$

In other words, the *ion longitudinal motion effectively suppresses the long-wavelength instability* of the *electron drift branch* $\omega=\omega_3$ (see (14.3)) with sufficiently large k_z, but *does not lead to stabilization of the drift-Alfven branch* $\omega=\omega_1$ (see (14.2)).

14.1.4 Stabilization by a cold electron admixture
In the presence of an admixture of low-density cold electrons $(n_c/n_h\ll 1, T_c/T_h\to 0)$, it is sufficient for the modification of (14.1) to take into account only the *longitudinal inertia* of these electrons. Similarly to (14.15), we then have

$$\left[\left(1-\frac{\omega_{ne}}{\omega}\right)\left(1+i\pi^{1/2}\frac{\omega}{|k_z|v_{Te}}\right)-\frac{n_c}{n_h}\frac{k_z^2 T}{M_e\omega^2}\right]$$
$$\times\left[1-\frac{\omega^2}{c_A^2 k_z^2}\left(1+\frac{\omega_{ne}}{\omega}\right)\right]+z_i\left(1+\frac{\omega_{ne}}{\omega}\right)=0. \tag{14.18}$$

Here $T\equiv T_h$.

In the limit $M_e\to 0$ and for finite n_c/n_h, two roots of (14.18) become formally infinite. In neglecting these roots, (14.18) reduces to (14.2), corresponding to the *ideal longitudinal conductivity approximation*, $\epsilon_{33}\to\infty$, considered in Chapter 5 (see section 5.3). It can be seen that in this approximation the instability does not develop. Recall that in section 5.3, in finding the growth rate, we allowed for the *compressibility*, i.e., an effect of order β. Such an effect is not contained in (14.18).

Let us now analyse (14.18) supposing M_e to be finite. Then we find that it is a necessary condition for instability of the *electron drift branch* $\omega = \omega_3$ (see (14.3)) far from $k_z = k_z^{(0)}$ that (cf. (11.16))

$$z_i \gtrsim \beta n_e M_i / n_h M_e. \tag{14.19}$$

At the same time, we assumed $z_i \lesssim 1$. Therefore, the long-wavelength instability considered is suppressed if

$$n_e / n_h \gtrsim M_e / M_i \beta. \tag{14.20}$$

As for the drift-Alfven branch $\omega = \omega_1$, according to (14.18), its growth rate far from the branch intersection point contains an additional small factor $(M_e / M_i \beta) n_h / n_e$.

14.2   Resistive drift-Alfven instability

When $A_z \neq 0$, $A_\perp = 0$, the equation of electron longitudinal motion is of the form (cf. (9.20) and (10.27))

$$
\begin{aligned}
e_e(-ik_z \phi + i\omega A_z / c) + e_e V_{ne} \tilde{B}_x / c \\
- ik_z(\tilde{p}_e / n_0 + s\tilde{T}_e) - \nu_e M_e(V_{ze} - V_{zi}) = 0.
\end{aligned}
\tag{14.21}
$$

In the *electrostatic approximation* $(A_z = 0)$ we have instead of this:

$$-ie_e k_z \phi - ik_z(\tilde{p}_e / n_0 + s\tilde{T}_e) - \nu_e M_e(V_{ze} - V_{zi}) = 0. \tag{14.22}$$

According to Maxwell's equations (cf. (10.31) and (10.32))

$$k_\perp^2 A_z = (4\pi / c) e_e n_0 (V_{ze} - V_{zi}). \tag{14.23}$$

Therefore, from (14.21) we obtain

$$V_{ze} - V_{zi} = \frac{ik_z}{\nu_e M_e \lambda}(e_e \phi + \tilde{p}_e / n_0 + s\tilde{T}_e) \tag{14.24}$$

where

$$\lambda = 1 - i(\omega_{pe} / ck_\perp)^2 (\omega - \omega_{ne}) / \nu_e. \tag{14.25}$$

From (14.22) we find an expression for $V_{ze} - V_{zi}$ which differs from (14.24) by the substitution $\lambda \to 1$. This means that the *electrostatic approximation for resistive drift instability* ([2], section 5.2) is valid only if

$$(\omega_{pe}/ck_\perp)^2 \, |\, \omega - \omega_{ne}\, |\, /\nu_e \ll 1. \tag{14.26}$$

For perturbations with $k_\perp \simeq 1/a$, condition (14.26) means qualitatively

$$\beta < \nu_e/\Omega_e. \tag{14.27}$$

When (14.26) is unsatisfied, the resistive drift instability should be considered to be an *electromagnetic* one. We will proceed to the study of such an electromagnetic instability, i.e., the *resistive drift-Alfven instability*.

14.2.1 Starting equations
The *resistive drift-Alfven instability* can be studied using the same approach as the electrostatic resistive drift instability (see [2], Chapters 4 and 5) if one introduces the factor λ in the denominator of the right-hand side of (14.24).

Let us assume β to be in the range $\nu_e/\Omega_e \ll \beta \ll 1$ and consider electromagnetic perturbations neglecting the ion perturbed longitudinal motion, $V_{zi} = 0$, as well as the electron and ion perturbed temperatures, $\tilde{T}_e = \tilde{T}_i = 0$. Using (14.24) and the following linearized electron continuity equation

$$-i\omega\tilde{n}_e + c(E_y/B_0)\partial n_0/\partial x + in_0 k_z V_{ze} = 0 \tag{14.28}$$

we find the electron perturbed density

$$\tilde{n}_e = -\phi n_0 \left(\frac{ck_y \kappa_n}{B_0 \omega} + \frac{i e_e k_z^2 (1 - \omega_{ne}/\omega)}{M_e \nu_e (\lambda\omega + i k_z^2 T/M_e \nu_e)} \right). \tag{14.29}$$

Since, as mentioned above, we neglect the ion longitudinal motion, the behaviour of the ion component is determined only by the field $E_\perp$ (not by E_z). Hence the ion density $\tilde{n}_i$ turns out to be the same as in the electrostatic approximation. To find $\tilde{n}_i$ we use the ion continuity equation

$$-i\omega\tilde{n}_i + \frac{cE_y}{B_0}\frac{\partial n_0}{\partial x} + i\mathbf{k}_\perp \cdot \mathbf{V}_{\perp i}^{(1)}(1 + \omega_{ne}/\omega)n_0 = 0. \tag{14.30}$$

Here $\mathbf{V}_{\perp i}^{(1)}$ means the *inertial part* of the ion velocity

$$\mathbf{V}_{\perp i}^{(1)} = -i\omega e_i \mathbf{E}_\perp/M_i \Omega_i^2 \tag{14.31}$$

and the term with ω_{ne}/ω corresponds to the *magnetic viscosity* (see in detail [2], Chapters 4 and 5). As a result, it follows from (14.30) that

$$\tilde{n}_i = -\phi n_0 \left(\frac{ck_y \kappa_n}{B_0 \omega} + \frac{e_i k_\perp^2 (1 + \omega_{ne}/\omega)}{M_i \Omega_i^2} \right). \tag{14.32}$$

Equating the right-hand sides of (14.29) and (14.32), we arrive at the dispersion relation

$$(\omega - \omega_{ne})(\omega^2 + \omega\omega_{ne} - k_z^2 c_A^2)$$
$$- z_i c_A^2 k_z^2 (\omega + \omega_{ne})(1 - i\omega\nu_e M_e/k_z^2 T) = 0 \tag{14.33}$$

which is a *generalization* of the following dispersion relation for the *electrostatic resistive drift instability*

$$\omega - \omega_{ne} + z_i(\omega + \omega_{ne})(1 - i\omega\nu_e M_e/k_z^2 T) = 0 \tag{14.34}$$

(cf. equation (14.34) with equation (4.107) of [2]).

14.2.2 Analysis of the dispersion relation
Let us assume (cf. (14.26))

$$(\omega_{pe}/ck_\perp)^2 \, |\, \omega - \omega_{ne} \,| \, /\nu_e \gg 1. \tag{14.35}$$

Then the terms with z_i in (14.33) are small, so that (14.33) can be solved by the method of successive approximations, by analogy with section 14.1.1. Assuming $\omega = \omega^{(0)} + \omega^{(1)}$, in the zeroth approximation with respect to z_i, we arrive at precisely the same result as in the case of the collisionless plasma considered in section 14.1.1. Then we have the three oscillation branches with real frequencies $\omega_{1,2,3}$ defined by (14.2) and (14.3). The complex parts of the oscillation frequencies due to the finiteness of z_i have the maximum at $\omega_1 = \omega_3$, i.e., at $k_z = k_z^{(0)}$, where $k_z^{(0)}$ is defined by (14.4). Then it follows from (14.33) that

$$(\omega^{(1)})^2 = (4/3) z_i \omega_{ne}^2 (1 - 2i\nu_{ie}/\beta\omega_{ne}) \tag{14.36}$$

where $\nu_{ie} \equiv (M_e/M_i)\nu_e$ is the ion–electron collision frequency (cf. section 10.2.3).

When $\beta \ll \nu_{ie}/\omega_{ne}$, we have from (14.36) that the growth rate is of order

$$\gamma \simeq z_i^{1/2} (\omega_{ne}\nu_{ie}/\beta)^{1/2} \tag{14.37}$$

whilst for $\beta \gg \nu_{ie}/\omega_{ne}$ we have

$$\gamma \simeq z_i^{1/2} \nu_{ie}/\beta. \tag{14.38}$$

It can be seen that *the growth rate of the resistive drift-Alfven instability decreases with increasing β.*

14.3   Use of the model collisional terms for electromagnetic density-gradient instabilities

We followed in section 14.1 the approximation of negligible collisions and in section 14.2 the approximation of frequent collisions. In these two limiting cases the dispersion relations were found by essentially different methods. To obtain a unified picture of the *electromagnetic density-gradient instabilities* for arbitrary collisionality, one can use the *kinetic equation with a model collisional term.* A similar approach was used in [2], section 4.7, when the *electrostatic drift instabilities* were considered.

In section 14.3.1 we shall derive a general expression for the *perturbed distribution function.* The *electron permittivity* will be calculated in section 14.3.2. As for the *ion permittivity*, it is more complicated. Therefore, we shall initially calculate the *transverse ion permittivity*, neglecting the longitudinal motion (section 14.3.3) and then the *longitudinal ion permittivity* neglecting the transverse motion (section 14.3.4). Thereby, we shall obtain all the necessary for constructing the dispersion relation. The dispersion relation will be written and its structure will be analysed in section 14.3.5. Finally, we shall use this dispersion relation to study long-wavelength perturbations (section 14.3.6) and short-wavelength perturbations (section 14.3.7).

14.3.1 General expression for the perturbed distribution function
We start with the standard form of the linearized kinetic equation for the perturbed distribution function $\tilde{f}_\alpha$ of particle species α (electrons, ions) (cf. (1.49) and (9.1))

$$\frac{\partial \tilde{f}_\alpha}{\partial t} + \boldsymbol{v}\cdot\nabla\tilde{f}_\alpha + \boldsymbol{v}\times\Omega_\alpha\cdot\frac{\partial \tilde{f}_\alpha}{\partial \boldsymbol{v}} + \frac{e_\alpha}{M_\alpha}\left(\boldsymbol{E} + \frac{1}{c}\boldsymbol{v}\times\tilde{\boldsymbol{B}}\right)\cdot\frac{\partial f_{0\alpha}}{\partial \boldsymbol{v}} = \tilde{C}_\alpha. \quad (14.39)$$

Note—the tilde denotes perturbations.

The *collisional term* $\tilde{C}_\alpha$ allows for collisions between particles of both the same and different species. Therefore

$$\tilde{C}_\alpha = \sum_\beta \tilde{C}_{\alpha\beta} \qquad (\alpha,\beta = \text{e},\text{i}). \quad (14.40)$$

Instead of precise expressions for the Coulomb collisional term [9.4], we take the following approximate expression corresponding to [14.22]

$$\tilde{C}_{\alpha\beta} = -\nu_{\alpha\beta}\left[\tilde{f}_\alpha - \left(\frac{\tilde{n}_\alpha}{n_0} + \frac{M_\alpha}{T_\alpha}\boldsymbol{v}\cdot\tilde{\boldsymbol{V}}_\beta\right)f_{0\alpha}\right]. \qquad (14.41)$$

Here, as in Chapter 10, $\tilde{n}_\alpha$ and $\tilde{\boldsymbol{V}}_\beta$ are the perturbed density and velocity of the corresponding particle species, and $\nu_{\alpha\beta}$ is the frequency of collisions of particles of species α with particles of species β. Note that a similar expression in [2] (see equation (4.104) of [2]) contains an erratum.

A method of studying the *gradient instabilities* based on using a collisional term of the form of (14.40) and (14.41) was initially suggested in [14.22]. Further development of this method was given in [14.12]. In particular, in [14.12] an expression for the collisional term allowing for finiteness of the equilibrium velocities of the corresponding plasma components is given. As mentioned in [14.22] (see also [14.12] and section 4.7 of [2]) the collisional term of the form of (14.40) and (14.41) satisfies the *conservation laws of particle number and momentum*. However, it does not take into account perturbations of the temperature, and effects such as heat conductivity and other effects mentioned in section 4.7 of [2]. As in section 4.7 of [2], we shall use this collisional term in the case of low-frequency ($\omega \ll \Omega_i$) perturbations in a plasma with an inhomogeneous density and a homogeneous temperature ($\nabla n_0 \neq 0$, $\nabla T = 0$), at a not too high frequency of ion-electron collision, $\nu_{ie} < \omega^*$.

The collisional term of the form of (14.40) and (14.41) enables one to find the solution of (14.39) by the *method of trajectory integration* (for more information on this method see in detail [1], sections 5.1 and 7.2; [2], section 1.7; and the present book, section 1.3). We then have

$$\tilde{f}_\alpha = -\int_{-\infty}^{t} \mathrm{d}t' \exp[-\nu_\alpha(t-t')]\left[\frac{e_\alpha}{M_\alpha}\left(\boldsymbol{E} + \frac{1}{c}\boldsymbol{v}\times\tilde{\boldsymbol{B}}\right)\cdot\frac{\partial f_{0\alpha}}{\partial\boldsymbol{v}}\right.$$
$$\left. - \sum_\beta \nu_{\alpha\beta}\left(\frac{\tilde{n}_\alpha}{n_0} + \frac{M_\alpha}{T_\alpha}\boldsymbol{v}\cdot\tilde{\boldsymbol{V}}_\beta\right)f_{0\alpha}\right].$$

$$(14.42)$$

Here $\nu_\alpha = \sum_\beta \nu_{\alpha\beta}$. The integral is taken along trajectories described by (1.48). After integrating, for the case of *low-frequency perturbations* ($\omega \ll \Omega_i$) in the *low-β plasma approximation* ($A_2 \to 0$, $\omega_d \to 0$), we obtain

$$\tilde{f}_\alpha = f_\alpha^{(1)} + f_\alpha^{(2)} \qquad (14.43)$$

where (cf. (1.52))

$$f_\alpha^{(1)} = -\frac{e_\alpha f_{0\alpha}}{T_\alpha}\left(\phi - \frac{\exp[i\xi\sin(\alpha-\chi)]J_0}{\omega + i\nu_\alpha - k_z v_z}\right.$$
$$\left. \times [(\omega + i\nu_\alpha - \omega_{n\alpha})\phi - (v_z/c)(\omega - \omega_{n\alpha})A_z]\right) \quad (14.44)$$

$$f_\alpha^{(2)} = i\sum_\beta \frac{\nu_{\alpha\beta}\exp[i\xi\sin(\alpha-\chi)]}{\omega + i\nu_\alpha - k_z v_z}$$
$$\times [(\tilde{n}_\alpha/n_0)J_0 + (M_\alpha/T_\alpha)(iv_\perp J_0'\tilde{V}_{2\beta} + v_z J_0 \tilde{V}_{z\beta})].\,(14.45)$$

Here $\tilde{V}_{2\beta} = \boldsymbol{e}_z \cdot \boldsymbol{k}\times\tilde{\boldsymbol{V}}_{2\beta}/k_\perp$. In the right-hand side of (14.45) we have omitted the contribution of $\tilde{V}_{1\beta} \equiv \boldsymbol{k}_\perp\cdot\tilde{\boldsymbol{V}}_\beta/k_\perp$ as small for small ω/Ω (see in detail in [14.12]).

Note also that, conforming with [14.23, 14.24], when $\xi \gg 1$, the collision frequency ν_α should be replaced by $\nu_{\alpha\text{eff}} \simeq \xi^2\nu_\alpha$.

14.3.2 Electron permittivity

In calculating the electron contribution to the permittivity, we assume the electron Larmor radius to be negligibly small, $\xi_e \to 0$. In addition, the electron transverse inertia is neglected, $\omega/\Omega_e \to 0$. For convenience, we allow for only electron–ion collisions, $\nu_{ei} \neq 0$, but not electron–electron collisions, $\nu_{ee} \to 0$; this is, of course, a modelling assumption. The effect of the ion longitudinal motion on the electron permittivity is neglected, $\tilde{V}_{zi} = 0$. In addition, we allow for the contribution of $\tilde{V}_{2i}$ to the electron expression (14.45) to be negligibly small when $\xi_e \to 0$. As a result, we find from (14.43)–(14.45)

$$\tilde{f}_e = \frac{e_e F_e}{T_e(\omega + i\nu_{ei} - k_z v_z)}$$
$$\times \left((k_z v_z + \omega_{ne})\phi - \frac{v_z}{c}(\omega - \omega_{ne})A_z + i\nu_{ei}\frac{T_e}{e_e}\frac{\tilde{n}}{n_0}\right). \quad (14.46)$$

Here $F_e \equiv f_{0e}$. Note that (14.46) can also be obtained by means of the *drift kinetic equation* (cf. (9.2)) *with the collisional term*

$$\tilde{C}_{ei} = -\nu_{ei}(\tilde{f}_e - \tilde{n}_e F_e/n_0). \quad (14.47)$$

By means of (14.46) we express the electron perturbed charge density $\rho_e \equiv e_e\tilde{n}_e$ and the electron longitudinal current $j_{ze} \equiv e_e n_0 \tilde{V}_{ze}$ in terms of the potentials ϕ and A_z. Then, following the approach of sections 1.1 and 2.4, we find the *electron contribution* to $\epsilon_\parallel, \epsilon_\perp$:

$$\epsilon_\parallel^{(e)} = (k_z\lambda_{De})^{-2}(1 - \omega_{ne}/\omega)(1 - \omega\alpha_1/\alpha_2) \quad (14.48)$$
$$\epsilon_\perp^{(e)} = \kappa_n k_y\omega_{pe}^2/\omega\Omega_e k^2. \quad (14.49)$$

Here

$$\alpha_1 = \langle (\omega + i\nu_{ei} - k_z v_z)^{-1} \rangle$$
$$\alpha_2 = \left\langle \frac{\omega - k_z v_z}{\omega + i\nu_{ei} - k_z v_z} \right\rangle. \qquad (14.50)$$

In the explicit form

$$\alpha_1 = -Z/(\omega + i\nu_{ei})$$
$$\alpha_2 = 1 + i\nu_{ei}Z/(\omega + i\nu_{ei}) \qquad (14.51)$$

where $Z = Z(x)$ is defined by (2.25).

14.3.3 Transverse ion permittivity

When calculating the ion contribution to the permittivity tensor, we allow for only *ion–ion collisions* so that $\nu_{ie} \to 0$. We neglect the ion longitudinal inertia assuming $k_z \to 0$, $A_z \to 0$, $\tilde{V}_{zi} \to 0$ in (14.44) and (14.45). We then have from (14.43)–(14.45)

$$
\tilde{f}_i = -\frac{e_i F_i}{T_i}\phi\left(1 - \frac{\omega + i\nu_i - \omega_{ni}}{\omega + i\nu_i}\exp[i\xi\sin(\alpha - \chi)]J_0\right)
$$
$$
+ \frac{i\nu_i F_i}{\omega + i\nu_i}\exp[i\xi\sin(\alpha - \chi)] \qquad (14.52)
$$
$$
\times \left(\frac{\tilde{n}_i}{n_0}J_0 + i\frac{M_i}{T_i}v_\perp J_0' \tilde{V}_{2i}\right) \qquad F_i \equiv f_{0i}, \ \nu_i \equiv \nu_{ii}.
$$

By means of (14.52) we calculate the ion perturbed density

$$\tilde{n}_i = -\frac{e_i\phi}{T_i}[1 - (\omega - \omega_{ni})G_1/G_2]. \qquad (14.53)$$

Here

$$
G_1 = (\omega + i\nu_i)I_0\exp(-z_i) - i\nu_i z_i(I_0^2 - I_1^2)\exp(-2z_i)
$$
$$
G_2 = (\omega + i\nu_i)^2 - i\nu_i(\omega + i\nu_i)[I_0 + 2z_i(I_0 - I_1)] \qquad (14.54)
$$
$$
\times \exp(-z_i) - \nu_i^2 z_i(I_0^2 - I_1^2)\exp(-2z_i)
$$

where $I_0 = I_0(z_i)$ and $I_1 = I_1(z_i)$ are, as usual, the Bessel functions of imaginary argument.

Using (14.53), we find, similarly to section 14.3.2

$$\epsilon_\perp^{(i)} = (k_\perp \lambda_{Di})^{-2}[1 - (\omega - \omega_{ni})G_1/G_2]. \qquad (14.55)$$

It is clear that, in the assumed approximation, $\epsilon_\parallel^{(i)} = 0$. In section 14.3.4 we shall take into account the ion longitudinal motion and thereby find a finite $\epsilon_\parallel^{(i)}$.

14.3.4 Longitudinal ion permittivity

We shall calculate $\epsilon_\parallel^{(i)}$ only in the approximation $z_i \to 0$. Then, by analogy with (14.46), it follows from (14.43)–(14.45) that

$$\tilde{f}_i(E_z) = \frac{iF_i}{\omega + i\nu_i - k_z v_z}\left[\frac{e_i v_z E_z}{T_i}\left(1 - \frac{\omega_{ni}}{\omega}\right) + \nu_i\left(\frac{\tilde{n}}{n_0} + \frac{M_i}{T_i}v_z \tilde{V}_{zi}\right)\right].$$

(14.56)

Next, using (14.56) and following the standard procedure, we obtain

$$\epsilon_\parallel^{(i)} = -(k_z \lambda_{\mathrm{Di}})^{-2}(1 - \omega_{ni}/\omega)\delta_1/\delta_2.$$ (14.57)

Here

$$\delta_1 = \left\langle \frac{k_z v_z}{\omega + i\nu_i - k_z v_z}\right\rangle$$

$$\delta_2 = \left\langle \frac{\omega - k_z v_z(1 + i\nu_i\omega M_i/T_i k_z^2)}{\omega + i\nu_i - k_z v_z}\right\rangle.$$

(14.58)

By analogy with (14.51), these quantities can be expressed in terms of the function Z.

14.3.5 Dispersion relation

Neglecting the ion longitudinal motion, from (2.76), (14.48), (14.49) and (14.55) we obtain the following dispersion relation

$$\left(1 - \frac{\omega_{ne}}{\omega}\right)\left(1 - \omega\frac{\alpha_1}{\alpha_2}\right)\left[1 - \frac{\omega(\omega - \omega_{ni})}{k_z^2 c_A^2 z_i}\left(1 - \omega\frac{G_1}{G_2}\right)\right]$$
$$+ \frac{T_e}{T_i}\left(1 - \frac{\omega_{ni}}{\omega}\right)\left(1 - \omega\frac{G_1}{G_2}\right) = 0.$$

(14.59)

In the limiting case of negligibly weak collisions, equation (14.59) coincides with (2.76) and (2.77). In the opposite limiting case of sufficiently frequent collisions (and for $z_i \ll 1$), it yields physically the same results as the *two-fluid approach*, see section 14.2.

Let us comment in detail on the *long-wavelength approximation*, $z_i \ll 1$. In this approximation

$$1 - \omega\frac{G_1}{G_2} = z_i\left[1 - \frac{3}{4}z_i + iz_i\frac{\nu_i}{\omega}\left(\frac{1}{4} + \frac{\omega}{\omega + i\nu_i}\right)\right].$$ (14.60)

It can be seen that, for both large and small ν_i/ω, the ion dissipative contribution to the dispersion relation is characterized by the parameter $z_i\nu_i/\omega$. This is in agreement with the *two-fluid approach*, see Chapters 9 and 10. Physically, the terms of order $z_i\nu_i/\omega$ correspond to the *ion transverse viscosity*.

Note that the following collisional term corresponds to expressions (14.55) and (14.60) (cf. (14.41))

$$\tilde{C}_{ii} = -\nu_{ii}\left[\tilde{f}_i - \left(\frac{\tilde{n}_i}{n_0} + \frac{M_i}{T_i}v_2\tilde{V}_{2i}\right)f_{0i}\right]. \tag{14.61}$$

Neglecting $\tilde{V}_{2i}$ one obtains, instead of (14.60), a physically incorrect result in which the dissipative terms yield the contribution to the dispersion relation of order ν_i/ω (a friction force type effect). Such an error had been made in [14.25].

In taking the *ion longitudinal motion* into account, by means of (14.57), we find the following dispersion relation (instead of (14.59))

$$\left[\left(1 - \frac{\omega_{ne}}{\omega}\right)\left(1 - \omega\frac{\alpha_1}{\alpha_2}\right) - \frac{T_e}{T_i}\left(1 - \frac{\omega_{ni}}{\omega}\right)\frac{\delta_1}{\delta_2}\right]$$
$$\times \left[1 - \frac{\omega(\omega - \omega_{ni})}{k_z^2 c_A^2 z_i}\left(1 - \omega\frac{G_1}{G_2}\right)\right] \tag{14.62}$$
$$+ \frac{T_e}{T_i}\left(1 - \frac{\omega_{ni}}{\omega}\right)\left(1 - \omega\frac{G_1}{G_2}\right) = 0.$$

Since δ_1/δ_2 was calculated in the approximation $z_i \to 0$, the quantity $1 - \omega G_1/G_2$ should be taken from (14.60).

When $(\omega, \nu_i) \gg k_z v_{Ti}$, it follows from (14.58) that

$$\frac{\delta_1}{\delta_2} = \frac{k_z^2 T_i}{M_i\omega^2}\left[1 + \frac{k_z^2 T_i}{M_i\omega^2}\left(1 + \frac{2\omega}{\omega + i\nu_i}\right)\right]. \tag{14.63}$$

It can be seen that in the collisionless limit, $\nu_i/\omega \to 0$, we have in the square brackets the standard term $1 + 3k_z^2 T_i/M_i\omega^2$. In the opposite limiting case, $\nu_i/\omega \gg 1$, equation (14.63) contains a dissipative term of order $k_z^2 v_{Ti}^2/\omega\nu_i$. This is in accordance with the *two-fluid approach* of section 10.3.3.

Note that (14.63) corresponds to a collisional term of the form

$$\tilde{C}_{ii} = -\nu_{ii}\left[\tilde{f}_i - \left(\frac{\tilde{n}_i}{n_0} + \frac{M_i}{T_i}v_z\tilde{V}_{zi}\right)f_{0i}\right]. \tag{14.64}$$

The author of [14.25] has neglected the term with $\tilde{V}_{zi}$ and has obtained an erroneous result whereby the dissipative contribution in (14.63) is of order ν_i/ω.

Thus, it is clear that, in order to calculate correctly the ion contribution into the dispersion relation, it is necessary to use a collisional term taking into account the *law of conservation of momentum*.

14.3.6 Long-wavelength drift-Alfven instabilities

Allowing for (14.60), we reduce (14.59) to the form

$$\left(1 - \frac{\omega_{ne}}{\omega}\right)\left\{1 - \frac{\omega^2}{c_A^2 k_z^2}\left(1 - \frac{\omega_{ni}}{\omega}\right)\left[1 + iz_i\frac{\nu_i}{\omega}\left(\frac{1}{4} + \frac{\omega}{\omega + i\nu_i}\right)\right]\right\}$$
$$+ z_i\frac{T_e}{T_i}\frac{1 - \omega_{ni}/\omega}{1 - \omega\alpha_1/\alpha_2} = 0.$$

$$(14.65)$$

Hence it follows that the long-wavelength oscillation branches (14.2) and (14.3) exist for arbitrary degree of the plasma collisionality (within the scope of the adopted approximation $\nu_{ie} < \omega^*$). The *ion transverse viscosity* (an effect of order $z_i\nu_i/\omega$) favours a damping of these oscillation branches. In neglecting such a viscosity, we have instead of (14.65),

$$\left(1 - \frac{\omega_{ne}}{\omega}\right)\left[1 - \frac{\omega^2}{c_A^2 k_z^2}\left(1 - \frac{\omega_{ni}}{\omega}\right)\right] + z_i\frac{T_e}{T_i}\frac{1 - \omega_{ni}/\omega}{1 - \omega\alpha_1/\alpha_2} = 0. \quad (14.66)$$

This dispersion relation was analysed in [14.12]. As limiting cases, the respective results of sections 14.1 and 14.2 are obtained from (14.66).

14.3.7 Electromagnetic short-wavelength resistive drift instability

Let us consider the perturbations with $z_i \gg 1$ assuming the electrons to be collisional. We suppose $\omega \ll \nu_{ei} \ll k_z v_{Te}$ and $T_e \equiv T_i$. It then follows from (14.59) that (cf. (3.100) of [14.16])

$$\omega = \frac{\omega_{ne}}{(2\pi z_i)^{1/2}}\left(2 + \beta\frac{\kappa_n^2 k_y^2}{k_\perp^2 k_z^2} - 2i\frac{\omega_{ne}\nu_{ei}}{k_z^2 v_{Te}^2}\right)^{-1}. \quad (14.67)$$

This expression describes the *electromagnetic short-wavelength resistive drift instability* similar to the kinetic instability characterized by (14.9) and (14.10).

14.4 Suppression of density-gradient instabilities in a finite-β plasma

14.4.1 Suppression of the kinetic drift-Alfven instability due to ion Landau damping

In contrast to section 14.1.1, we now allow for the interaction between perturbations and *resonant ions*.

At first let us estimate the order of magnitude of this effect.

Let us assume $z_i \ll 1$. Then, neglecting the resonant ions, we have the growth rate (14.5) attainable for $\omega \approx 2^{-1/2} k_z c_A \approx \omega_{ne}$. The number of ions moving along the equilibrium magnetic field with velocity close to ω/k_z is characterized by an exponential of the form

$$\exp(-\omega^2/k_z^2 v_{Ti}^2) \simeq \exp(-1/2\beta_i). \tag{14.68}$$

This value is small for small β. However, the ratio of the growth rate (14.5) to the oscillation frequency ω_{ne} is also small. Therefore if β is not too small (although smaller than unity), the resonant ions can become important.

The exponential of type (14.68) is obtained under the assumption that the resonant ions are those with $v_z = \omega/k_z$. This is a consequence of the so-called *longitudinal resonance*. However, the particles move not only along the field lines but also across them. The transverse motion of the particles is characterized by the magnetic-drift velocity (1.27). Due to the magnetic drift the so-called *transverse resonance* (*magneto-drift resonance*) occurs (cf., e.g., section 4.2). Then the resonant particles are those with $V_d(v_\perp) = \omega/k_y$. When $\omega \approx \omega_{ne}$, the number of such particles is proportional to the exponential

$$\exp(-\omega/k_y V_D) \simeq \exp(-1/\beta_i). \tag{14.69}$$

It can be seen that, in allowance for the longitudinal ion resonance (14.68), it is, generally speaking, necessary to take into account also the transverse resonance. This is most important for $z_i \gg 1$ when $\omega \ll \omega_{ne}$ (see (14.9)] and, as a consequence, $\exp(-\omega/k_y V_D) \gg \exp(-1/\beta_i)$.

Quantitatively, the role of the ion resonances can be studied by means of the following dispersion relation

$$2 - I_0 \exp(-z_i) \left(1 + \frac{\omega_{ne}}{\omega} \right) - \frac{\omega^2 - \omega_{ne}^2}{k_z^2 c_A^2 z_i} [1 - I_0 \exp(-z_i)]$$

$$+ i\pi^{1/2} S \frac{\omega}{|k_z| v_{Te}} \left[1 - \frac{\omega_{ne}}{\omega} + \left(\frac{M_i}{M_e} \right)^{1/2} \left(1 + \frac{\omega_{ne}}{\omega} \right) Q \right] = 0. \tag{14.70}$$

Here

$$S = 1 - \left(\frac{\omega}{k_z c_A} \right) \frac{1 - I_0 \exp(-z_i)}{z_i} \left(1 + \frac{\omega_{ne}}{\omega} \right) \tag{14.71}$$

$$Q = \int_0^\infty J_0^2[(2\lambda z_i)^{1/2}] \exp\left(-\lambda - \frac{(\omega - \omega_{Di}\lambda)^2}{k_z^2 v_{Ti}^2} \right) d\lambda. \tag{14.72}$$

We assume $T_e = T_i$, so that $\omega_{ne} = -\omega_{ni}$.

Note that, in neglecting the ion magnetic drift, $\omega_{Di} \to 0$,

$$Q = I_0(z_i) \exp(-z_i - \omega^2/k_z^2 v_{Ti}^2). \tag{14.73}$$

In this approximation equation (14.70) can be obtained by means of (2.75) and (2.76). Substitution of (14.73) by (14.72) taking into account the magnetic drift is a model to a degree. Note also that the structure of the integral (14.72) is similar to that of integrals (4.35). Details of obtaining (14.70) are in [1.7].

The imaginary terms in (14.70) correspond to the resonant inter-action of the waves with electrons (the term with $1 - \omega_{ne}/\omega$) and with ions (the term with the integral Q). The real part of this equa-tion coincides with the real part of (14.7). Therefore, the picture of the oscillation branches considered in section 14.1.1 holds. The sign of the growth rate is opposite to the sign of the expression in the square brackets of the last term of (14.70). Hence it follows that, in accordance with the above, the resonant ions result in damping of the oscillations (since $\omega_{ne}/\omega > 0$).

Assuming $\text{Im}\,\omega = 0$ in (14.70) and separating the real and imag-inary terms, we obtain the following equations for the *instability boundary*:

$$2 - I_0 \exp(-z_i)\left(1 + \frac{\omega_{ne}}{\omega}\right) - \frac{\omega^2 - \omega_{ne}^2}{k_z^2 c_A^2 z_i}[1 - I_0 \exp(-z_i)] = 0 \tag{14.74}$$

$$1 - \frac{\omega_{ne}}{\omega} + \left(\frac{M_i}{M_e}\right)^{1/2}\left(1 + \frac{\omega_{ne}}{\omega}\right)Q = 0. \tag{14.75}$$

These equations were analysed in [1.7]. According to [1.7], the instability is suppressed for all z_i if

$$\beta > 0.13. \tag{14.76}$$

The authors of [14.26] analysed (14.74) and (14.75) neglecting the magneto-drift effect, i.e., taking Q in the form of (14.73). In addition to [1.7], in [14.26] arbitrary values of k_z were considered. According to [14.26], the marginal value of β for plasma stability is $\beta = 0.15$ which is close to (14.76). Other results of [14.26] seem to be doubtful due to having neglected the magneto-drift effect. In addition, when $k_z > k_z^{(0)}$, the *ion longitudinal inertia* should be taken into account (see section 14.1.3); this was neglected in [14.26].

The magneto-drift effect was taken into account in [14.27] in the electrostatic approximation. In accordance with [2.7], the authors of [14.27] have concluded that the magneto-drift effect favours stabilization.

Subsequent analysis of the problem discussed is given in [14.28, 14.29].

14.4.2 Suppression of the resistive drift-Alfven instability
In the case of *collisionless ions* the *resistive drift-Alfven instability* (see section 14.2) for finite β can be suppressed by *ion Landau damping*. Such an effect was considered in [14.25]. Although the general conclusion of [14.25] concerning the possibility of the suppression of this instability seems to be undoubted, the specific results of [14.25] do not seem to be sufficiently precise due to the incorrect calculation of the electron contribution to the dispersion relation (cf. the calculations of [14.25] with those in section 14.3).

Suppression of the same instability in the regime of *collisional ions* was studied in [14.14]. However, there is now some doubt concerning the correctness of the respective results of [14.14].

Therefore, further studies of both these approaches seem to be necessary. For this, some of the starting equations of section 14.3 can be of use. At the same time, for this problem, it seems to be impossible to use the general results of [14.25] especially due to the fact that the ion collisions have been considered incorrectly in [14.25] (see the explanations in section 14.3).

14.5   Electromagnetic short-wavelength temperature-gradient instabilities

In this section we shall consider the *short-wavelength perturbations*, $z_i \gg 1$. In contrast to section 14.1.1, we now completely neglect the effects associated with the finiteness of the value $z_i^{-1/2}$. The remaining assumptions are standard.

Then equations (2.77) yield

$$\epsilon_{\parallel} = (k_z \lambda_{\mathrm{De}})^{-2} b(\omega)$$
$$\epsilon_{\perp} = (k_{\perp} \lambda_{\mathrm{Di}})^{-2}(1 + T_i \omega_{ne}/T_e \omega). \tag{14.77}$$

Here

$$b(\omega) = 1 - \frac{\omega_{ne}}{\omega}(1 + \eta_e x^2) + Z(x)\left(1 - \frac{\omega_{ne}}{\omega} + \frac{\eta}{2}\frac{\omega_{ne}}{\omega}(1 - 2x^2)\right) \tag{14.78}$$

where $x = \omega/ \mid k_z \mid v_{Te}$. In this case equation (2.76) takes the form

$$1 + T_e/T_i + \omega_{ne}/\omega + b(\omega)$$
$$= 2(\omega_{pe}/ck_{\perp})^2 b(\omega)(1 + T_i\omega_{ne}/T_e\omega)x^2 T_e/T_i. \tag{14.79}$$

Here, as usual, $\omega_{pe}^2 = 4\pi e^2 n_0/M_e$ is the square of the electron plasma frequency.

14.5.1 Oscillation branches

In the main, we shall deal with perturbations with $\omega/\mid k_z \mid v_{Te} \ll 1$. In this approximation, neglecting small imaginary terms, from (14.79) we find the dispersion relation (cf. (6.78))

$$(\omega - \omega_{ni})(\omega - \omega_{ne}) - (1 + T_e/T_i)k_z^2 c_A^2 z_i = 0. \qquad (14.80)$$

Similarly to (6.78), this dispersion relation describes the *short-wavelength drift-Alfven (SDA) perturbations*. In contrast to (6.78), equation (14.80) concerns the case where $\beta \ll 1$ and $T_e \neq T_i$.

14.5.2 The results of instability analysis

An analysis of dispersion relation (14.79) was made in [6.39]. We here present only the simplest results of this analysis.

Instability is possible only if (cf. section 6.6.3)

$$\eta < 0 \qquad \text{or} \qquad \eta > 2(1 + T_i/T_e). \qquad (14.81)$$

The characteristic wave numbers of growing perturbations are of order

$$k_\perp \simeq \omega_{pe}/c \qquad (14.82)$$
$$k_z \simeq \beta^{1/2}/a. \qquad (14.83)$$

The growth rate is of the order of the oscillation frequency and both are of order

$$\gamma \simeq \operatorname{Re}\omega \simeq v_{Ti}/a. \qquad (14.84)$$

Note that the *instability boundary* obtained from (14.79) is *doubly connected* for positive η, as illustrated in figure 14.1. The instability regions are separated by a gap at

$$k_\perp = k_\perp^* \equiv (T_e/T_i)^{1/2}\omega_{pe}/c. \qquad (14.85)$$

The doubly-connected nature of the instability boundary appears to be a reflection of the fact that equation (14.79), in fact, describes not one, but *two types of instabilities*, one of which lies in the range $k_\perp < k_\perp^*$ and the other in the range $k_\perp > k_\perp^*$. The *electromagnetic short-wavelength instability* discussed here corresponds to perturbations with $k_\perp < k_\perp^*$, i.e., to the left upper shaded range in figure 14.1. The so-called *electron temperature-gradient instability* (see [2], section 3.2.2) corresponds to perturbations with $k_\perp > k_\perp^*$, i.e., to the

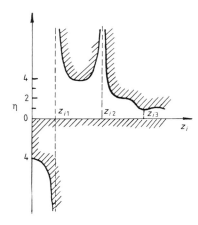

Figure 14.1   Instability ranges (shaded). Typical values of
z_i: $z_{i1} = 0.97$, $z_{i2} = M_i\beta_e/2M_e$, $z_{i3} = 1.76M_i/M_e$.

right upper shaded range in figure 14.1. Such an instability is usually studied in the *electrostatic approximation* which is valid only when $k_\perp \gg \omega_{pe}/c$. Therefore, it is clear that equation (14.79) and figure 14.1 take into account not only the electrostatic variety of the electron temperature-gradient instability but also the *electromagnetic range* of this instability ($k_\perp \simeq \omega_{pe}/c$).

Note that when $k_\perp \ll \omega_{pe}/c$, equation (14.79) describes an *electromagnetic modification of the electron-acoustic oscillation branches.* It then reduces to (cf. (14.80))

$$\omega(\omega - \omega_{ni}) - k_z^2 c_A^2 z_i = 0. \tag{14.86}$$

It should be mentioned here that the second inequality of (14.81) concerns only perturbations with $k_\perp < k_\perp^*$. As for perturbations with $k_\perp > k_\perp^*$, they are unstable only for smaller η (see section 3.2 of [2] and figure 14.1).

Let us also note that c/ω_{pe} is assumed to be large compared with the electron Larmor radius and small compared with the ion Larmor radius. This means that we consider a plasma with $M_e/M_i \ll \beta \ll 1$.

In the lower part of figure 14.1 the perturbations with $z_i \lesssim 1$ are also allowed for. It can be seen that when $\eta < 0$, the *electromagnetic short-wavelength instability turns smoothly into the long-wavelength one.* The marginal value $\eta = -4$ corresponding to perturbations with $z_i \to 0$ is simply the result following from (14.14).

14.6 Hydrodynamic drift-Alfven instability in a plasma with $\beta \ll M_e/M_i$

In the case of a *highly rarefied plasma*, $\beta \ll M_e/M_i$, when the Alfven velocity is greater than the electron thermal velocity, $c_A \gg v_{Te}$, the frequency range $\omega \gg k_z v_{Te}$ can be important in addition to the range $k_z v_{Ti} \ll \omega \ll k_z v_{Te}$. If $\omega \gg k_z v_{Te}$, from (2.75) and (2.76) we find the dispersion relation

$$(\omega^2 - \omega\omega_{ni} - k_z^2 c_A^2)(\omega - \omega_{ne}) = -(ck_\perp/\omega_{pe})^2\omega^2(\omega - \omega_{ni}). \quad (14.87)$$

We have taken $\nabla T = 0$ for simplicity.

If $k_\perp$ is sufficiently small, $k_\perp \ll \omega_{pe}/c$, the right-hand side of (14.87) is also small. Therefore, in the zeroth approximation with respect to $k_\perp c/\omega_{pe}$ we again arrive at (14.2) and (14.3). Taking finite $k_\perp$ into account, we find that near $k_z = k_z^{(0)}$ (see (14.4)) the frequencies of the branches $\omega_{1,3}$ are complex, the growth rate being given by (assuming $T_e = T_i$)

$$\gamma = (2/3)\omega_{ne}ck_\perp/\omega_{pe}. \quad (14.88)$$

This corresponds to the *hydrodynamic drift-Alfven instability*.

It is clear that such an instability is possible only when the longitudinal plasma length is sufficiently large compared with its transverse one. This situation can be realized in the *ionospheric plasma* [14.30]. As a rule, *ionospheric instabilities* are treated as electrostatic ones (refer to the literature, in particular, [14.30]). The *electromagnetic effects* in those instabilities were considered in [14.31] where the *interchange instability* was studied.

15 Allowance for Magnetic-field Curvature Effects

In the present chapter we shall derive the basic equations of the theory of small-scale instabilities in a finite-β plasma allowing for *magnetic-field curvature*. We shall restrict our analysis to the simplest case of a curvilinear magnetic field: a *shearless field of cylindrical geometry*. Thereby we shall neglect the effects of a longitudinal inhomogeneity of the magnetic field.

Firstly we shall consider the *equilibrium* of a collisionless plasma in the above field, section 15.1. Then we shall explain the *local permittivity* of such a plasma, section 15.2. Thereby we shall generalize the approach of Chapters 1 and 2 to the case of the above geometry of the curvilinear magnetic field. In contrast to the case of a rectilinear magnetic field, in the case of a curvilinear magnetic field, there exist branches of *hydromagnetic perturbations* with frequencies higher than the drift frequencies. Based on the results of section 15.2, in section 15.3 we shall find some *kinetic dispersion relations* for such perturbations and analyse them.

In the approximation of a rectilinear magnetic field discussed in Chapters 1 and 2, the problem of *low-frequency long-wavelength perturbations* could be essentially simplified. Such a simplification has been based on the *splitting* of the general dispersion relation into two dispersion relations, one of which corresponds to the Alfven (inertial) perturbations and the second to the magnetoacoustic (inertialess) perturbations. When the magnetic-field curvature is sufficiently strong, the above *splitting does not take place*, so that the picture of the branches of low-frequency long-wavelength perturbations changes. This is discussed in section 15.4 where we show that in this case it is also possible to introduce the concept of inertial and inertialess perturbations. However now, in contrast to Chapter 2, these types of perturbations are characterized by *oscillation frequencies of essentially different scale*. More complicated situations are also possible when, just as in the presence of a magnetic-field curvature, both types of perturbations have frequencies of the order of the drift frequencies. Such situations are also discussed in section 15.4.

In section 15.5 we consider the *two-fluid description* of small-scale *interchange perturbations* in a plasma cylinder.

The problem of allowing for the curvature effects in the theory of plasma instabilities has a *long history*. Rosenbluth and Longmire [15.1] have suggested that the physical effects of gravitational and magnetic drift are identical. Such a suggestion has initiated a development of the *theory of plasma instabilities in a gravitational field*. The results of this theory were presented in [2], Chapters 6 and 9. A more simplified approach to the theory of plasma instabilities in a curvilinear magnetic field is the *theory of hydromagnetic (single-fluid) stability of a plasma cylinder*. The results of this theory have been presented in Kadomtsev's review [15.2] and in [2], Chapter 13.

The problem of a *kinetic description* of plasma instabilities in the presence of magnetic field curvature was initially considered by Tserkovnikov [1.4]. He has allowed for both *drift and magneto-drift effects*, as well as the *inertial and magneto-viscosity effects*. Paper [1.4] has initiated the work by Rudakov and Sagdeev [1.5], where the *drift kinetic equation* has been used, and by Kadomtsev [3.1] using the CGL *approach*. We have already commented on these papers in Chapter 3. We should now note that a *key role* in the subsequent development of the theory of plasma instabilities has been played by the well-known paper by Rosenbluth *et al* [15.3]. It has been clear since [15.3] that some non-trivial physical effects exist beyond the scope of both the drift kinetic equation approach and CGL's approach. This circumstance has been a starting point for papers [15.4, 1.6] where a *general approach* to the theory of plasma instabilities in a curvilinear magnetic field allowing for *finite-β effects* has been developed (for the given example of cylindrical geometry). The results of these papers are used in the present chapter. One more step in the development of such a theory was made in [15.5] where a *two-energy component plasma* in a curvilinear cylindrically symmetric magnetic field was considered. A *two-fluid description* of the curvature effects was developed, in particular, in [2.1]. Below we shall use some results of these papers.

15.1 Equilibrium

15.1.1 Equilibrium magnetic field
We assume that the equilibrium has circular cylindrical symmetry. In other words, we suppose that in the cylindrical coordinate system

r, θ, z all the equilibrium parameters of the plasma and the magnetic field depend only on the radial coordinate r. Then the vector of this field $\boldsymbol{B}_0$ will be characterized by the components

$$\boldsymbol{B}_0 = (0, \, B_\theta, \, B_z). \tag{15.1}$$

The subscript zero in B_θ, B_z is omitted for simplicity.

When $B_z = 0$, the magnetic field lines are concentric circles. When $B_\theta = 0$, the field lines are straight and parallel to the cylinder axis. (Such a field was considered in the preceding chapters.) If $B_\theta \neq 0$ and $B_z \neq 0$ they have helical spiral form.

The modulus of the magnetic field is expressed in terms of B_θ, B_z by

$$B = (B_\theta^2 + B_z^2)^{1/2}. \tag{15.2}$$

The unit vector along the field $\boldsymbol{e}_0 = \boldsymbol{B}/B$ is associated with the unit vectors $\boldsymbol{e}_\theta$, $\boldsymbol{e}_z$ by

$$\boldsymbol{e}_0 = \boldsymbol{e}_\theta h_\theta + \boldsymbol{e}_z h_z \tag{15.3}$$

where $h_\theta \equiv B_\theta/B$, $h_z \equiv B_z/B$. In addition to the unit vectors $\boldsymbol{e}_r = \nabla r/r$ and $\boldsymbol{e}_0$, we shall also use the unit vector $\boldsymbol{e}_b$ defined by

$$\boldsymbol{e}_b = \boldsymbol{e}_0 \times \boldsymbol{e}_r \equiv h_z \boldsymbol{e}_\theta - h_\theta \boldsymbol{e}_z \tag{15.4}$$

where the subscript 'b' means the binormal to the field line.

15.1.2 Constants of particle motion

Since the equilibrium is not dependent on z and θ, besides the energy per unit mass $\mathcal{E} = v^2/2$ (see section 1.2.1), there exist two further precise constants of particle motion. As these constants one can take the axial component of the *generalized momentum* per unit mass, denoted by V_z, and the azimuthal component of *generalized angular momentum* per unit mass, denoted by L. Let us write these constants of motion in a form similar to (1.25)

$$V_z = v_z - \int_{r_g}^{r} \Omega h_\theta \, \mathrm{d}r' \qquad L = rv_\theta + \int_{r_g}^{r} r'\Omega h_z \, \mathrm{d}r' \tag{15.5}$$

where r_g means the r-coordinate of the guiding centre of the particle.

The value r_g is connected to V_z and L by the condition (cf. section 1.2.1)

$$v_b(r_g) = 0 \tag{15.6}$$

where $v_b = \boldsymbol{v} \cdot \boldsymbol{e}_b$. Expressing v_b in terms of v_θ and v_z and allowing for (15.5), we find that such a condition can be presented in the form

$$Lh_z(r_g) - r_g V_z h_\theta(r_g) = 0. \tag{15.7}$$

Using (15.5), one can also construct the constant of motion

$$V_\| = h_z(r_g)V_z + Lh_\theta(r_g)/r_g \tag{15.8}$$

denoting the longitudinal velocity of the guiding centre. Therefore, we can operate with both the totality of the constants of motion $\mathcal{E}$, V_z, L and similar totality $\mathcal{E}$, r_g, $V_\|$.

If the magnetic field is weakly inhomogeneous in the Larmor radius, using (15.6), we find the following expression for the radial coordinate of the guiding centre (cf. (1.24))

$$r_g = r + v_b/\Omega. \tag{15.9}$$

In this approximation the constant of motion $V_\|$, defined by (15.8), means

$$V_\| = v_\|(1 - v_b h_\theta^2/r\Omega) \tag{15.10}$$

where $v_\| = \boldsymbol{v} \cdot \boldsymbol{e}_0$ is the particle velocity along the magnetic field.

Since $v_\perp^2 + v_\|^2 = \text{const} = 2\mathcal{E}$, according to (15.10), the quantity

$$V_\perp = v_\perp + \frac{v_\|^2 h_\theta^2}{r v_\perp \Omega} v_b \tag{15.11}$$

is also a constant of motion. Here $v_\perp$ is the modulus of the transverse particle velocity $\boldsymbol{v}_\perp$ defined by the relation $\boldsymbol{v}_\perp = \boldsymbol{v} - v_\| \boldsymbol{e}_0$.

In the case of a weakly inhomogeneous magnetic field one can construct one more constant of motion, namely, the magnetic moment (cf. (1.26)). In detail this constant of motion will be discussed below.

15.1.3 Transverse equilibrium condition

Let us represent the equilibrium distribution function $f_0(\boldsymbol{v}, \boldsymbol{r})$ in the form (cf. (1.29))

$$f_0 = F(\mathcal{E}, V_\|, r_g). \tag{15.12}$$

For r_g and $V_\|$ of the form of (15.9) and (15.10) this means approximately (cf. (1.37))

$$f_0 = F(\mathcal{E}, v_\|, r) + \frac{v_b}{\Omega}\left(\frac{\partial F}{\partial r} - \frac{v_\| h_\theta^2}{r}\frac{\partial F}{\partial v_\|}\right). \tag{15.13}$$

Transform from the variables $(\mathcal{E}, v_\|)$ to $(\mathcal{E}_\perp, v_\|)$ where, similarly to section 1.2.1, $\mathcal{E}_\perp = v_\perp^2/2$ is the transverse particle energy per unit mass. Then

$$\left(\frac{\partial F}{\partial v_\|}\right)_\mathcal{E} = \left(\frac{\partial F}{\partial v_\|}\right)_{\mathcal{E}_\perp} - v_\|\frac{\partial F}{\partial \mathcal{E}_\perp}. \tag{15.14}$$

Allowing for (15.13) and (15.14), we calculate the transverse equilibrium current (cf. (1.38) and (1.39))

$$j_b \equiv \boldsymbol{j} \cdot \boldsymbol{e}_b = \sum e \int v_b f_0 \, \mathrm{d}\boldsymbol{v} = \frac{c}{B} \left(\frac{\partial p_\perp}{\partial r} + \frac{h_\theta^2}{r} (p_\perp - p_\parallel) \right). \quad (15.15)$$

Here $p_\perp$ is the transverse plasma pressure defined by (1.40) while $p_\parallel$ is the longitudinal plasma pressure equal to

$$p_\parallel = \sum M \int v_\parallel^2 f_0 \, \mathrm{d}\boldsymbol{v}. \quad (15.16)$$

It is clear that (15.15) can also be obtained by expressing the function f_0 in the form (cf. (1.32))

$$f_0 = F_1(\hat{\mathcal{E}}_\perp, V_\parallel, r_g) \quad (15.17)$$

where $\hat{\mathcal{E}}_\perp = V_\perp^2/2$. Then the right-hand side of (15.17) is expanded as a series of small corrections in all three arguments of F_1.

According to Maxwell's equations

$$\begin{aligned} j_b &= \frac{c}{4\pi} \boldsymbol{e}_b \cdot \nabla \times \boldsymbol{B} = \frac{c}{4\pi} (h_z \, \mathrm{curl}_\theta \boldsymbol{B} - h_\theta \, \mathrm{curl}_z \boldsymbol{B}) \\ &= -\frac{c}{4\pi} \left(\frac{\partial B}{\partial r} + \frac{B}{r} h_\theta^2 \right). \end{aligned} \quad (15.18)$$

It follows from (15.15) and (15.18) that

$$\frac{\partial}{\partial r} \left(p_\perp + \frac{B^2}{8\pi} \right) + \frac{B_\theta^2}{4\pi r} \left(1 + \frac{4\pi(p_\perp - p_\parallel)}{B^2} \right) = 0 \quad (15.19)$$

which is the *transverse equilibrium condition*. This result generalizes (1.42) for the case of a curvilinear cylindrically symmetric magnetic field. At the same time, (15.19) is a generalization of the well-known hydromagnetic equilibrium condition (see, e.g., [15.2]) for the case of a plasma with anisotropic pressure, $p_\parallel \neq p_\perp$ [15.6].

15.1.4 Equilibrium trajectories
Similarly to section 1.2.4, we express the particle velocity $\boldsymbol{v}$ in terms of $v_\perp$, $v_\parallel$, α:

$$\boldsymbol{v} = v_\parallel \boldsymbol{e}_0 + v_\perp (\boldsymbol{e}_r \cos \alpha + \boldsymbol{e}_b \sin \alpha). \quad (15.20)$$

Note also that in considering the particle motion in a magnetic field of *arbitrary geometry*, the particle velocity can be presented in a form similar to (15.20)

$$\boldsymbol{v} = v_\parallel \boldsymbol{e}_0 + v_\perp (\boldsymbol{e}_1 \cos \alpha + \boldsymbol{e}_2 \sin \alpha) \quad (15.21)$$

where the meaning of the values $v_\parallel$, $v_\perp$, α, e_0 is the same while e_1 and e_2 are unit vectors similar to e_r and e_b, i.e., they form a right-hand orthogonal triad with e_0. We shall present starting equations for $v_\perp$, $v_\parallel$, α in the case of general geometry. Then we shall explain how these equations will be simplified in the above case of cylindrical symmetry.

We multiply scalarly the first equation of (1.43) by $e_1 \cos \alpha + e_2 \sin \alpha$, e_0, and $-e_1 \sin \alpha + e_2 \cos \alpha$, successively. Allowing for $de_i/dt = (v \cdot \nabla)e_i$ we then find

$$\begin{aligned}
dv_\perp/dt &= -v_\parallel v_\perp \nabla \cdot e_0/2 + \delta f_\perp \\
dv_\parallel/dt &= v_\perp^2 \nabla \cdot e_0/2 + \delta f_\parallel \\
d\alpha/dt &= -\Omega + \delta f_\alpha.
\end{aligned} \tag{15.22}$$

where

$$\begin{aligned}
\delta f_\perp - (v_\parallel/v_\perp)\delta f_\parallel &= -v_\perp^2 (e_1 \cos \alpha + e_2 \sin \alpha) \cdot (e_0 \cdot \nabla)e_0 + A_2 \\
\delta f_\alpha &= \frac{v_\perp^2}{v_\parallel}(e_1 \sin \alpha - e_2 \cos \alpha) \cdot (e_0 \cdot \nabla)e_0 \\
&\quad + v_\perp[e_1 \cdot (e_1 \cdot \nabla)e_2 \cos \alpha - e_2 \cdot (e_2 \cdot \nabla)e_1 \sin \alpha] + C_0 + C_2.
\end{aligned} \tag{15.23}$$

Here A_2 and C_2 are terms containing $\cos 2\alpha$ and $\sin 2\alpha$, and C_0 is an addition to the cyclotron frequency of order $\rho \nabla$ non-oscillating with α. An explicit form of these quantities is not required by us since the effects connected with them are unimportant.

For the field of the form of (15.1), equations (15.22) reduce to

$$\begin{aligned}
dv_\perp/dt &= (v_\parallel^2/r)h_\theta^2 \cos \alpha \\
dv_\parallel/dt &= -(v_\perp v_\parallel/r)h_\theta^2 \cos \alpha \\
d\alpha/dt &= -\Omega - (1/r)[(v_\parallel^2/v_\perp)h_\theta^2 + v_\perp h_z^2] \sin \alpha.
\end{aligned} \tag{15.24}$$

In changing from (15.22) to (15.24), we have taken $e_1 = e_r$, $e_2 = e_b$ and have allowed for

$$\begin{aligned}
\nabla \cdot e_0 &= 0 \\
(e_0 \cdot \nabla)e_0 &= -e_r h_\theta^2/r \\
e_b \cdot (e_b \cdot \nabla)e_r &= h_z^2/r.
\end{aligned} \tag{15.25}$$

Multiplying the second equation of (1.43) by e_r, e_θ, e_z, we find

$$\begin{aligned}
dr/dt &= v_r = v_\perp \cos \alpha \\
d\theta/dt &= v_\theta/r = (h_\theta v_\parallel + h_z v_\perp \sin \alpha)/r \\
dz/dt &= v_z = h_z v_\parallel - h_\theta v_\perp \sin \alpha.
\end{aligned} \tag{15.26}$$

Using (15.9), we present $\Omega(r)$ in the form

$$\Omega(r) = \Omega(r_{\mathrm{g}}) - \kappa_B v_\perp \sin \alpha \qquad (15.27)$$

where κ_B is defined similarly to (1.27). It then follows from the last equation of (15.24) in the zeroth approximation in κ_B and ρ/r that

$$\alpha^{(0)}(t) = \alpha_0 - (t - t_0)\Omega(r_{\mathrm{g}}) \qquad (15.28)$$

where α_0 and t_0 mean the same as in section 1.2.4.

Allowing for (15.28), we obtain that the first two equations of (15.24) reduce to (15.10) and (15.11), respectively. Thereby $v_\perp$ and $v_\parallel$ can be expressed in terms of the constants of motion $V_\parallel$ and $V_\perp$. Then taking into account small terms in the last equation of (15.24), expression (15.28) is replaced by

$$\alpha(t) = \alpha^{(0)}(t) + \delta\alpha(t) \qquad (15.29)$$

where

$$\delta\alpha(t) = \frac{1}{\Omega V_\perp}[\kappa_B V_\perp^2 - \frac{1}{r}(V_\parallel^2 h_\theta^2 + V_\perp^2 h_z^2)][\cos\alpha^{(0)}(t) - \cos\alpha_0]. \quad (15.30)$$

It is assumed that all the functions of r in the right-hand side of (15.30) are taken at $r = r_{\mathrm{g}}$.

We now solve equations (15.26). The first of them has already been solved, see (15.9). From the remaining two equations of (15.26) we find

$$\theta = \theta_0 + \frac{t - t_0}{r_{\mathrm{g}}}(h_\theta V_\parallel + h_z V_{\mathrm{d}}) + \frac{h_z V_\perp}{r_{\mathrm{g}}\Omega}[\cos\alpha^{(0)}(t) - \cos\alpha_0]$$

$$z = z_0 + (t - t_0)(h_z V_\parallel - h_\theta V_{\mathrm{d}}) - \frac{h_\theta V_\perp}{\Omega}[\cos\alpha^{(0)}(t) - \cos\alpha_0].$$
$$(15.31)$$

Here V_{d} is the magnetic-drift velocity of a particle as defined by (cf. (1.27))

$$V_{\mathrm{d}} = \frac{\kappa_B V_\perp^2}{2\Omega} - \frac{V_\parallel^2 h_\theta^2}{r\Omega}. \qquad (15.32)$$

Note that in obtaining (15.31), we have used the identity $h_z h_z' = -h_\theta h_\theta'$ and have omitted the terms of order $h_z' V_\perp^2/2\Omega V_\parallel h_\theta$ (the prime means the derivative with respect to r).

Introducing vectors $\boldsymbol{V}_{\mathrm{d}} \equiv \boldsymbol{e}_{\mathrm{b}} V_{\mathrm{d}}$ and $\tilde{\boldsymbol{v}}_\perp \equiv \boldsymbol{v}_\perp - \boldsymbol{V}_{\mathrm{d}}$ and defining the magnetic moment μ by (1.26), one can ensure that $\mathrm{d}\mu/\mathrm{d}t = 0$, i.e., μ is a constant of motion.

15.2   Local permittivity

We assume in this section that in a plasma cylinder there are small-scale perturbations, and that, by analogy with Chapter 1, they can be described within the scope of the WKB approximation by means of the *local dispersion relation*. Let us consider how such a dispersion relation looks and calculate the *permittivity tensor components* contained in it.

15.2.1 Derivation of starting equations for the permittivity
Allowing for the equilibrium to be symmetric with respect to θ and z, we take the dependence of the perturbations on these coordinates in the form $\exp(im\theta + ik_z z)$, where m and k_z are called the *cylindrical wave numbers*. By means of these wave numbers, we construct the 'quasi-slab' wave numbers k_b and $k_\parallel$ defined by

$$k_b = h_z m/r - h_\theta k_z \qquad\qquad k_\parallel = h_z k_z + h_\theta m/r. \qquad (15.33)$$

Physically, the wave numbers k_b and $k_\parallel$ are similar to k_y and k_z in the case where $\boldsymbol{B}_0 \parallel \boldsymbol{z}$.

By analogy with Chapter 1, we characterize the radial dependence of the perturbations by the exponential $\exp(i \int k_r \, \mathrm{d}r)$. The derivatives of k_r with respect to r will be neglected. Then one can introduce the wave vector $\boldsymbol{k}$, characterizing it by the components

$$\boldsymbol{k} = (k_r, \, k_b, \, k_\parallel). \qquad (15.34)$$

Similarly to Chapter 1, we introduce the scalar and vector potentials ϕ, A_2, A_3, the perturbed charge density and electric current densities ρ, j_2, j_3, and the permittivity tensor $\epsilon_{ik}(i, k = 0, 2, 3)$. As a result, we arrive at the *local dispersion relation* of the form of (1.4).

Expressions for the tensor ϵ_{ik} can be found, as in Chapter 1, by means of the perturbed distribution function calculated by the method of trajectory integration. This function can be constructed by *re-designations* in the formulas of section 1.3. Consequently, no additional calculations are required. Among such re-designations, we first note those concerning the wave numbers

$$k_x \rightarrow k_r \qquad\qquad k_y \rightarrow k_b \qquad\qquad k_z \rightarrow k_\parallel. \qquad (15.35)$$

In section 1.3 we dealt with a few expressions for the perturbed distribution function in terms of different totalities of the constants of motion, namely expressions (1.52), (1.57) and (1.60). Remembering the analysis of section 15.1, we conclude, generalizing equation (1.52) to the case of the geometry considered, that we should make the substitutions

$$v_z \rightarrow v_\parallel \qquad X \rightarrow r_g. \qquad (15.36)$$

Similarly, equation (1.57) is generalized by means of the substitutions

$$\mathcal{E}_\perp \rightarrow \hat{\mathcal{E}}_\perp \qquad v_z \rightarrow V_\parallel \qquad X \rightarrow r_g \qquad (15.37)$$

while in equation (1.60) we should re-designate $X \rightarrow r_g$. In addition, under all the above generalizations, the magnetic-drift velocity V_d should be taken in the form of (15.32).

As a result, we again arrive at the expressions for ϵ_{ik} $(i, k = 0, 2, 3)$ of the form of (1.64) or at the corresponding modifications of these expressions as explained in section 1.4.

However, the situation becomes essentially changed under the *transition* from ϵ_{ik} $(i, k = 0, 2, 3)$ to $\epsilon_{\alpha\beta}$ $(\alpha, \beta = 1, 2, 3)$. As in section 7.3, where an additional transverse current was taken into account, in the case considered, we deal with the *transverse equilibrium condition* different from that in Chapter 1 (cf. (15.19) with (1.42)). In addition, the expression for the *magnetic-drift velocity* (15.32) *differs* from the similar expression in Chapter 1 (see (1.27)). Thereby, the above transition from ϵ_{ik} to $\epsilon_{\alpha\beta}$ should be done anew. It can be realized in the following manner.

As in section 2.1, we assume the perturbations to be of low frequency, $(\omega, k_\parallel v_\parallel) \ll \Omega$. We start with dispersion relation (1.4). We replace the first row of the determinant in (1.4) by its sum with the second row multiplied by $(ik_b/k_\perp^2)(\kappa_B + h_\theta^2/r)$ and with the third row multiplied by $(-k_\parallel/k_\perp)$. Then we replace the first column by its sum with the second column multiplied by $(-ik_b/k_\perp^2)(\kappa_B + h_\theta^2/r)$ and with the third column multiplied by $(-k_\parallel/k_\perp)$. As a result, we obtain the dispersion relation of the form of (1.1) in which [15.5]

$$\epsilon_{11} = -\sum \frac{4\pi e^2}{M k_\perp^2} \left\langle \tilde{G} \left[1 - J_0^2 - \frac{\omega_d}{\omega} J_0^2 + \frac{\omega_\Delta}{\omega} \left(\frac{4 J_0 J_1}{\xi} - 1 \right) \right] \right.$$
$$\left. - \frac{G_0}{\omega} \zeta_0 H^2 + \frac{1}{B} \frac{\partial F}{\partial \mu} \alpha_{11} \right\rangle$$

$$\epsilon_{12} = -\epsilon_{21} = i \sum \frac{4\pi e^2}{M\omega} \left\langle \frac{v_\perp^2}{2\Omega} \left[\tilde{G} \left(1 - \frac{2 J_0 J_1}{\xi} \right) \right. \right. \qquad (15.38)$$
$$\left. \left. - G_0 \zeta_0 \frac{2 J_1}{\xi} H + \frac{1}{B} \frac{\partial F}{\partial \mu} \alpha_{12} \right] \right\rangle$$

$$\epsilon_{13} = \epsilon_{31} = \sum \frac{4\pi e^2}{M\omega k_\perp} \left\langle v_\parallel \left(G_0 \zeta_0 J_0 H + \frac{1}{B} \frac{\partial F}{\partial \mu} \alpha_{13} \right) \right\rangle.$$

Here

$$\tilde{G} = \frac{\partial F}{\partial \mathcal{E}} + \frac{1}{B}\frac{\partial F}{\partial \mu} + \frac{k_b}{\omega\Omega}\frac{\partial F}{\partial r}$$

$$G_0 = \frac{\partial F}{\partial \mathcal{E}} + \frac{k_b}{\omega\Omega}\frac{\partial F}{\partial r}$$

$$\alpha_{11} = \frac{1}{\omega}\left(\frac{k_{\parallel}^2 v_{\parallel}^2}{\omega}(1 - J_0^2) + \omega_d J_0^2 - \omega_M - \frac{\omega_\Delta^2}{\omega}\frac{4J_1^2}{\xi^2} + \frac{\omega_d\omega_\Delta}{\omega}\right)$$

$$\alpha_{12} = \frac{1}{\omega}[\omega_\Delta(4J_1^2/\xi^2 - 1) - (\tilde{\omega}_R + \omega_R)] \qquad (15.39)$$

$$\alpha_{13} = k_{\parallel}v_{\parallel}(1 - J_0^2)/\omega$$

$$H = J_0\omega_d - (2J_1/\xi)\omega_\Delta$$

$$\omega_d = \omega_M + \omega_R$$

$$\omega_\Delta = \omega_M - \tilde{\omega}_R$$

$$\omega_M = k_b v_{\perp}^2 \kappa_B/2\Omega$$

$$\omega_R = k_b v_{\parallel}^2/\Omega R$$

$$\tilde{\omega}_R = k_b v_{\perp}^2/2\Omega R$$

where $R \equiv -r/h_\theta^2$ is the radius of field-line curvature (which is negative since it is directed toward the centre of the cylinder).

Note that in contrast to (2.1), we use here the equilibrium distribution function depending on $(r, \mathcal{E}, \mu)$, i.e., F in (15.38) and (15.39) is the same as F_2 in section 1.2.2.

The remaining components of $\epsilon_{\alpha\beta}$ have a form similar to (2.2). According to section 1.4, in terms of $(r, \mathcal{E}, \mu)$ these components become [15.5]

$$\epsilon_{22} = \sum \frac{4\pi e^2}{M\omega^2}\left\langle\frac{v_{\perp}^2}{2B}\frac{\partial F}{\partial \mu} + \omega G_0\zeta_0 J_1^2\right\rangle$$

$$\epsilon_{23} = i\sum \frac{4\pi e^2}{M\omega}\langle v_{\parallel}v_{\perp}G_0\zeta_0 J_0 J_1\rangle \qquad (15.40)$$

$$\epsilon_{33} = \sum \frac{4\pi e^2}{M\omega}\langle v_{\parallel}^2 G_0\zeta_0 J_0^2\rangle.$$

Let us also present relationships for transforming the derivatives of the equilibrium distribution function (cf. (15.14))

$$\left(\frac{\partial F}{\partial \mathcal{E}}\right)_{r,\mu} = \frac{1}{v_\parallel}\frac{\partial F}{\partial v_\parallel}$$

$$\frac{1}{B}\left(\frac{\partial F}{\partial \mu}\right)_{r,\mathcal{E}} = \frac{\partial F}{\partial \mathcal{E}_\perp} - \frac{1}{v_\parallel}\frac{\partial F}{\partial v_\parallel} \tag{15.41}$$

$$\left(\frac{\partial F}{\partial r}\right)_{\mathcal{E},\mu} = \left(\frac{\partial F}{\partial r}\right)_{\mathcal{E}_\perp,v_\parallel} + \kappa_B \mathcal{E}_\perp \left(\frac{\partial F}{\partial \mathcal{E}_\perp} - \frac{1}{v_\parallel}\frac{\partial F}{\partial v_\parallel}\right).$$

These relationships allow one to transform from (15.38) and (15.40) to (2.1) and (2.2) when the curvature vanishes.

15.2.2 *The approximation of low-frequency long-wavelength perturbations*

Equations (15.38) and (15.40) determine the permittivity of a collisionless plasma in the cylindrically symmetric magnetic field with respect to *low-frequency perturbations* ($\omega \ll \Omega$) with arbitrary ratio of the Larmor radius of particles to the transverse wavelength (arbitrary ξ). We shall simplify these equations for $\xi \ll 1$, corresponding to the *long-wavelength approximation*.

As a result, similarly to (2.3), we find

$$\epsilon_{11} = -\sum \frac{4\pi e^2}{Mk_\perp^2 \omega}\left\langle \omega\tilde{G}\left[\frac{\xi^2}{2}\left(1 - \frac{\omega_M}{2\omega}\right) - \frac{\omega_R + \tilde{\omega}_R}{\omega}\right]\right.$$

$$- G_0\zeta_0(\omega_R + \tilde{\omega}_R - \frac{\xi^2}{8}\omega_M)^2 + \frac{1}{B}\frac{\partial F}{\partial \mu}\left[\frac{\xi^2}{2}\left(\frac{k_\parallel^2 v_\parallel^2}{\omega} + \omega_M - \frac{\omega_\Delta\omega_M}{2\omega}\right)\right.$$

$$\left.\left. + \omega_R + \frac{\omega_\Delta(\omega_R + \tilde{\omega}_R)}{\omega}\right]\right\rangle$$

$$\epsilon_{12} = i\sum \frac{4\pi e^2}{M\omega^2}\left\langle \frac{v_\perp^2}{2\Omega}\left(\frac{3}{8}\xi^2\omega\tilde{G} - \omega G_0\zeta_0(\omega_R + \tilde{\omega}_R - \frac{\xi^2}{8}\omega_M)\right.\right.$$

$$\left.\left. - \frac{1}{B}\frac{\partial F}{\partial \mu}(\omega_R + \tilde{\omega}_R - \frac{\xi^2}{4}\omega_M)\right)\right\rangle \tag{15.42}$$

$$\epsilon_{13} = \sum \frac{4\pi e^2}{M\omega^2 k_\perp}\left\langle v_\parallel\left(\omega G_0\zeta_0(\omega_R + \tilde{\omega}_R - \frac{\xi^2}{8}\omega_M) + \frac{k_\parallel v_\parallel}{B}\frac{\partial F}{\partial \mu}\frac{\xi^2}{2}\right)\right\rangle$$

$$\epsilon_{22} = \sum \frac{4\pi e^2 k_\perp^2}{M\omega^2\Omega^2}\left\langle \mathcal{E}_\perp^2\left(\frac{1}{B}\frac{\partial F}{\partial \mu} + \omega G_0\zeta_0\right)\right\rangle$$

$$\epsilon_{23} = i\sum \frac{4\pi e^2 k_\perp}{M\omega\Omega}\langle v_\parallel \mathcal{E}_\perp G_0\zeta_0\rangle$$

$$\epsilon_{33} = \sum \frac{4\pi e^2}{M\omega}\langle v_\parallel^2 G_0\zeta_0\rangle.$$

For the transformation to (2.3), one should take $(\omega_R, \tilde{\omega}_R) \to 0$, $\omega_M = \omega_\Delta = \omega_d$.

15.3 Kinetic theory of local hydromagnetic perturbations in a finite-β plasma

Using (1.1) and (15.42), in the present section we shall obtain and analyse the local dispersion relations in the limit of infinite longitudinal conductivity, $\epsilon_{33} \to 0$, and of zeroth Larmor radius of particles, $\xi^2 \to 0$. This limit corresponds to *ideal hydromagnetic perturbations* (see [15.2] and Chapter 12 of [2]). Thereby we shall find out what follows from the kinetic approach for such perturbations.

According to Chapter 5, in the limit $\epsilon_{33} \to \infty$ the dispersion relation (1.1) reduces to (5.1). Consequently, we should calculate only the components ϵ_{11}, ϵ_{12}, ϵ_{22}. Using the above assumptions, from (15.42) we find

$$
\begin{aligned}
\epsilon_{11} =& \frac{c^2}{c_A^2} + \frac{c^2 k_\parallel^2}{\omega^2}\sigma_- + \frac{4\pi k_b^2 c^2}{\omega^2 k_\perp^2 R^2 B^2}\Bigg[R(p_\parallel' + p_\perp') \\
& - \frac{B^2}{4\pi}(\sigma_- + \sigma_\perp) + \sum M\left\langle \frac{\omega}{\omega - k_\parallel v_\parallel}\left(v_\parallel^2 + \frac{v_\perp^2}{2}\right)^2 \frac{\partial F}{\partial \mathcal{E}}\right\rangle \Bigg] \\
\epsilon_{12} =& -\mathrm{i}\frac{4\pi k_b c^2}{\omega^2 R B^2}\Bigg[\frac{B^2}{4\pi}(\sigma_- + \sigma_\perp) \\
& + \sum M\left\langle \frac{\omega}{\omega - k_\parallel v_\parallel}\frac{v_\perp^2}{2}\left(v_\parallel^2 + \frac{v_\perp^2}{2}\right)\frac{\partial F}{\partial \mathcal{E}}\right\rangle \Bigg] \\
\epsilon_{22} =& \frac{4\pi c^2 k_\perp^2}{\omega^2 B^2}\left(-\sigma_\perp \frac{B^2}{4\pi} + \sum M\left\langle \frac{\omega}{\omega - k_\parallel v_\parallel}\frac{v_\perp^4}{4}\frac{\partial F}{\partial \mathcal{E}}\right\rangle\right)
\end{aligned}
$$

(15.43)

where

$$
\begin{aligned}
\sigma_- &= \frac{4\pi(p_\parallel - p_\perp)}{B^2} \\
\sigma_\perp &= -\frac{4\pi}{B^2}\sum M\left\langle \frac{v_\perp^4}{4}\frac{1}{B}\frac{\partial F}{\partial \mu}\right\rangle
\end{aligned}
$$

(15.44)

and a prime denotes the radial derivative. Note that the designations σ_- and $\sigma_\perp$ were introduced in [15.7].

15.3.1 Interchange perturbations
For the interchange perturbations, $k_\parallel = 0$, from (5.1) and (15.43) the following dispersion is obtained

$$\omega^2 = \frac{c_A^2}{2R^2} \frac{k_b^2}{k_\perp^2} \{ -R(\kappa_{p\perp}\beta_\perp + \kappa_{p\parallel}\beta_\parallel) \tag{15.45}$$
$$+ [1/2(1+\beta_\perp)][2(3\beta_\perp + 4\beta_\parallel) + 2\beta_\perp^2 + 4\beta_\perp\beta_\parallel - \beta_\parallel^2]\}.$$

Here $\beta_{\perp,\parallel} = 8\pi p_{\perp,\parallel}/B^2$, $\kappa_{p\perp,\parallel} = \partial \ln p_{\perp,\parallel}/\partial r$.

In the case of an isotropic plasma pressure, $p_\perp = p_\parallel = p$, (15.45) means [3.1]

$$\omega^2 = \frac{2p}{nM_iR^2}\frac{k_b^2}{k_\perp^2}\left(-\kappa_p R + \frac{14+5\beta}{4(1+\beta)}\right). \tag{15.46}$$

In the limit $| \kappa_p R | \gg 1$ and when $\kappa_p R > 0$, it hence follows that interchange perturbations are unstable with growth rate

$$\gamma = \frac{| k_b |}{k_\perp}\left(\frac{2p'}{M_i nR}\right)^{1/2}. \tag{15.47}$$

According to (15.46), at finite $\kappa_p R$ *the interchange instability is suppressed* if

$$\kappa_p R < (14+5\beta)/4(1+\beta). \tag{15.48}$$

This result concerns a *collisionless plasma*. In the case of a *collisional plasma*, (15.48) is replaced by [15.2] (see below, section 15.5)

$$\kappa_p R < \frac{2\gamma_0}{(1+\gamma_0\beta/2)} \tag{15.49}$$

where $\gamma_0 = 5/3$. It can be seen that in a collisionless plasma the stabilization effect due to finite a/R (where $a \simeq | \partial \ln p/\partial r |^{-1}$) takes place at both large and small β, while in a collisional plasma this effect weakens if $\beta \gg 1$.

In the presence of a *plasma pressure anisotropy*, $\beta_\parallel \neq \beta_\perp$, the terms of order a/R are, as before, stabilizing if $\beta_\perp > \beta_\parallel$. In contrast to this, when $\beta_\parallel$ is large compared with $\beta_\perp$, a destabilization effect is possible. In particular, when $\beta_\perp = 0$, (15.45) takes the form

$$\omega^2 = \frac{c_A^2\beta_\parallel}{2R^2}\frac{k_b^2}{k_\perp^2}(-R\kappa_{p\parallel} + 4 - \beta_\parallel/2). \tag{15.50}$$

It can be seen that the terms of order a/R are destabilizing if

$$\beta_\parallel > 8. \tag{15.51}$$

In this case an instability is possible even in the absence of the pressure gradient, i.e., at $\kappa_{p\parallel} = 0$. The growth rate of such an instability is equal to

$$\gamma = \frac{c_A}{2R} \frac{k_b}{k_\perp} \beta_\parallel^{1/2} (\beta_\parallel - 8)^{1/2}. \tag{15.52}$$

This instability is similar to the *fire-hose instability* [1.1] for which it is required that $\beta_\parallel > 2$ (see section 15.3.2). Due to a *higher threshold* on $\beta_\parallel$ and a lower growth rate, the above instability in a homogeneous plasma in a curvilinear magnetic field is, probably, less important than the fire-hose instability.

15.3.2 Oblique perturbations with $\omega \ll k_\parallel v_{Ti}$

When $\omega \ll k_\parallel v_{Ti}$, it follows from (5.1) and (15.43) that

$$\omega^2 = -\frac{p_\parallel' + p_\perp'}{nM_i R} \frac{k_b^2}{k_\perp^2} + (1 - \sigma_-)c_A^2 \left(k_\parallel^2 + \frac{k_b^2}{k_\perp^2 R^2} \frac{\sigma_\perp + \sigma_-}{1 + \sigma_\perp} \right). \tag{15.53}$$

Hence we find that the plasma is unstable even when neglecting the magnetic-field curvature and the plasma inhomogeneity if

$$\sigma_- > 1. \tag{15.54}$$

This is the condition for the *fire-hose instability* [1.1].

We shall now consider $\sigma_- < 1$. In addition, we shall take $1 + \sigma_\perp > 0$ since in the opposite case the *mirror instability* takes place [1.1]. Then we find that the plasma pressure anisotropy in the presence of a magnetic-field curvature has a destabilizing effect if

$$\sigma_- + \sigma_+ < 0. \tag{15.55}$$

When the velocity distribution of the particles is bi-Maxwellian with longitudinal and transverse temperatures $T_\parallel$ and $T_\perp$, we have

$$\sigma_\perp = \beta_\perp (1 - \beta_\perp/\beta_\parallel). \tag{15.56}$$

Then (15.55) means

$$T_\perp > T_\parallel. \tag{15.57}$$

Consequently, in contrast to section 15.3.1, a plasma in a curvilinear magnetic field is less stable against oblique perturbations when $T_\perp > T_\parallel$ and more stable when $T_\parallel > T_\perp$.

In the absence of plasma pressure anisotropy, $p_\parallel = p_\perp = p$, $\sigma_- = \sigma_\perp = 0$, equation (15.53) yields

$$\omega^2 = -\frac{2p'}{nM_i R} \frac{k_b^2}{k_\perp^2} + c_A^2 k_\parallel^2. \tag{15.58}$$

Precisely the same result is obtained within the scope of the single-fluid approach [15.2]. Equation (15.58) describes a suppression of the flute instability at finite $k_\parallel$ (cf. [2]). The stabilization criterion is of the form

$$k_\parallel^2 k_\perp^2 / k_b^2 > \beta \kappa_p / R. \tag{15.59}$$

This criterion is in accordance with the well-known *Suydam stability condition* concerning perturbations in a plasma in a sheared magnetic field [2].

15.3.3 *Oblique perturbations with $v_{Ti} \ll \omega/k_\parallel \ll v_{Te}$*

As mentioned above, when $p_\parallel = p_\perp$, the limiting cases of perturbations considered in section 15.3.2 are qualitatively the same as in the single-fluid description. We shall now study the perturbations with $v_{Ti} \ll \omega/k_\parallel \ll v_{Te}$. In the case of an isotropic plasma pressure, from (5.1) and (15.43) we find the dispersion relation

$$\omega^2 = k_\parallel^2 c_A^2 + \frac{2k_b^2}{nM_i Rk_\perp^2}\left(-(p_e' + p_i') + \frac{p_i(14 + 5\beta_i)}{4R(1 + \beta_i)}\right). \tag{15.60}$$

This dispersion relation has no analogue under the single-fluid description of plasma. However, one can show that a result similar to (15.60) is obtained under the *two-fluid description.*

The perturbations with the above range of longitudinal phase velocities are of interest in the case when the stability criterion (15.59) is satisfied. Then equation (15.60) gives a dispersion law of the oscillation branches with real frequencies, which can be excited under interaction with *resonant particles* or due to *collisional dissipative effects.*

15.4   General analysis of long-wavelength perturbations at finite magnetic-field curvature

According to section 2.1, in the absence of magnetic-field curvature, the dispersion relation for low-frequency long-wavelength perturbations *splits* into two dispersion relations corresponding to the inertial (Alfven) and inertialess (magnetoacoustic) oscillation branches. In this case, *coupling* between such branches is of order $\xi_i^2 \equiv k_\perp^2 \rho_i^2$. In the presence of a curvature, the picture of the oscillation branches generally changes. Let us consider how the dispersion relation is modified by increasing the curvature.

15.4.1 *Field of weak curvature and plasma with* $\beta \simeq 1$

Assuming $a/R \ll 1$, we shall first consider the case of a plasma with $\beta \simeq 1$. Then $\omega_M \simeq \omega_\Delta \approx \omega_D \simeq \omega^*$, where ω_D is the same as in Chapter 2. Turning to (15.42), we note that the role of the curvature is unimportant if

$$k_\perp^2 \rho_i^2 \gg a/R. \qquad (15.61)$$

Under this condition we have a picture of the oscillation branches considered in section 2.1.2, and the expression for the growth rate of the Alfven perturbations (2.33), obtained by allowing for the correction terms of order ξ^2, holds. Under the condition opposite to (15.61):

$$a/R \gg k_\perp^2 \rho_i^2 \qquad (15.62)$$

it is necessary to allow for perturbations of *two characteristic frequencies*: $\omega \simeq \omega^*$ and $\omega \gg \omega^*$.

When $\omega \simeq \omega^*$, the component ϵ_{11} is large, of order $a/R\xi^2$. Therefore, the dispersion relation for perturbations of such frequencies can be found by equating the coefficient at ϵ_{11} to zero. Then we again arrive at dispersion relation (2.8) with D_2 of the form of (2.6). It is also clear that under condition (15.62), Alfven perturbations with $\omega \simeq \omega^*$ are not realized. However these perturbations *do not disappear* as a whole: they transit in the above frequency range $\omega \gg \omega^*$. By means of (1.1) and (15.42) we find that the dispersion relation for these branches again reduces to (2.7) with $D_1^{(0)}$ of the form of (2.5). However, expression (2.9) for $\epsilon_{11}^{(0)}$ contained in (2.5) is now replaced by (cf. (15.43))

$$\epsilon_{11}^{(0)} = \frac{c^2}{c_A^2} + \frac{8\pi c^2 k_b^2 p'}{\omega^2 k_\perp^2 B^2 R}. \qquad (15.63)$$

Here we also have assumed for simplicity that the plasma pressure is isotropic. The role of the pressure anisotropy can be taken into account by means of (15.43).

Note also that, under condition (15.62), the above conclusion about the Alfven branches with $\omega \gg \omega^*$ is valid only *in the absence of a cold plasma*. We shall consider how this conclusion will be modified in the presence of a cold plasma in section 15.4.4.

Above we have discussed two limiting cases: (15.61) and (15.62). It is also of interest to study an intermediate case

$$k_\perp^2 \rho_i^2 \simeq a/R. \qquad (15.64)$$

Then the dispersion relation for Alfven branches takes the form of (2.7) with $D_1^{(0)}$ of the form of (2.5) in which expression (15.63) is replaced by

$$\epsilon_{11}^{(0)} = \bar{\epsilon}_{11}^{(0)} + \frac{8\pi c^2 k_b^2 p'}{\omega^2 k_\perp^2 B^2 R} \tag{15.65}$$

where $\bar{\epsilon}_{11}^{(0)}$ means the right-hand side of the first equality of (2.9).

15.4.2 Field of weak curvature and plasma with $\beta \ll 1$
It is clear from (15.42) that, for $\beta \ll 1$ and $a/R \ll 1$, the case

$$a/R \ll k_\perp^2 \rho_i^2 \ll a/R\beta \tag{15.66}$$

is possible. In this case the oscillation branches are defined by the relationships of section 2.1.2. However, the expression for the growth rate of the Alfven branches becomes different: instead of (2.33), we now have

$$\gamma = -\frac{\operatorname{Im} \epsilon_{11}^{(1)}}{\partial D_1^{(0)}/\partial\omega} \tag{15.67}$$

where, according to the first equation of (15.42)

$$\operatorname{Im} \epsilon_{11}^{(1)} = -\sum \frac{4\pi e^2 k_b^2}{M k_\perp^2 \omega R^2 \Omega^2} \tag{15.68}$$
$$\times \left\langle \delta(\omega - k_\| v_\| - \omega_d)(v_\|^2 + v_\perp^2/2)^2 \left(\frac{\partial F}{\partial \mathcal{E}} + \frac{k_b}{\omega\Omega}\frac{\partial F}{\partial r}\right)\right\rangle$$

and ω_d is the same as in Chapter 1. By comparing (15.68) with (2.28) it can be seen that the *role of the curvature* consists qualitatively in this case in the following substitution in the expression for the growth rate of Alfven branches:

$$\xi^2 \to (a/R\beta\xi)^2. \tag{15.69}$$

The picture considered can be generalized simply for the case

$$a/R \simeq k_\perp^2 \rho_i^2 \ll a/R\beta. \tag{15.70}$$

Then, in addition to the expression for the growth rate of Alfven branches, the quantity $\epsilon_{11}^{(0)}$ should also be modified. Such a modification is given by (cf. (2.9), (15.63))

$$\epsilon_{11}^{(0)} = \frac{c^2}{c_A^2}\left(1 - \frac{\omega_{pi}^*}{\omega}\right) + \frac{8\pi c^2 k_b^2 p'}{\omega^2 k_\perp^2 B^2 R}. \tag{15.71}$$

When the parameters β and a/R, being small, satisfy the relation

$$\beta \gg a/R \qquad (15.72)$$

a class of perturbations is also possible with (cf. (15.66), (15.70))

$$a/R\beta \ll k_\perp^2 \rho_i^2 \ll 1. \qquad (15.73)$$

All the results for the approximation of straight magnetic field lines also hold in this case.

15.4.3 Field of strong curvature

Let $R \simeq a$. This corresponds to the case of a magnetic field of *strong curvature*. Then condition (15.61) cannot be satisfied, since, by our assumption, $k_\perp^2 \rho_i^2 \ll 1$. Therefore, the results of the approximation of straight field lines become invalid. In particular, in this case one cannot talk about splitting the oscillation branches into inertial and inertialess branches in the meaning of section 2.1.2. However, the concepts of the *inertial and inertialess branches* become fruitful in this case also. They can be introduced in the following manner.

In accordance with section 15.3, in the case of a curvilinear magnetic field, the general dispersion relation of low-frequency long-wavelength perturbations has solutions with $\omega \gg \omega^*$. The simplest examples of such solutions are given by (15.46) and (15.58). Such perturbations can be called *perturbations with large frequencies*. The *transverse plasma inertia* is important for them, so that they can also be called *Alfven perturbations*.

When $R \simeq a$, equation (1.1) with $\epsilon_{\alpha\beta}$ of the form of (15.42), generally speaking, does not exclude solutions with $\omega \simeq \omega^* \simeq \omega_R$. As can be seen, for instance, from the structure of the first equation of (15.42), for such solutions, the terms with ξ^2 are negligibly small. At the same time, these terms correspond to transverse plasma inertia. Hence it is clear that, when $R \simeq a$, the perturbations with $\omega \simeq \omega^*$, if they exist, should be *a priori inertialess*.

Thus, when $R \simeq a$, the inertial perturbations are those with $\omega \gg \omega^*$, while the inertialess perturbations correspond to the case $\omega \simeq \omega^*$.

It is clear that the problem of inertialess perturbations at $R \simeq a$ is, generally speaking, more complicated than that at $R \to \infty$. In particular, for the case of inertial perturbations with $k_\parallel = 0$, a relatively simple dispersion relation (2.14) with ϵ_{22} of the form of (2.16) is now replaced by

$$\epsilon_{11}(\epsilon_{22} - N^2) + \epsilon_{12}^2 = 0. \qquad (15.74)$$

Here, in accordance with (15.42)

$$\epsilon_{11} = \frac{4\pi k_b^2 c^2}{\omega^2 k_\perp^2 R^2 B^2}$$

$$\times \left[2Rp' + \sum M \left\langle \frac{\omega}{\omega - \omega_d} \left(v_\parallel^2 + \frac{v_\perp^2}{2} \right)^2 \left(\frac{\partial F}{\partial \mathcal{E}} + \frac{k_b}{\omega \Omega} \frac{\partial F}{\partial r} \right) \right\rangle \right]$$

$$\epsilon_{12} = -i \frac{4\pi k_b c^2}{\omega R B^2} \sum M \left\langle \frac{v_\perp^2/2}{\omega - \omega_d} \left(v_\parallel^2 + \frac{v_\perp^2}{2} \right) \left(\frac{\partial F}{\partial \mathcal{E}} + \frac{k_b}{\omega \Omega} \frac{\partial F}{\partial r} \right) \right\rangle$$

$$\epsilon_{22} = \frac{4\pi c^2 k_\perp^2}{\omega B^2} \sum M \left\langle \frac{v_\perp^4/4}{\omega - \omega_d} \left(\frac{\partial F}{\partial \mathcal{E}} + \frac{k_b}{\omega \Omega} \frac{\partial F}{\partial r} \right) \right\rangle.$$

$$(15.75)$$

For simplicity we assume the plasma to be isotropic, $\partial F/\partial \mu = 0$.

For the general case, analysis of (15.74) and (15.75) is difficult. Hence we shall below restrict ourselves to the analysis of the case where $\beta \ll 1$. Then (15.74) is replaced by

$$\epsilon_{11} = 0 \tag{15.76}$$

where ϵ_{11} is defined by the first equation of (15.75) with ω_d calculated at $\beta = 0$, i.e., with

$$\omega_d = \frac{k_b}{R\Omega} \left(v_\parallel^2 + \frac{v_\perp^2}{2} \right). \tag{15.77}$$

Allowing for (15.77), for the case of a Maxwellian velocity distribution, the expression for ϵ_{11} reduces to

$$\epsilon_{11} = \frac{1}{k_\perp^2 \lambda_D^2} \left(1 - \sum \hat{i} \left\langle \frac{\omega}{\omega - \omega_d} \right\rangle \right) \tag{15.78}$$

where the operator $\hat{i}$ is defined by (2.11) while $\langle \ldots \rangle$ is given by (2.12).

It is clear that ϵ_{11} of the form of (15.78) is the same as ϵ_{00} defined by an equation of the form of (1.64). Therefore, dispersion relation (15.76) corresponds to the *electrostatic approximation*.

15.4.4 *The role of a cold plasma*
One of the effects connected with the presence of a cold plasma is an *increase of the longitudinal plasma conductivity*, $\epsilon_{33} \to \infty$ (cf. the introduction to Chapter 5). As a result, the general dispersion relation (1.1) reduces to (5.1) not only in the hydromagnetic approximation, considered in section 15.3, but also in the case where $\omega \simeq \omega^*$. In addition, as can be seen from section 5.3, at a high density of a cold plasma the Alfven velocity is significantly decreased. It is clear from

the structure of the first equation of (15.43) that, as a result of this effect, the frequencies of hydromagnetic perturbations can be essentially lower than those defined by (15.46) and (15.58). In particular, the frequencies of such perturbations can become comparable to the characteristic drift and magneto-drift frequencies of hot plasma components. In this case, if there are no additional small parameters, such as β or a/R, it is impossible, generally speaking, to reduce the general problem to more particular problems of inertial and inertialess perturbations. However, the general problem is simplified at $\beta \ll 1$ since in this case equation (5.1) reduces to

$$\epsilon_{11} - N^2 \cos^2 \theta = 0 \qquad (15.79)$$

corresponding to the *inertial (Alfven) branches*. Then the real part of ϵ_{11} is defined by a relationship of type (15.71) with a renormalized value ω_{pi}^*

$$\omega_{\mathrm{pi}}^* \to \omega_p^* n_{\mathrm{h}}/n_{\mathrm{c}}. \qquad (15.80)$$

Here, as in Chapter 5, n_{h} and n_{c} are the densities of the hot and cold components, and ω_p^* is the drift frequency of the hot component.

If the oscillation frequency defined by the real terms of (15.79) is real, the growth rate of perturbations is defined by (15.67) with Im $\epsilon_{11}^{(1)}$ equal to (15.68) in which ω_{d} is defined by (15.77).

15.5   Two-fluid description of small-scale interchange perturbations in a cylindrical finite-β plasma

Consider the interchange perturbations, $k_{\parallel} = 0$, in a cylindrical plasma using the *two-fluid approach*.

Similarly to (9.32) and (9.33), by means of (9.20) and (9.29), we find approximate expressions for the transverse velocities and heat fluxes of each particle species:

$$\begin{aligned}
\boldsymbol{V}_{\perp} &= \boldsymbol{h} \times (\nabla p/n - e\boldsymbol{E})/M\Omega \\
\boldsymbol{q}_{\perp} &= \frac{5}{2}\frac{p}{M\Omega}\boldsymbol{h} \times \nabla T.
\end{aligned} \qquad (15.81)$$

Substituting (15.81) into (9.19) and (9.21), we obtain (cf. (9.3) and (9.4))

$$\frac{\partial n}{\partial t} + \boldsymbol{V}_E \cdot \nabla n + n \operatorname{div} \boldsymbol{V}_E + \boldsymbol{h} \cdot \frac{\nabla p}{M} \times \nabla \frac{1}{\Omega} + \frac{\nabla p}{M\Omega}\operatorname{curl}\boldsymbol{h} = 0$$

$$\frac{\partial p}{\partial t} + \boldsymbol{V}_E \cdot \nabla p + \gamma_0 \operatorname{div} \boldsymbol{V}_E \qquad (15.82)$$

$$+ \gamma_0 \boldsymbol{h} \cdot \frac{\nabla(pT)}{M} \times \nabla \frac{1}{\Omega} + \gamma_0 \frac{\nabla(pT)}{M\Omega}\operatorname{curl}\boldsymbol{h} = 0.$$

Linearizing (15.82), we take into account $E_\parallel = 0$ and use relations

$$\text{div } \boldsymbol{V}_{\text{E}} = \frac{i\omega \tilde{B}_\parallel}{B_0} - \frac{cE_b}{B_0}(\kappa_B + 1/R)$$

$$\nabla p \cdot \text{curl} \boldsymbol{h} = ik_b \tilde{p}/R \qquad (15.83)$$

$$\nabla(pT)\text{curl}\boldsymbol{h} = ik_b(\tilde{p}T_0 + p_0\tilde{T})/R$$

where, as in (15.39), $R = -r/h_\theta^2$ is the radius of field-line curvature. We present equations (15.82) in linearized form (cf. (10.1), (10.2))

$$-i\omega\frac{\tilde{n}}{n_0} + i\omega_{\text{D}}^*\frac{\tilde{p}}{p_0} + \frac{cE_b}{B_0}(\kappa_n - \kappa_B - 1/R) + \frac{i\tilde{B}_\parallel}{B_0}(\omega - \omega_p^*) = 0$$

$$-i\frac{\tilde{p}}{p_0}(\omega - 2\gamma_0\omega_{\text{D}}^*) - i\gamma_0\omega_{\text{D}}^*\frac{\tilde{n}}{n_0} + \frac{cE_b}{B_0}[\kappa_p - \gamma_0(\kappa_B + 1/R)]$$

$$+ i\gamma_0\frac{\tilde{B}_\parallel}{B_0}(\omega - 2\omega_p^* + \omega_n) = 0.$$

$$(15.84)$$

Here

$$\omega_{\text{D}}^* = \frac{k_b T}{M\Omega}(\kappa_B + 1/R). \qquad (15.85)$$

Combining the electron and ion continuity equations (15.84) and using the *equilibrium condition* (15.19) for $p_\parallel = p_\perp \equiv p$, we find the perturbed-pressure balance equation similar to (10.3)

$$4\pi(\tilde{p}_e + \tilde{p}_i) + \tilde{B}_\parallel B_0 = 0. \qquad (15.86)$$

Then following (10.4), we introduce the *canonical variables* X, Y, Z. From (15.84), by analogy with (10.5), we arrive at the equation set

$$\omega_{\text{Di}}^* X + (\kappa_B + 1/R - \kappa_n)Y - \omega Z - \frac{2(\omega - \omega_{\text{pi}}^*)}{1 + \kappa_B R}\frac{\tilde{B}_\parallel}{B_0} = 0$$

$$\omega X + \gamma_0\omega_{\text{Di}}^* Z - \frac{2\gamma_0(\omega_{\text{pi}}^* - \omega_{ni})}{1 + \kappa_B R}\frac{\tilde{B}_\parallel}{B_0} = 0$$

$$-2\gamma_0\omega_{\text{Di}}^* X + [\kappa_p - \gamma_0(\kappa_B + 1/R)]Y + \frac{2\gamma_0(\omega - \omega_{\text{pi}}^*)}{1 + \kappa_B R}\frac{\tilde{B}_\parallel}{B_0} = 0.$$

$$(15.87)$$

Here we assume that $T_{0e} = T_{0i}$.

In contrast to (10.5), equation set (15.87) is not complete. This is a consequence of the fact that the *splitting* of the oscillation branches into inertial and inertialess branches *does not take place* in the case of a curvilinear magnetic field (cf. section 15.4). An additional equation completing the set (15.87) can be obtained by analogy with section 9.6. Then we have (cf. (9.49))

$$\boldsymbol{e}_0 \cdot \text{curl}(M_i n \, d\boldsymbol{V}_i/dt + \nabla \cdot \boldsymbol{\pi}) + ik_b B_0 \tilde{B}_\parallel / 2\pi R = 0. \qquad (15.88)$$

Since we adopt the WKB approximation, to calculate the viscosity tensor components we can use expression (9.57) with the substitutions in the subscripts $x \to r$ and $y \to b$, and in the derivatives $\partial/\partial x \to \partial/\partial r$ and $\partial/\partial y \to ik_b$. The velocity and heat flux are taken from (15.81). After linearizing and transforming to the variables X, Y, Z, equation (15.88) reduces to

$$\frac{\tilde{B}_\parallel}{B_0} D_I - \omega_{\text{Di}}^*(\omega - 3\omega_{\text{Di}})X - (\kappa_B + 1/R)(\omega - \frac{3}{2}\omega_{\text{Di}}^*)Y - \frac{3}{2}\omega_{\text{Di}}^*\omega_{\text{Di}} Z = 0. \qquad (15.89)$$

Here, as in section 2.1, $\omega_{\text{Di}} = k_b \kappa_B T_i / M_i \Omega_i$ is the averaged magneto-drift frequency due to the magnetic-field inhomogeneity

$$D_I = \omega^2 - \omega(\omega_{pi}^* + \omega_{\text{Di}}) + \omega_{pi}^* \omega_{\text{Di}} \sigma + \frac{2(p_{0i}' + p_{0e}')}{M_i n_0 R} \frac{k_b^2}{k_\perp^2} \qquad (15.90)$$

and σ is defined by (2.10).

Let us consider how, using (15.87) and (15.89), one can obtain stabilization criterion (15.49). For this, we should take $\omega \gg (\omega^*, \omega_D)$. Then, from (15.87) we find approximately

$$X = 0$$

$$Y = \frac{2\gamma_0 \omega}{(1 + \kappa_B R)[\kappa_p - \gamma_0(\kappa_B + 1/R)]} \frac{\tilde{B}_\parallel}{B_0} \qquad (15.91)$$

$$Z = -\frac{2(\kappa_p + \gamma_0 \kappa_n)}{(1 + \kappa_B R)[\kappa_p - \gamma_0(\kappa_B + 1/R)]} \frac{\tilde{B}_\parallel}{B_0}.$$

Neglecting the drift and magneto-drift frequencies, equation (15.89) is also simplified to

$$\left(\omega^2 + \frac{2(p_{0i}' + p_{0e}')}{M_i n_0 R} \frac{k_b^2}{k_\perp^2}\right) \frac{\tilde{B}_\parallel}{B_0} + \omega(\kappa_B + 1/R)Y = 0. \qquad (15.92)$$

Substituting Y here from (15.91), we arrive at (cf. (15.46))

$$\omega^2 = \frac{2p}{n_0 M_i R^2} \frac{k_b^2}{k_\perp^2}\left(-\kappa_p R + \frac{2\gamma_0}{1 + \gamma_0 \beta/2}\right). \qquad (15.93)$$

Hence stabilization criterion (15.49) follows.

Assuming the curvature to be strong, $1/R \simeq (\kappa_p, \kappa_B)$, let us now consider how a dispersion relation for the inertial oscillation branches can be obtained from (15.87) and (15.89). The frequencies of these oscillation branches are supposed to be of the order of the magneto-drift frequencies. Therefore, D_I in (15.89) is a large quantity of order $(k_\perp \rho_i)^{-2}$. It then follows from (15.89) that

$$\tilde{B}_\| = 0. \tag{15.94}$$

Obviously, equation set (15.87) becomes complete in this approximation. Equating the determinant of (15.87) to zero, we find

$$\omega^2 = -\gamma_0 \omega_{\mathrm{D}}^{*2} \frac{\kappa_p - 2\gamma_0 \kappa_n + \gamma_0(\kappa_B + 1/R)}{\kappa_p - \gamma_0(\kappa_B + 1/R)}. \tag{15.95}$$

This equation differs from (10.6) with $T_i = T_e$ only by the formal substitution $\kappa_B \to \kappa_B + 1/R$.

We shall return to the discussion of equation (15.95) in section 16.4.

It is clear from (15.87) and (15.89) that a generalization of (15.93) to the case where $\kappa_p R \gg 1$ and $\omega \simeq \omega^*$ is given by the dispersion relation

$$D_I = 0. \tag{15.96}$$

This dispersion relation coincides with that obtained by the kinetic approach (cf. (2.5), (2.7), (2.9), (15.63)).

16 Magnetic-field Curvature Effects in Kinetic and Two-fluid Instabilities

In the present chapter magnetic-field curvature effects in the kinetic and two-fluid instabilities will be considered.

In Chapter 2 we have shown that in the approximation of a rectilinear magnetic field and of zero Larmor radius, there takes place a splitting of low-frequency oscillation branches into inertial (Alfven) and inertialess branches. In this approximation the *Alfven oscillation branches* prove to be insensitive to *resonant wave-particle interaction*. In section 15.4 we have shown that such a situation is *essentially different* in the presence of a magnetic-field curvature, so that even in the approximation of zero Larmor radius the Alfven oscillation branches prove to be dependent on this interaction. In section 16.1 we shall give relationships describing the resonant interaction in the simplest case where $k_\parallel = 0$.

In section 3.2 we discussed the *transverse drift waves* in the approximation of a rectilinear magnetic field. In this approximation, when $k_\parallel = 0$ and $\nabla T = 0$, their growth rate was small, of order M_e/M_i, due to the absence of resonant interaction with electrons. In section 16.2 we shall show that this result will be *essentially modified* in the presence of a magnetic-field curvature.

In Chapter 5 *Alfven instabilities* in a *two-energy component plasma* in a magnetic field with straight field lines were discussed. Using the results of section 15.4.4, in section 16.3 we shall study the *magnetic-field curvature effects* on these instabilities.

In section 15.5 we have found the starting equations of the *two-fluid approach* for the *interchange oscillation branches*, $k_\parallel = 0$, and have given a preliminary analysis of the structure of these branches. This analysis will be continued in section 16.4 where we shall formulate the physical consequences of the *curvature effects* on the above-mentioned perturbations of a *collisional plasma*. This will be a generalization of the results of section 11.1 concerning *magneto-drift entropy instabilities*.

In section 11.2 we considered the instability in a *collisional finite-β plasma* with $\nabla T = 0$ due to finite *electron heat conductivity*. In section 16.5 we shall show that an analogous type of instability can take place even in the *zero-β plasma* if the *magnetic-field curvature is finite*. It is then important to note that the magnetic-field curvature can lead to the mentioned instability even when the *direction of curvature is favourable* with the viewpoint of the problem of the interchange instability.

In Chapter 12 we studied instabilities in *force-free* and *almost force-free high-β plasmas* neglecting the magnetic-field curvature. In section 16.6 we shall consider the *role of the curvature* in the problem of instabilities in such plasmas.

In the present chapter we use the results of [1.6, 2.1, 15.5, 16.1, 16.2]. The concept of residual instability was introduced in [15.3]. Note also the paper [16.3] devoted to a heuristic analysis of two-fluid and kinetic stability of the *tandem mirror*.

16.1   Residual interchange instability

Consider perturbations with $k_\parallel = 0$ in a plasma with $\beta \to 0$. Combining (15.71) and (15.78), we arrive at the dispersion relation [1.6]

$$\omega^2 - \omega\omega_{ni} + \gamma_H^2 + i\pi\omega^3\Delta/z_i = 0 \qquad (16.1)$$

where

$$\gamma_H^2 = 2\kappa_p p k_b^2 / M_i n_0 R k_\perp^2 \qquad (16.2)$$

$$\Delta = \sum (1 - \omega_n/\omega)\langle\delta(\omega - \omega_d)\rangle. \qquad (16.3)$$

We assumed for simplicity that $\nabla T = 0$ and $T_i = T_e \equiv T$.

Neglecting the terms with ω_{ni} and Δ in (16.1), we find

$$\omega^2 = -\gamma_H^2. \qquad (16.4)$$

In the case where $\kappa_p R > 0$ this dispersion relation corresponds to the *hydromagnetic interchange instability* with the growth rate equal to γ_H (the subscript 'H' denotes hydromagnetic).

Taking into account the term with ω_{ni} in (16.1), equation (16.4) is replaced by (cf. (4.52))

$$\omega^2 - \omega\omega_{ni} + \gamma_H^2 = 0. \qquad (16.5)$$

When

$$\omega_{ni}^2 > 4\gamma_H^2 \tag{16.6}$$

equation (16.5) describes the *finite-Larmor-radius stabilization effect* initially pointed out by Rosenbluth, Krall and Rostoker [15.3].

Let condition (16.6) be satisfied, so that the hydromagnetic interchange instability is suppressed. Moreover, let us suppose for simplicity that

$$\omega_{ni}^2 \gg 4\gamma_H^2 \tag{16.7}$$

Then we have two stable oscillation branches with frequencies

$$\omega_1 = \omega_{ni} - \omega_2 \tag{16.8}$$

$$\omega_2 = \omega_H^2/\omega_{ni}. \tag{16.9}$$

The branch $\omega = \omega_1$ can be called the *fast interchange branch*, while the branch $\omega = \omega_2$ is the *slow interchange branch*.

Let us now allow for the term with Δ in (16.1). Then we find that the above branches have growth rates defined by

$$\gamma_1 = (2\pi\kappa_p R)^{1/2} \frac{|\omega_2|}{z_i} \exp(-\kappa_p R/2) \tag{16.10}$$

$$\gamma_2 = -2(2\pi)^{1/2} \frac{|\omega_2|}{z_i^{3/2}} \exp(-2/z_i). \tag{16.11}$$

It follows from (16.10) and (16.11) that the fast interchange branch is growing, while the slow interchange branch is damped. The growth of the fast interchange branch corresponds to the *residual interchange instability*.

16.2 Transverse drift waves in a curvilinear magnetic field

According to (2.57) and the last equation of (15.40), transverse drift waves in a plasma with $\nabla T = 0$ and $\nabla n_0 \neq 0$ in the presence of a magnetic-field curvature are described by the local dispersion relation

$$c^2 k_\perp^2 + \sum \frac{4\pi e^2 n_0}{T}(\omega - \omega_n)\left\langle \frac{v_\parallel^2 J_0^2}{\omega - k_b V_d} \right\rangle = 0 \tag{16.12}$$

where V_d is defined by (15.32).

We assume for simplicity that $(\beta, a/R) \ll 1$. Then for $\omega \simeq \omega_n$, one can consider $\omega \gg k_b V_D$ where $V_D \equiv \langle V_d \rangle$. In this approximation the resonant interaction of particles with perturbations can be considered as a correctional effect. Although the growth rate of such an interaction will be exponentially small, we shall allow for it in order to extrapolate the results into the range $(\beta, a/R) \simeq 1$ (cf. the similar procedure in section 3.2).

We also use the simplifying suppositions $k_\perp \rho_e \to 0$, $k_\perp \rho_i \gg 1$. In addition, we mention that the ion contribution into (16.12) is small compared with the electron contribution, of order M_e/M_i. Therefore, the ion contribution should be allowed for only in the imaginary part of (16.12) (cf. section 3.2). As a result, (16.12) is transformed to

$$1 + (ck_\perp/\omega_{pe})^2 - \omega_{ne}/\omega - i\pi\omega S = 0 \tag{16.13}$$

where

$$S = M_e \sum \frac{1}{T}(1 - \omega_n/\omega)\langle J_0^2 v_\parallel^2 \delta[\omega - k_b(V_\beta + V_R)]\rangle \tag{16.14}$$

and V_β and V_R are the parts of the magnetic-drift velocity associated with finite β and $1/R$

$$V_\beta = -\kappa_n v_\perp^2 \beta/4\Omega \qquad V_R = (v_\parallel^2 + v_\perp^2/2)/R\Omega. \tag{16.15}$$

For the real part of the oscillation frequency $\omega_0 \equiv \mathrm{Re}\,\omega$, from (16.13) we find the same expression as in the rectilinear magnetic field approximation, i.e., expression (6.66) (see also (3.16)), while the growth rate is defined by

$$\gamma = \pi(\omega^3/\omega_{ne})S. \tag{16.16}$$

It is assumed in the right-hand side of equation (16.16) that $\omega = \omega_0$.

It follows from (16.14), (6.66) and (16.16) that the contribution in γ is due to particles for which

$$\frac{\Omega}{\Omega_e}\frac{T}{M_e} = \left(1 + \frac{c^2 k_\perp^2}{\omega_{pe}^2}\right)\left[\frac{1}{\kappa_n R}\left(v_\parallel^2 + \frac{v_\perp^2}{2}\right) - \frac{\beta v_\perp^2}{4}\right]. \tag{16.17}$$

Consequently, when $R \to \infty$, the contribution to (16.16) is due to only the ions. In this case the growth rate is positive (an instability takes place) and is defined by (cf. (3.18))

$$\gamma = \frac{M_e}{M_i}\omega\left(1 + \frac{c^2 k_\perp^2}{\omega_{pe}^2}\right)^{-3/2}\frac{2 + c^2 k_\perp^2/\omega_{pe}^2}{(\beta z_i)^{1/2}}$$

$$\times \exp\left[-\frac{2}{\beta}\left(1 + \frac{c^2 k_\perp^2}{\omega_{pe}^2}\right)^{-1}\right]. \tag{16.18}$$

If the curvature is sufficiently strong, so that (cf. (15.72))

$$\beta \kappa_n R \ll 1 \tag{16.19}$$

and $\kappa_n R > 0$ (unfavourable curvature, cf. section 16.1), the contribution to the growth rate is due to only the electrons. However, according to (6.66) and (16.14), the sign of the electron contribution is opposite to the sign of the ion one, so that the former contains the negative factor $1 - \omega_{ne}/\omega = -(ck_\perp/\omega_{pe})^2$. Consequently, in this case the perturbations prove to be damped. Their decay rate, $-\gamma$, is defined by

$$
\gamma = - (2\pi)^{1/2}\omega_0 \left(\frac{ck_\perp}{\omega_{pe}}\right)^2 \left(1 + \frac{c^2 k_\perp^2}{\omega_{pe}^2}\right)^{-5/2} \left(1 + \frac{\kappa_n}{2}\beta R\right)^{-1}
$$
$$
\times \exp\left[-\frac{\kappa_n R}{2}\left(1 + \frac{c^2 k_\perp^2}{\omega_{pe}^2}\right)^{-1}\right]. \tag{16.20}
$$

Thus the *transverse drift waves are stable for a sufficiently strong unfavourable curvature.*

16.3 Alfven instabilities in a two-energy component plasma

The general expression for the growth rate of Alfven waves excited by high-energy ions can be found similarly to (5.3), starting with the relationships given in section 15.2. However, we shall not use that general expression but will restrict ourselves to qualitative estimates and some limiting cases.

First of all, we note that the curvature can be totally neglected when the first inequality of (15.73) is satisfied. (We keep in mind the perturbations with $k_\perp \rho_i \ll 1$, so that the second inequality of (15.73) is automatically satisfied.) In this case the Alfven instabilities in a two-energy component plasma are described by the relationships given in Chapter 5.

We now assume that the curvature increases and becomes so strong that both sides of the first inequality of (15.73) prove to be of the same order, i.e.,

$$a/R \simeq \beta k_\perp^2 \rho_i^2. \tag{16.21}$$

In this case the expression for the growth rate can be presented in the form (cf. (5.34), (5.35))

$$
\gamma = -\frac{1 - \omega_n/\omega}{(k_\perp \lambda_D)^2 \omega \partial Q/\partial \omega} \langle [(\omega_R + \omega_\beta \xi^2/8)^2 + O(\omega_\beta^2 \xi^4)]\delta(\omega - k_\parallel v_\parallel - \omega_\beta)\rangle. \tag{16.22}
$$

Here $\lambda_{\rm D}$ is the Debye length of high-energy ions, ω_n is their density-gradient drift frequency, $\omega_R = k_{\rm b}V_R$, $\omega_\beta = k_{\rm b}V_\beta$, where V_R and V_β are defined by (16.15); $O(\omega_\beta^2\xi^4)$ means the terms of order $\omega_\beta^2\xi^4$. The quantity Q is of the form

$$Q = \frac{c^2}{c_{\rm A}^2}\left(1 - \frac{k_\parallel^2 c_{\rm A}^2}{\omega^2} + \frac{n_{\rm h}}{n_{\rm c}}\frac{\omega_n\langle\omega_\beta\rangle}{\omega^2} + \frac{n_{\rm h}}{n_{\rm c}}\frac{k_{\rm b}^2}{k_\perp^2}\frac{\kappa_n v_T^2}{R\omega^2}\right) \qquad (16.23)$$

where v_T is the thermal velocity of the high-energy ions. As in Chapter 5, we assume the high-energy ions to be Maxwellian with $\nabla n \neq 0$ and $\nabla T = 0$. It is assumed that the real part of the oscillation frequency satisfies the equation

$$Q = 0 \qquad (16.24)$$

and that the roots of (16.24) are real.

It follows from (16.22) that, under condition (16.21), the resonant condition $\omega - k_\parallel v_\parallel - \omega_\beta = 0$ holds, i.e., it is the same as in the case of a rectilinear magnetic field. However, the factor at the δ-function is modified. The oscillation branches described by (16.23) and (16.24) are also modified.

Expression (16.22) holds also in the case of an intermediate curvature when

$$\beta k_\perp^2 \rho_{\rm i}^2 \ll a/R \ll \beta. \qquad (16.25)$$

It then reduces to

$$\gamma = -\frac{1 - \omega_n/\omega}{(k_\perp\lambda_{\rm D})^2\omega\partial Q/\partial\omega}\langle\omega_R^2\delta(\omega - k_\parallel v_\parallel - \omega_\beta)\rangle. \qquad (16.26)$$

Finally, in the case of a strong curvature, $a/R \gtrsim \beta$, the resonant factor, i.e., the δ-function, could be modified by replacing ω_β by ω_R. Then

$$\gamma = -\frac{1 - \omega_n/\omega}{(k_\perp\lambda_{\rm D})^2\omega\partial Q/\partial\omega}\langle\omega_R^2\delta(\omega - k_\parallel v_\parallel - \omega_R)\rangle. \qquad (16.27)$$

In this case expression (16.23) for Q is also simplified:

$$Q = \frac{c^2}{c_{\rm A}^2}\left(1 - \frac{k_\parallel^2 c_{\rm A}^2}{\omega^2} + \frac{n_{\rm h}}{n_{\rm c}}\frac{k_{\rm b}^2}{k_\perp^2}\frac{\kappa_n v_T^2}{R\omega^2}\right). \qquad (16.28)$$

Now suppose that $n_{\rm h}/n_{\rm c}$ is sufficiently small so that the last term in the brackets of (16.28) is insignificant. Then

$$\omega^2 = k_\parallel^2 c_{\rm A}^2. \qquad (16.29)$$

Allowing for (16.29) and assuming $v_T \gg c_A$, we find that the term ω in the argument of the δ-function in (16.27) can be neglected (cf. section 5.3), so that the resonant condition now means

$$v_\parallel = -\omega_R/k_\parallel. \tag{16.30}$$

In addition, keeping in mind the case where $a/R \ll 1$, we find that $\omega_n \gg \omega_R$. Therefore, when ω/ω_R is small, the ratio ω_n/ω is large. As a result, (16.27) reduces to

$$\gamma = \frac{n_h}{n_c} \frac{k_b^2}{k_\perp^2} \frac{1}{R^2 v_T^2} \frac{\omega_n}{\omega} \langle (v_\parallel^2 + v_\perp^2/2)^2 \delta(k_\parallel v_\parallel + \omega_R) \rangle. \tag{16.31}$$

In order of magnitude, this means

$$\gamma \simeq \frac{n_h}{n_c} \frac{\omega_n}{k_\perp^2 \rho^2} \frac{v_T}{c_A}. \tag{16.32}$$

Note that, in neglecting the curvature, instead of (16.32), we have the estimate (see (5.37))

$$\gamma \simeq \frac{n_h}{n_c} (k_\perp \rho)^2 \frac{v_T}{c_A} \omega_n. \tag{16.33}$$

Comparing (16.32) with (16.33), we see that, under transition from the case of a rectilinear magnetic field to the case of a field with a strong curvature, the *growth rate increases by a factor of* $(k_\perp \rho)^{-4}$. Such a result will not appear unexpected if one remembers the structure of (16.22) (see also (15.69)).

16.4 Magneto-drift entropy instabilities in a curvilinear magnetic field

The *entropy waves* in a curvilinear magnetic field are described by dispersion relation (15.95). It follows from (15.95) that, when the stability criterion of the interchange perturbations (15.49) is satisfied, an instability with growth rate of order ω_D is possible if

$$\frac{\kappa_T}{\kappa_p} > 1 - \frac{1}{R\kappa_p} - \frac{1}{2\gamma_0} + \frac{\beta}{4}. \tag{16.34}$$

When $R \to \infty$, this inequality reduces to (11.1). If, in addition, $\beta \ll 1$ and allowing for $\gamma_0 = 5/3$, from (16.34) we find instability conditions (11.2).

We now assume $\beta \ll 1$ while the product $R\kappa_p$ is supposed to be of the order of unity. Then, according to (16.34), the first inequality of (11.2) is replaced by the following instability condition of a plasma with positive η:

$$\eta > \frac{0.7 - 1/R\kappa_p}{0.3 + 1/R\kappa_p}. \tag{16.35}$$

It can be seen that in the case of a favourable curvature, $R\kappa_p < 0$, the *instability boundary is shifted* into a range of greater η while in the case of an unfavourable curvature, $R\kappa_p > 0$, it is shifted into a range of smaller η.

16.5   The destabilization effect of finite electron heat conductivity in a curvilinear magnetic field

We assume $\beta \ll a/R$. Thereby we suppose that the magnetic-field inhomogeneity is due to the field curvature but not to the diamagnetic current. We describe the electron component by the continuity and heat-balance equations allowing for the longitudinal (collisional) electron heat flux. Corresponding equations were given in Chapter 10. We neglect the temperature inhomogeneity, $\nabla T_0 = 0$, and consider the curvature to be small, $a/R \ll 1$. The perturbations are assumed to be electrostatic. As a result, we obtain the following expression for the perturbed electron density

$$\frac{\tilde{n}_e}{n_0} = -\frac{ic\kappa_n E_b}{B_0} \frac{\omega - (2\gamma_0 - 1)\omega_{De} + i\Delta}{\omega(\omega - 2\gamma_0\omega_{De} + i\Delta) + \omega_{De}(\gamma_0\omega_{De} - i\Delta)}. \tag{16.36}$$

Here, as in section 10.3.1, $\Delta = (2/3)3.16 k_\parallel^2 T_e / M_e \nu_e$. The expression for the perturbed electron density $\tilde{n}_i$ differs from (16.36) by replacing electron subscripts by ion ones and by neglecting the terms with Δ.

Using the quasi-neutrality condition $\tilde{n}_e = \tilde{n}_i$, we find the dispersion relation (cf. (11.11))

$$\omega^2 - \gamma_0\omega\omega_{De}(1-\tau) - \tau\gamma_0(2\gamma_0-1)\omega_{De}^2 + i\Delta\left(\omega - \omega_{Di}\frac{2\gamma_0 - 1 + \gamma_0\tau}{1+\tau}\right) = 0 \tag{16.37}$$

where $\tau = T_{0i}/T_{0e}$. In the limiting case $\tau \ll 1$ it hence follows that

$$\mathrm{Re}\,\omega = (2\gamma_0 - 1)\omega_{Di}$$

$$\gamma = \frac{(\gamma_0 - 1)^2\omega_{Di}^2\Delta}{\gamma_0^2\omega_{De}^2 + \Delta^2}. \tag{16.38}$$

In the opposite limiting case, $\tau \gg 1$, instead of this, we have

$$\mathrm{Re}\,\omega = \gamma_0 \omega_{\mathrm{Di}}$$
$$\gamma = \frac{(\gamma_0 - 1)^2 \omega_{\mathrm{De}}^2 \Delta}{\gamma_0^2 \omega_{\mathrm{Di}}^2 + \Delta^2}. \tag{16.39}$$

It can be seen that in both cases $\gamma > 0$, i.e., an instability takes place. These results hold qualitatively when $\tau \simeq 1$.

The *excitation of perturbations is related to the presence* of the terms with Δ in dispersion relation (16.37), i.e., to *finite longitudinal electron heat conductivity*.

16.6 Influence of the curvature on instabilities in force-free and almost force-free high-β plasmas

In studying the role of the curvature in the problem of instabilities in force-free and almost force-free high-β plasmas, we shall restrict ourselves to the *hydrodynamic inertialess instability* (see section 12.3.2). In the absence of a curvature such an instability is described by dispersion relation (12.32). It follows from (12.32) that the perturbations are growing when the parameter $b \equiv \partial \ln B / \partial \ln n$ lies in the range of (12.33).

In the presence of curvature, one can find a dispersion relation coinciding formally with (12.32) but with a different value of b

$$b = \partial \ln B / \partial \ln n + 1/\kappa_n R. \tag{16.40}$$

Allowing for equilibrium condition (15.19) for $p_\parallel = p_\perp$, we conclude that in the case of an almost force-free plasma, $\kappa_p \neq 0$, the curvature should be taken into account if

$$| \kappa_p R | \lesssim 1/\beta. \tag{16.41}$$

In the case of a strictly force-free plasma ($\kappa_p = 0$, $\kappa_B = 1/R$)

$$b = 2/\kappa_n R. \tag{16.42}$$

Then it follows from inequalities opposite to (12.33) that the *curvature yields the stabilization* if

$$-6.7 < \kappa_n R < 0.9. \tag{16.43}$$

It can be seen that it is necessary for stabilization that the radius of curvature should be of the order of the characteristic length of the density gradient.

References

[1]   Mikhailovskii A B 1974 *Instabilities of a Homogeneous Plasma (Theory of Plasma Instabilities vol 1)* (New York: Consultants Bureau)

[2]   Mikhailovskii A B 1974 *Instabilities of an Inhomogeneous Plasma (Theory of Plasma Instabilities vol 2)* (New York: Consultants Bureau)

[3]   Mikhailovskii A B 1993 *Instabilities in a Confined Plasma* (Bristol: Institute of Physics Publishing)

[1.1]   Mikhailovskii A B 1975 *Reviews of Plasma Physics* vol 6 ed M A Leontovich (New York: Consultants Bureau) p 77

[1.2]   Mikhailovskii A B 1962 *Nucl. Fusion* **2** 162

[1.3]   Mikhailovskii A B 1967 *Reviews of Plasma Physics* vol 3 ed M A Leontovich (New York: Consultants Bureau) p 159

[1.4]   Tserkovnikov Yu A 1958 *Sov. Phys.–JETP* **5** 58

[1.5]   Rudakov L I and Sagdeev R Z 1960 *Sov. Phys.–JETP* **10** 952

[1.6]   Mikhailovskaya L V and Mikhailovskii A B 1963 *Nucl. Fusion* **3** 276

[1.7]   Mikhailovskaya L V and Mikhailovskii A B 1964 *Sov. Phys.–JETP* **18** 1077

[1.8]   Mikhailovskii A B and Fridman A M 1967 *Sov. Phys.–JETP* **24** 965

[1.9]   Mikhailovskii A B and Fridman A M 1968 *Sov. Phys.–Tech. Phys.* **12** 1305

[1.10] Mikhailovskii A B and Pokhotelov O A 1978 *Sov. Phys.–Tech. Phys.* **22** 779

[1.11] Mikhailovskii A B 1977 *Instabilities of an Inhomogeneous Plasma (Theory of Plasma Instabilities vol 2)* 2nd edn (Moscow: Atomizdat) (in Russian)

[1.12] Krall N A and Rosenbluth M N 1963 *Phys. Fluids* **6** 254

[1.13] Mikhailovskii A B and Onishchenko O G 1987 *J. Plasma Phys.* **37** 15

[1.14] Fo R A C, Schneider R S and Ziebell L F 1989 *J. Plasma Phys.* **42** 165

[2.1]   Mikhailovskaya L V 1968 *Sov. Phys.–Tech. Phys.* **12** 1451

[2.2]   Mikhailovskaya L V 1971 *Sov. Phys.–Tech. Phys.* **16** 183

[2.3]   Pearlstein L D and Krall N A 1966 *Phys. Fluids* **9** 2231

[2.4]   Krall N A and Pearlstein L D 1966 *Plasma Physics and Controlled Nuclear Fusion Research* vol 1 (Vienna: International Atomic Energy Agency) p 735

[2.5]   Mikhailovskii A B 1963 *Sov. Phys.–JETP* **17** 1043

[2.6]   Mikhailovskii A B and Fridman A M 1971 *Plasma Phys.* **13** 1163

[2.7]   Mikhailovskii A B and Klimenko V A 1980 *J. Plasma Phys.* **24** 385

[2.8]   Mikhailovskii A B 1971 *Nucl. Fusion* **11** 323
[2.9]   Mikhailovskii A B 1980 *Sov. J. Plasma Phys.* **6** 31–3
[2.10] Mikhailovskii A B and Rudakov L I 1963 *Sov. Phys.–JETP* **17** 621
[2.11] Lakhin V P, Mikhailovskii A B and Novakovskii S V 1986 *Sov. J. Plasma Phys.* **12** 326
[2.12] Mikhailovskii A B and Onishchenko O G 1987 *J. Plasma Phys.* **37** 29

[3.1]   Kadomtsev B B 1960 *Sov. Phys.–JETP* **10** 780
[3.2]   Ishihara T 1973 *J. Plasma Phys.* **9** 389
[3.3]   Chamberlain J W 1963 *J. Geophys. Res.* **68** 5667
[3.4]   Krall N A 1967 *Phys. Fluids* **10** 2263
[3.5]   Krall N A 1968 *Advadnces in Plasma Physics* vol 1 ed A Simon and W B Thompson (New York: Interscience) p 153
[3.6]   Mikhailovskii A B 1971 *Sov. Phys.–Tech. Phys.* **15** 1597
[3.7]   Wu C S 1971 *J. Geophys. Res.* **76** 4454
[3.8]   Wu C S 1971 *J. Geophys. Res.* **76** 6961
[3.9]   Buti B and Lakhina G S 1973 *Phys. Rev.* **A 7** 319
[3.10] Nambu M 1974 *Phys. Fluids* **17** 1855
[3.11] Fridman A M 1970 *Sov. Phys.–Tech. Phys.* **14** 1019
[3.12] Mikhailovskii A B 1970 *Sov. Phys.–Dokl.* **15** 471
[3.13] Bodin H A B and Newton A A 1969 *Phys. Fluids* **12** 2175
[3.14] Bodin H A B 1977 *Rep. Prog. Phys.* **40** 1415
[3.15] Kadomtsev B B and Timofeev A V 1962 *Sov. Phys.–Dokl.* **7** 826
[3.16] Krall N A and Rosenbluth M N 1965 *Phys. Fluids* **8** 1488
[3.17] Buchelnikova N S 1964 *Nucl. Fusion* **4** 165
[3.18] D'Angelo N 1969 *Advances in Plasma Physics* vol 2 ed A Simon and W B Thompson (New York: Interscience) p 1
[3.19] Unti T, Atkinson G, Wu C S and Neugebauer M 1972 *J. Geophys. Res.* **77** 2250
[3.20] Unti T, Neugebauer M and Wu C S 1973 *J. Geophys. Res.* **78** 7237
[3.21] Buti B and Lakhina G S 1973 *J. Plasma Phys.* **10** 249
[3.22] Rudakov L I and Sagdeev R Z 1961 *Sov. Phys.–Dokl.* **6** 415
[3.23] Galeev A A, Oraevskii V N and Sagdeev R Z 1963 *Sov. Phys.–JETP* **17** 615
[3.24] Mikhailovskaya L V and Mikhailovskii A B 1964 *Sov. Phys.–Tech. Phys.* **8** 896

[4.1]   Gurovich V T, Naidorf B Yu and Fridman A M 1970 *Sov. Phys.–Tech. Phys.* **15** 337
[4.2]   Mikhailovskii A B and Tsypin V S 1965 *Nucl. Fusion* **4** 240
[4.3]   Bajaj N K and Tandon J N 1968 *Nucl. Fusion* **8** 297
[4.4]   Pearlstein L D and Freidberg J P 1978 *Phys. Fluids* **21** 1218
[4.5]   Tamao T 1969 *Phys. Fluids* **12** 1458
[4.6]   Chance M S, Coroniti F V and Kennel C F 1973 *J. Geophys. Res.* **78** 7521
[4.7]   Lin C S and Parks G K 1978 *J. Geophys. Res.* **83** 2628
[4.8]   Tamao T 1978 *Planet. Space Sci.* **26** 1141

[5.1]   Hasegawa A 1969 *Phys. Fluids* **12** 2042
[5.2]   Hasegawa A 1971 *Phys. Rev. Lett.* **27** 11
[5.3]   Hasegawa A 1971 *Rev. Geophys. Space Phys.* **9** 703
[5.4]   Hasegawa A 1975 *Plasma Instabilities and Nonlinear Effects* (Berlin:

Springer)
[5.5] Kozhevnikov A A, Mikhailovskii A B and Pokhotelov O A 1976 *Planet. Space Sci.* **24** 465
[5.6] Mikhailovskii A B 1975 *Sov. Phys.–JETP* **41** 890
[5.7] Pilipenko V A and Pokhotelov O A 1975 *Geomagn. Aeron.* **15** 784
[5.8] Pokhotelov O A and Pilipenko V A 1976 *Geomagn. Aeron.* **16** 296
[5.9] Brown W L, Cahill L J, Davis L R, McIlwain C E and Roberts C S 1968 *J. Geophys. Res.* **73** 153
[5.10] Lanzerotti L J, Hasegawa A and Maclennan C G 1969 *J. Geophys. Res.* **74** 5565
[5.11] Lanzerotti L J and Hasegawa A 1975 *J. Geophys. Res.* **80** 1019
[5.12] Pokhotelov O A, Pilipenko V A and Amata E 1985 *Planet. Space Sci.* **33** 1229
[5.13] Pokhotelov O A, Nezlina Yu M and Pilipenko V A 1986 *Trans. (Doklady) USSR Acad. Sci., Earth Sci. Sect.* **286** 18
[5.14] Pokhotelov O A, Pilipenko V A, Nezlina Yu M, Koch I, Kremser G, Korth A and Amata E 1986 *Planet. Space Sci.* **34** 695
[5.15] Ryzhov N M, Sizonenko V L, Suprunenko V A, Sukhomlin E A and Gutarev Yu V 1974 *Sov. Phys.–Tech. Phys.* **19** 722–5
[5.16] Mikhailovskii A B, Pokhotelov O A, Ryzhov N M and Suprunenko V A 1976 *Sov. J. Plasma Phys.* **2** 46
[5.17] Southwood D J 1976 *J. Geophys. Res.* **81** 3340
[5.18] Mikhailovskii A B 1978 *Instabilities of Plasma in Magnetic Traps* (Moscow: Atomizdat) (in Russian)
[5.19] Pokhotelov O A 1978 *Geomagn. Aeron.* **18** 735
[5.20] Patel V L and Migliuolo S 1980 *J. Geophys. Res.* **85** 1736
[5.21] Migliuolo S and Patel V L 1981 *J. Geophys. Res.* **86** 667
[5.22] Migliuolo S 1983 *J. Geophys. Res.* **88** 2065
[5.23] Patel V L, Ng P H and Cahill L J 1983 *J. Geophys. Res.* **88** 5677
[5.24] Migliuolo S 1983 *J. Geophys. Res.* **88** 5783
[5.25] Ng P H and Patel V L 1983 *J. Geophys. Res.* **88** 10 035
[5.26] Patel V L, Ng P H and Ludlow G R 1984 *J. Geophys. Res.* **89** 8851
[5.27] Migliuolo S 1984 *J. Geophys. Res.* **89** 11023

[6.1] Mikhailovskii A B and Timofeev A V 1963 *Sov.Phys.– JETP* **18** 626
[6.2] Mikhailovskii A B 1965 *Nucl. Fusion* **5** 125
[6.3] Ohkawa T and Yoshikawa M 1966 *Phys. Rev. Lett.* **17** 685
[6.4] Makowski M A and Emmert G A 1985 *Phys. Fluids* **28** 2838
[6.5] Krall N A and Rosenbluth M N 1962 *Phys. Fluids* **5** 1435
[6.6] Dnestrovskii Yu N 1965 *Sov. Phys.–Tech. Phys.* **9** 1649
[6.7] Bujaj N K and Krall N A 1972 *Phys. Fluids* **15** 657
[6.8] Higuchi Y 1974 *Can. J. Phys.* **52** 1265
[6.9] Gladd N T and Huba J D 1979 *Phys. Fluids* **22** 911
[6.10] Gladd N T and Krall N A 1982 *Phys. Fluids* **25** 1241
[6.11] Higuchi Y 1973 *Nature Phys. Sci.* **246** 117
[6.12] Zelenyi L M 1975 *Sov. J. Plasma Phys.* **1** 319
[6.13] Spalding I J, Eden M J, Phelps A D R and Allen T K 1969 *Plasma Physics and Controlled Nuclear Fusion Research* vol 2 (Vienna: International Atomic Energy Agency) p 639
[6.14] Dimov G I, Zakaidakov V V and Kishinevskii M E 1976 *Sov. J. Plasma Phys.* **2** 326
[6.15] Fowler T K and Logan B G 1977 *Comment. Plasma Phys. Nucl. Fusion*

2 167

[6.16]   Davidson R C and Gladd N T 1975 *Phys. Fluids* **18** 1327

[6.17]   Davidson R C, Gladd N T, Wu C S and Huba J D 1976 *Phys. Rev. Lett.* **37** 750

[6.18]   Davidson R C, Gladd N T, Wu C S and Huba J D 1977 *Phys. Fluids* **20** 301

[6.19]   Davidson R C, Gladd N T, Huba J D, Hui B H, Liewer P C, Liu C S, Mondt J P, Ogden J M, Wu C S, Hamasaki S, Krall N A and Wagner C E 1977 *Plasma Physics and Controlled Nuclear Fusion Research* vol 3 (Vienna: International Atomic Energy Agency) p 113

[6.20]   Huba J D, Gladd N T and Papadopoulos K 1978 *J. Geophys. Res.* **83** 5217

[6.21]   Gurnett D A, Frank L A and Lepping R P 1976 *J. Geophys. Res.* **81** 6059

[6.22]   Gary S P and Eastman T E 1979 *J. Geophys. Res.* **84** 7378

[6.23]   Gurnett D A, Anderson R R, Tsurutani B T, Smith E J, Paschmann G, Haerendel G, Bame S J and Russel C T 1979 *J. Geophys. Res.* **84** 7043

[6.24]   Post R F and Rosenbluth M N 1966 *Phys. Fluids* **9** 730

[6.25]   Tang W M, Pearlstein L D and Berk H L 1972 *Phys. Fluids* **15** 1153

[6.26]   Berk H L and Gerver M J 1976 *Phys. Fluids* **19** 1646

[6.27]   Gerver M J 1976 *Phys. Fluids* **19** 1581

[6.28]   Byers J A, Cohen B I, Condit W C and Hanson J D 1978 *J. Comput. Phys.* **27** 363

[6.29]   Bhatia K G and Lakhina G S 1980 *Astrophys. Space Sci.* **70** 467

[6.30]   Ioffe M S, Kanaev B I, Pastukhov V P and Yushmanov E E 1975 *Sov. Phys.–JETP* **40** 1064

[6.31]   Coensgen F H, Cummings W F, Logan B G, Molvik A W, Nexen W E, Simonen T C, Stallard B W and Turner W C 1975 *Phys. Rev. Lett.* **35** 1501

[6.32]   Simonen T C 1976 *Phys. Fluids* **19** 1365

[6.33]   Berk H L, Fowler T K, Pearlstein L D, Post R F, Callen J D, Horton W C and Rosenbluth M N 1969 *Plasma Physics and Controlled Nuclear Fusion Research* vol 2 (Vienna: International Atomic Energy Agency) p 151

[6.34]   Ferron J R, Dimonte G, Wong A Y and Leikind B J 1983 *Phys. Fluids* **26** 2227

[6.35]   Ferron J R and Wong A Y 1984 *Phys. Fluids* **27** 1287

[6.36]   Koepke M, McCarrick M J, Majeski R P and Ellis R F 1986 *Phys. Fluids* **29** 3439

[6.37]   Mense A T, Emmert G A and Callen J D 1975 *Nucl. Fusion* **15** 703

[6.38]   Perraut S, Gendrin R, Robert P, Roux A, Devilledary C and Jones D 1978 *Space Sci. Rev.* **22** 347

[6.39]   Mikhailovskii A B 1968 *Sov. Phys.–Tech. Phys.* **12** 993

[6.40]   Pyatak A I, Sizonenko V L and Stepanov K N 1973 *Sov. Phys.–Tech. Phys.* **18** 309

[6.41]   Pyatak A I and Sizonenko V L 1974 *Sov. Phys.–Tech. Phys.* **19** 635

[6.42]   Pyatak A I and Sizonenko V L 1974 *Sov. Phys.–Tech. Phys.* **19** 640

[6.43]   Kadomtsev B B and Pogutse O P 1984 *JETP Lett.* **39** 269

[6.44]   Kadomtsev B B and Pogutse O P 1984 *Nonlinear and Turbulent Processes in Physics* vol 1 ed R Z Sagdeev (New York: Harwood) p 257

[6.45] Petviashvili V I and Pogutse I O 1986 *JETP Lett.* **43** 343
[6.46] Shukla P K, Pecseli H L and Rasmussen J J 1986 *Phys. Scr.* **34** 171
[6.47] Novakovskii S V, Mikhailovskii A B and Onishchenko O G 1988 *Phys. Lett.* **132** 33

[7.1]   Sagdeev R Z 1966 *Reviews of Plasma Physics* vol 4 ed M A Leontovich (New York: Consultants Bureau) p 23
[7.2]   Stepanov K N 1965 *Sov. Phys.-Tech. Phys.* **9** 1633
[7.3]   Mikhailovskii A B and Tsypin V S 1966 *JETP Lett.* **3** 158
[7.4]   Pyatak A I, Sizonenko V L and Stepanov K N 1973 *Sov. Phys.-Tech. Phys.* **18** 304
[7.5]   Goeldbloed J P, Pyatak A I and Sizonenko V L 1973 *Sov. Phys.-JETP* **37** 1051
[7.6]   Pyatak A I and Vrba P 1975 *Czech. J. Phys* B **25** 979
[7.7]   Krall N A and Liewer P C 1971 *Phys. Rev.* A **4** 2094
[7.8]   Gladd N T 1976 *Plasma Phys.* **18** 27
[7.9]   Krall N A and Book D 1969 *Phys. Rev. Lett.* **23** 574
[7.10] Lominadze D G and Sizonenko V L 1974 *Sov. Phys.-Tech. Phys.* **19** 1145
[7.11] Parail V V 1971 *Sov. Phys.-Tech. Phys.* **16** 376
[7.12] Lakhina G S and Sen A 1973 *Nucl. Fusion* **13** 913
[7.13] McBride J B and Ott E 1972 *Phys. Lett.* **39A** 363
[7.14] Aref'ev V I, Gordeev A V and Rudakov L I 1969 *Plasma Physics and Controlled Nuclear Fusion Research* vol 2 (Vienna: International Atomic Energy Agency) p 165
[7.15] Zelenyi L M 1977 *Sov. J. Plasma Phys.* **3** 38

[8.1]   Zelenyi L M and Mikhailovskii A B 1978 *Sov. J. Plasma Phys.* **4** 247
[8.2]   Ivanov A A, Meilikhov E Z, Parail V V and Frank-Kamenetskii D A 1970 *Sov. Phys.-Dokl.* **14** 652
[8.3]   Gary S P and Sanderson J J 1970 *J. Plasma Phys.* **4** 739
[8.4]   Davidson R C and Krall N A 1977 *Nucl. Fusion* **17** 1313
[8.5]   McBride J D, Ott E, Boris J P and Orens J H 1972 *Phys. Fluids* **15** 2367
[8.6]   McBride J D, Ott E and Wagner C E 1973 *Nucl. Fusion* **13** 151
[8.7]   Lemons D S and Gary S P 1977 *J. Geophys. Res.* **82** 2337
[8.8]   Eviatar A and Goldstein M L 1980 *J. Geophys. Res.* **85** 753
[8.9]   Detyna E and Wooding E R 1975 *Plasma Phys.* **17** 539
[8.10] Hsia J B, Chiu S M, Hsia M F, Chou R L and Wu C S 1979 *Phys. Fluids* **22** 1737
[8.11] Alfven H 1960 *Rev. Mod. Phys.* **32** 710
[8.12] Danielsson L 1970 *Phys. Fluids* **13** 2288
[8.13] Sherman J C 1973 *Astrophys. Space Sci.* **24** 487
[8.14] Raadu M A 1978 *Astrophys. Space Sci.* **55** 125
[8.15] Petelski E F, Fahr H J, Ripken H W, Brenning N and Axnas I 1980 *Astron. Astrophys.* **87** 20
[8.16] Huba J D and Wu C S 1976 *Phys. Fluids* **19** 988
[8.17] Sizonenko V L and Stepanov K N 1967 *Nucl. Fusion* **7** 131
[8.18] Sizonenko V L and Stepanov K N 1978 *Nucl. Fusion* **18** 1081
[8.19] Priest E R and Sanderson J J 1972 *Plasma Phys.* **14** 951
[8.20] Sanderson J J and Priest E R 1972 *Plasma Phys.* **14** 959
[8.21] Biskamp D 1970 *J. Geophys. Res.* **75** 4659

[8.22] Wong H V 1970 *Phys. Fluids* **13** 757
[8.23] Lashmore-Davies C N 1970 *J. Phys. A: Math. Gen.* **3** L40
[8.24] Lashmore-Davies C N 1971 *Phys. Fluids* **14** 1481
[8.25] Lashmore-Davies C N and Martin T J 1973 *Nucl. Fusion* **13** 193
[8.26] Forslund D W, Morse R L and Nielsen C W 1970 *Phys. Rev. Lett.* **25** 1266
[8.27] Lominadze D G 1973 *Sov. Phys.-JETP* **36** 686

[9.1] Mikhailovskii A B and Tsypin V S 1971 *Plasma Phys.* **13** 785
[9.2] Mikhailovskii A B 1967 *Sov. Phys.-JETP* **25** 623
[9.3] Braginskii S I 1958 *Sov. Phys.-JETP* **6** 358
[9.4] Braginskii S I 1965 *Reviews of Plasma Physics* vol 1 ed M A Leontovich (New York: Consultants Bureau) p 205
[9.5] Kaufman A 1960 *La Theorie des Gas Neutres et Ionises* (Paris: Hermann) p 293
[9.6] Nemov V V 1970 *Nucl. Fusion* **10** 19
[9.7] Grad H 1949 *Commun. Pure Appl. Math.* **2** 331
[9.8] Herdan R and Lilley B I 1960 *Rev. Mod. Phys.* **32** 731
[9.9] Mikhailovskii A B and Tsypin V S 1971 *Sov. Phys.-JETP* **32** 287
[9.10] Budker G I 1961 *Plasma Physics and the Problem of Controlled Thermonuclear Reactions* vol 1 ed M A Leontovich (Oxford: Pergamon) pp 145–52
[9.11] Schluter A 1957 *Z. Naturf.* a12 822
[9.12] Chew G F, Goldberger M L and Low F E 1956 *Proc. R. Soc.* A **236** 112
[9.13] Zhdanov S K and Trubnikov B A 1977 *Sov. Phys.-JETP* **45** 256
[9.14] Milantiev V P 1969 *Plasma Phys.* **11** 145

[10.1] Mikhailovskii A B and Tsypin V S 1971 *Sov. Phys.-JETP* **32** 922
[10.2] Mikhailovskii A B and Tsypin V S 1972 *Plasma Phys.* **14** 449
[10.3] Nemov V V *Nucl. Fusion* **12** 643
[10.4] Tsypin V S 1972 *Sov. Phys.-Tech. Phys.* **16** 1377
[10.5] Tsypin V S 1972 *Sov. Phys.-Tech. Phys.* **17** 543
[10.6] Mikhailovskii A B 1982 *J. Plasma Phys.* **28** 1
[10.7] Bolshov L I, Dreizin Yu A and Dykhne A M 1977 *JETP Lett.* **26** 4
[10.8] Dobrowolny M 1972 *Phys. Fluids* **15** 2263

[11.1] Lapshin V I and Nemov V V 1975 *Nucl. Fusion* **15** 1107
[11.2] Lapshin V I and Nemov V V 1975 *Nucl. Fusion* **15** 1113
[11.3] Lapshin V I and Nemov V V 1976 *Sov. Phys.-Tech. Phys.* **21** 37
[11.4] Dreizin Y A and Sokolov E P 1981 *Sov. J. Plasma Phys.* **7** 516
[11.5] Dreizin Y A and Sokolov E P 1982 *Sov. J. Plasma Phys.* **8** 319
[11.6] Ivanov V N and Mikhailovskii A B 1985 *Sov. J. Plasma Phys.* **11** 278
[11.7] Galeev A A, Moiseev S S and Sagdeev R Z 1964 *J. Nucl. Energy* C **6** 645
[11.8] Baikov I S 1966 *JETP Lett.* **4** 201
[11.9] Mikhailovskii A B 1967 *Sov. Phys.-JETP* **25** 831
[11.10] Moiseev S S 1966 *JETP Lett.* **4** 55

[12.1] Alfven H and Smars E 1960 *Nature* **188** 801
[12.2] Lehnert B 1968 *Nucl. Fusion* **8** 173
[12.3] Budker G I, Mirnov V V and Ryutov D D 1971 *JETP Lett.* **14** 212
[12.4] Alikhanov S G, Konkashbaev I K and Chebotaev P Z 1970 *Nucl. Fusion*

10 13
[12.5]   Brandt B and Braams C M 1974 *Nucl. Fusion* **S1974** 385
[12.6]   El-Nadi A and Rosenbluth M N 1973 *Phys. Fluids* **16** 2036
[12.7]   Waltz R E 1978 *Nucl. Fusion* **18** 901
[12.8]   Aydemir A Y, Berk H L, Mirnov V, Pogutse O P and Rosenbluth M N
         1987 *Phys. Fluids* **30** 3083
[12.9]   Byers J A 1977 *Phys. Rev. Lett.* **39** 1476
[12.10] Bodin H A B and Newton A A 1980 *Nucl. Fusion* **20** 1255
[12.11] Hasegawa A, Daido H, Fujita M, Mima K, Murakami M, Nakai S,
         Nishihara K, Terai K and Yamanaka C 1986 *Phys. Rev. Lett.* **56**
         139
[12.12] Hasegawa A, Nishihara K, Daido H, Fujita M, Ishisaki R, Miki F, Mima
         K, Murakami M, Nakai S, Terai K and Yamanaka C 1988 *Nucl. Fusion*
         28 369
[12.13] Kammash T and Galbraith D L 1989 *Nucl. Fusion* **29** 1079
[12.14] Mikhailovskii A B, Fridman A M and Tsypin V S 1971 *Nucl. Fusion* **11**
         123
[12.15] Chudin N V 1971 *Sov. Phys.–Tech. Phys.* **16** 37
[12.16] Saison R 1975 *Phys. Fluids* **18** 274
[12.17] Jablon C 1977 *Nucl. Fusion* **17** 787

[13.1]   Syrovatskii S I 1957 *Usp. Fiz. Nauk* **62** 247
[13.2]   Landau L D and Lifshits E M 1959 *Electrodynamics of Continuous Media*
         (Reading, MA: Addison-Wesley)
[13.3]   Chandrasekhar S 1962 *Hydrodynamic and Hydromagnetic Stability* (Ox-
         ford: Clarendon)
[13.4]   Ong R S B and Roderick N 1972 *Planet. Space Sci.* **20** 1
[13.5]   Lau Y Y and Liu C S 1980 *Phys. Fluids* **23** 939
[13.6]   Chen L and Hasegawa A 1974 *J. Geophys. Res.* **79** 1033
[13.7]   Ohsawa Y, Nosaki K and Hasegawa A 1976 *Phys. Fluids* **19** 1139
[13.8]   Fejer J A 1964 *Phys. Fluids* **7** 499
[13.9]   Sen A K 1964 *Phys. Fluids* **7** 1293
[13.10] Lerche I 1966 *J. Geophys. Res.* **71** 2365
[13.11] Sen A K 1968 *J. Geophys. Res.* **73** 5015
[13.12] Lerche I 1968 *J. Geophys. Res.* **73** 5017
[13.13] Talwar S P 1965 *Phys. Fluids* **8** 1295
[13.14] Duhau S and Gratton J 1975 *J. Plasma Phys.* **13** 451
[13.15] Fejer J A 1963 *Phys. Fluids* **6** 508
[13.16] Kadomtsev B B, Mikhailovskii A B and Timofeev A V 1965 *Sov. Phys.–
         JETP* **20** 1517
[13.17] McKenzie J F 1970 *Planet. Space Sci.* **18** 1
[13.18] Duhau S, Gratton F and Gratton J 1971 *Phys. Fluids* **14** 2067
[13.19] Jaggi R K and Wolf R A 1971 *Phys. Fluids* **14** 648
[13.20] Sperling J L and Bhadra D K 1979 *Plasma Phys.* **21** 225
[13.21] Achterberg A 1980 *Plasma Phys.* **22** 277
[13.22] Jokipii J R and Davis L 1969 *Astrophys J.* **156** 1101
[13.23] Parker E N 1964 *Astrophys J.* **139** 690
[13.24] Ershkovich A I Nusinov A A and Chernikov A A 1973 *Sov. Astron.–A.
         J.* **16** 705
[13.25] Breus T K and Krymskii A M 1987 *Planet Space Sci.* **35** 1213
[13.26] Dobrowolny M and D'Angelo N 1972 *Cosmic Plasma Physics* ed
         K Schindler (New York: Plenum) p 149

[13.27] Bavassano B, Dobrowolny M and Moreno G 1978 *Solar Phys.* **57** 445
[13.28] D'Angelo N 1973 *J. Geophys. Res.* **78** 1206
[13.29] Melander B G and Parks G K 1981 *J. Geophys. Res.* **86** 4697
[13.30] Huba J D 1981 *J. Geophys. Res.* **86** 3653
[13.31] Huba J D 1981 *J. Geophys. Res.* **86** 8991
[13.32] Migliuolo S 1988 *J. Geophys. Res.* **93** 867
[13.33] Stix T H 1973 *Phys. Fluids* **16** 1922
[13.34] Catto P H, Rosenbluth M N and Liu C S 1973 *Phys. Fluids* **16** 1719
[13.35] Suckewer S, Eubank H P, Goldston R J, McEnerney J, Sauthoff N R and Towner H H 1981 *Nucl. Fusion* **21** 1301
[13.36] Kadomtsev B B, Morozov D Kh and Pogutse O P 1984 *JETP Lett.* **39** 269
[13.37] Morozov D Kh and Pogutse O P 1986 *Sov. J. Plasma Phys.* **12** 381–6

[14.1] Kadomtsev B B 1965 *Plasma Turbulence* (London: Academic)
[14.2] Coroniti F V and Kennel C F 1970 *J. Geophys. Res.* **75** 1863
[14.3] Kennel C F 1969 *Rev. Geophys.* **7** 379
[14.4] McPherron R L and Coleman P J 1971 *J. Geophys. Res.* **76** 3010
[14.5] Hasegawa A 1971 *J. Geophys. Res.* **76** 5361
[14.6] Rostoker G, Samson J C and Higuchi Y 1972 *J. Geophys. Res.* **77** 4700
[14.7] Kikuchi H 1972 *Cosmic Plasma Physics* ed K Schindler (New York: Plenum) p 45
[14.8] Su S Y, Konradi A and Fritz T A 1977 *J. Geophys. Res.* **82** 1859
[14.9] Kintner P M and Gurnett D A 1978 *J. Geophys. Res.* **83** 39
[14.10] Patel V L 1978 *Geophys. Res. Lett.* **5** 291
[14.11] Maclennan C G, Lanzerotti L J, Hasegawa A, Berring E A, Bernbrock G R, Sheldon W R, Rosenberg T J and Matthews D L 1978 *Geophys. Res. Lett.***5** 403
[14.12] Mikhailovskii A B and Pogutse O P 1966 *Sov. Phys.–Tech. Phys.* **11** 153
[14.13] Bhatnagar P L, Gross E P and Krook M 1954 *Phys. Rev.* **94** 511
[14.14] Mikhailovskii A B 1972 *Nucl. Fusion* **12** 55
[14.15] Tang J T, Luhmann N C, Nishida Y and Ishi K 1975 *Phys. Rev. Lett.* **34** 70
[14.16] Kadomtsev B B and Pogutse O P 1970 *Reviews of Plasma Physics* vol 5 ed M A Leontovich (New York: Consultants Bureau) p 249
[14.17] Kadomtsev B B 1964 *Sov. Phys.–JETP* **18** 847
[14.18] Galeev A A and Rudakov L I 1964 *Sov. Phys.–JETP* **18** 444
[14.19] Sagdeev R Z, Shapiro V D and Shevchenko V I 1978 *Sov. J. Plasma Phys.* **4** 306
[14.20] Hasegawa A and Mima K 1978 *Phys. Fluids* **21** 87
[14.21] Sanuki H and Weiland J 1980 *J. Plasma Phys.* **23** 29
[14.22] Mikhailovskii A B and Pogutse O P 1964 *Sov. Phys.–Dokl.* **9** 379
[14.23] Pitaevskii L P 1963 *Sov. Phys.–JETP* **17** 658
[14.24] Rukhadze A A and Silin V P 1967 *Sov. Phys.–Dokl.* **11** 606
[14.25] Dominguez R R 1979 *Nucl. Fusion* **19** 105
[14.26] Berk H L and Dominguez R R 1977 *J. Plasma Phys.* **18** 31
[14.27] Winske D and Gary S P 1979 *Phys. Fluids* **22** 665
[14.28] Hastings D E and McCune J E 1982 *Phys. Fluids* **25** 509
[14.29] Huba J D and Gary S P 1982 *Phys. Fluids* **25** 1821
[14.30] Fejer B G and Kelley M C 1980 *Rev. Geophys. Space Phys.* **18** 401
[14.31] Hudson M K and Kennel C F 1975 *J. Plasma Phys.* **14** 121

[15.1]   Longmire C L and Rosenbluth M N 1957 *Ann. Phys.* **1** 120
[15.2]   Kadomtsev B B 1966 *Reviews of Plasma Physics* vol 2 ed M A
          Leontovich (New York: Consultants Bureau) p 153
[15.3]   Rosenbluth M N, Krall N A and Rostoker N 1962 *Nucl. Fusion Suppl.*
          1 143
[15.4]   Mikhailovskaya L V and Mikhailovskii A B 1963 *Nucl. Fusion* **3** 113
[15.5]   Meerson B I, Mikhailovskii A B and Pokhotelov O A 1977 *Plasma Phys.*
          19 1177
[15.6]   Shafranov V D 1966 *Reviews of Plasma Physics* vol 2 ed M A Leontovich
          (New York: Consultants Bureau) p 103
[15.7]   Taylor J B and Hastie R J 1965 *Phys. Fluids* **8** 1175

[16.1]   Borisov N D and Mikhailovskii A B 1972 *Sov. Phys.–Dokl.* **16** 553–5
[16.2]   Borisov N D 1973 *Sov. Phys.–Tech. Phys.* **17** 1615
[16.3]   Horton W 1980 *Nucl. Fusion* **20** 321

Index